This engaging collection provides compelling reasons for "starting close to home" when responding to the challenges of climate change. A range of fascinating case studies shows that thinking locally is vital for understanding the complex flows of people, power, and knowledge that shape environmental problems and solutions. A watershed can be both a place and a turning point; what a brilliant idea to showcase the diverse mobilizations of climate justice on the shores of the Great Lakes at this pivotal time for the planet.

—Sherilyn MacGregor, Reader in Environmental Politics, The University of Manchester, UK

A compilation of rich and deeply moving "stories" from young climate justice leaders and activists which make a compelling and truly inspirational read.

—Tahseen Jafry, Professor and Director, The Centre for Climate Justice, Glasgow Caledonian University, Glasgow, UK

This book shares stories from the frontlines of those fighting for climate justice. The often personal accounts inform and inspire, from the everyday politics of how we eat, work, and play, to labour movement organizing, community initiatives, and direct action to spark ideas for reconnecting through art and the sacred. This book serves as an important reminder and inspiration to all those concerned about climate justice of what we can do in our daily lives to make a difference.

—Leah Temper, Ecological Economist, Scholar Activist, and Filmmaker based at McGill University, Montreal, Canada, and the Autonomous University of Barcelona, Spain

How do we foster a compassionate response to the human dimensions of today's climate changes? This is the challenge that this volume meets head-on as it considers the global calls for dealing with climate injustices from a regional grounding, and thus shows us how local actions can scale up to a global response in the absence of meaningful political leadership. Drawing from actions in the areas of policy, education, and community-building, this edited book offers a diversity of case studies and a wonderful Action Glossary that can inspire each of us to re-think ways of growing into our time of climate justice.

—Timothy B. Leduc, Assistant Professor, Wilfrid Laurier University, Canada

LOCAL ACTIVISM FOR GLOBAL CLIMATE JUSTICE

This book will inspire and spark grassroots action to address the inequitable impacts of climate change, by showing how this can be tackled and the many benefits of doing so.

With contributions from climate activists and engaged young authors, this volume explores the many ways in which people are proactively working to advance climate justice. The book pays special attention to Canada and the Great Lakes watershed, showing how the effects of climate change span local, regional, and global scales through the impact of extreme weather events such as floods and droughts, with related economic and social effects that cross political jurisdictions. Examining examples of local-level activism that include organizing for climate-resilient and equitable communities, the dynamic leadership of Indigenous peoples (especially women) for water and land protection, and diaspora networking, *Local Activism for Global Climate Justice* also provides theoretical perspectives on how individual action relates to broader social and political processes.

Showcasing a diverse range of inspirational and thought-provoking case studies, this book will be of great interest to students and scholars of climate justice, climate change policy, climate ethics, and global environmental governance, as well as teachers and climate activists.

Patricia E. Perkins is Professor in the Faculty of Environmental Studies at York University, Canada, where she teaches ecological economics, community economic development, climate justice, and critical interdisciplinary research design.

Routledge Advances in Climate Change Research

A Critical Approach to Climate Change Adaptation
Discourses, Policies and Practices
Edited by Silja Klepp and Libertad Chavez-Rodriguez

Contemplating Climate Change
Mental Models and Human Reasoning
Stephen M. Dark

Climate Change, Moral Panics and Civilization
Amanda Rohloff
Edited by André Saramago

Climate Change and Social Inequality
The Health and Social Costs of Global Warming
Merrill Singer

Cities Leading Climate Action
Urban Policy and Planning
Sabrina Dekker

Culture, Space and Climate Change
Vulnerability and Resilience in European Coastal Areas
Thorsten Heimann

Communication Strategies for Engaging Climate Skeptics
Religion and the Environment
Emma Frances Bloomfield

Regenerative Urban Development, Climate Change and the Common Good
Edited by Beth Schaefer Caniglia, Beatrice Frank, John L. Knott Jr., Kenneth S. Sagendorf, Eugene A. Wilkerson

Local Activism for Global Climate Justice
The Great Lakes Watershed
Edited by Patricia E. Perkins

Political Responsibility for Climate Change
Ethical Institutions and Fact-Sensitive Theory
Theresa Scavenius

For more information about this series, please visit: https://www.routledge.com/Routledge-Advances-in-Climate-Change-Research/book-series/RACCR

LOCAL ACTIVISM FOR GLOBAL CLIMATE JUSTICE

The Great Lakes Watershed

Edited by Patricia E. Perkins

LONDON AND NEW YORK

First published 2020
by Routledge
2 Park Square, Milton Park, Abingdon, Oxon OX14 4RN

and by Routledge
52 Vanderbilt Avenue, New York, NY 10017

Routledge is an imprint of the Taylor & Francis Group, an informa business

British Library Cataloguing-in-Publication Data
A catalogue record for this book is available from the British Library

Library of Congress Cataloging-in-Publication Data
Names: Perkins, Patricia E., author.
Title: Local activism for global climate justice: the Great Lakes Watershed/Edited by Patricia E. Perkins.
Description: Abingdon, Oxon; New York, NY: Routledge, 2020. | Series: Routledge advances in climate change research | Includes bibliographical references and index.
Identifiers: LCCN 2019018713 (print) | LCCN 2019980447 (ebook) | ISBN 9780367335878 (hardback) | ISBN 9780367335892 (paperback) | ISBN 9780429320705 (ebook)
Subjects: LCSH: Climatic changes—Great Lakes Region (North America) | Great Lakes Watershed (North America)—Environmental conditions. | Environmental policy—Great Lakes Watershed (North America) | Water quality management—Great Lakes Watershed (North America)
Classification: LCC QC902.2.G74 P47 2020 (print) | LCC QC902.2.G74 (ebook) | DDC 363.738/740977—dc23
LC record available at https://lccn.loc.gov/2019018713
LC ebook record available at https://lccn.loc.gov/2019980447

ISBN: 978-0-367-33587-8 (hbk)
ISBN: 978-0-367-33589-2 (pbk)
ISBN: 978-0-429-32070-5 (ebk)

Typeset in Bembo
by Deanta Global Publishing Services, Chennai, India

CONTENTS

FIGURES

ABOUT THE COVER

The image on the cover of this book is a reproduction of a print by Samay Arcentales Cajas, a Toronto-born Kichwa graphic artist and filmmaker (samaycajas.com). It is entitled "No More Mercury!" and was created in support of the Grassy Narrows (Asubpeeschoseewagong) First Nation's decades-long struggle for environmental justice, following mercury contamination of the English-Wabigoon river system by a paper mill and by logging which has devastated the environment and fish, people's health, and the local economy since the 1960s. Climate change will increase toxic mercury exposure worldwide due to increased wind and precipitation, thawing permafrost, higher temperatures, and greater atmospheric CO_2—disproportionately harming children, whose development is permanently affected by mercury exposure. The image also evokes Turtle Island, the Indigenous conception of the world resting on a turtle's back over the water that gives life to all.

This image previously appeared on the cover of *Women and Environments International Magazine*, no. 98/99, https://www.yorku.ca/weimag/

CONTRIBUTORS

Ifrah Abdillahi is a PhD candidate in the Social and Behavioral Health Sciences division at the Dalla Lana School of Public Health (DLSPH), University of Toronto. Her dissertation poses critical questions at the interface of healthy city design, population health, and climate adaptation. Having worked extensively on social justice projects in Canada, Europe, and East Africa, her research is situated within the broader study of urban health equity. She holds a Master's in Gender and Political Science with technical expertise in gender equity and gender analysis, and is a fellow in the Collaborative Doctoral Specialization in Global Health at DLSPH. Prior to joining Dalla Lana, Ifrah was a research associate for Yale University on the *Africa and the International Criminal Court* project, and has also consulted in the private sector with monitoring and evaluation subcontractors for USAID and DFID on governance and health-related portfolios.

Alison Adams is a PhD student at the University of Vermont's Rubenstein School. She studies the nonmaterial connections between humans and nature, with a focus on how environmental change affects individuals' subjective well-being and cultural practices, and the justice and equity implications of these effects. She also explores how art can contribute to our understanding of the intangible ways people and nature interact. Previously, Alison worked in environmental advocacy on issues such as stopping uranium mining around the Grand Canyon and securing stronger protections for public lands. Alison has a BA in History of Art from Yale University, and a MS in Natural Resources from the University of Vermont.

Paul Baines is the Outreach and Education coordinator for Great Lakes Commons and founder of the Great Lakes Commons Map. He comes to this water reconciliation work with a background in critical pedagogy, democratic media, and environmental and cultural studies. Paul is at home in the Lake Ontario watershed, while often touring the Great Lakes by bike or van connecting people, issues, and perspectives.

Douglas Baxter recently completed his Master's in Environmental Studies at York University. He has a Bachelor's in Environmental Studies, also from York. He is interested in the relationships among firm efficiency, profitability, and sustainability in changing ecological systems.

Sam Bliss is a PhD student in ecological economics at the University of Vermont. His doctoral research and social life both revolve around non-market food systems. Sam is also the US correspondent of the scholar-activist collective Research & Degrowth. He is fomenting movements for gift-based food networks and economic degrowth in North America. Join him!

Natália Britto dos Santos is a PhD student in Environmental Studies at York University, Toronto. She has experience in protected areas conservation and management in Brazil, and as a volunteer environmental educator in both Brazil and Toronto. Her study interests include ecosystem benefits to human well-being, environmental values, and connectedness with nature.

JoEllen Calderara is an active community leader in the Central Vermont region. She was heavily involved in Vermont's recovery efforts from flooding events in 2011, including Tropical Storm Irene. She was a founding board member of the Vermont Disaster Relief Fund and Chair of the Central Vermont Long Term Recovery Committee—providing case management, overseeing allocations of recovery dollars, and spearheading recovery efforts for mobile home park communities.

Stephen M. Clare completed his Master's in Renewable Resources at McGill University's Department of Natural Resource Sciences, Montreal. He aspires to use his research to support sustainable resource management, and his thesis documented the struggle of Indigenous communities in Panama to gain tenure and management rights to traditional forest resources.

Alice Damiano is a PhD candidate in Renewable Resources, Department of Natural Resource Sciences, McGill University, Montreal. Originally from Italy, she is now doing interdisciplinary research on human-Earth relationships within the Economics for the Anthropocene (E4A) project. She is working on the idea of learning from Indigenous peoples about how to establish better relationships with the environment, especially in the context of natural and environmental disasters.

Monica Krista de Vera is a Master's in Environmental Studies candidate at York University, Toronto, specializing in Planning. She completed her undergraduate degree at the University of Toronto in Philosophy and Environmental Studies.

Her research interests include municipal governance and politics, urban planning, social movements, and activism.

Meagan Dellavilla completed her master's degree in Environmental Studies at York University, Toronto. She currently works as the Research and Evaluation Manager for Community Food Centres Canada. Originally from upstate New York (land traditionally occupied and cared for by the Seneca Nation of the Haudenosaunee Confederacy), Meagan is of mixed-European ancestry.

Emily Dyett is the Youth & Climate Justice Coordinator for the Western New York Environmental Alliance (WNYEA). She received her BS in Environmental Studies with a focus on Education and Policy from the University at Buffalo, New York. She has worked in community education, natural resource conservation, and outreach with organizations like Earth Spirit Educational Services (delivering programs focused on the unique ecology of the Buffalo region to students throughout Erie County), WNY Partnership for Regional Invasive Species Management (PRISM), and Alliance for the Great Lakes. Her work with WNYEA on the Youth & Climate Justice Initiative focuses on empowering the next generation of climate activists in WNY and engaging young people with the climate justice movement both regionally and nationally.

Laura Gilbert is a PhD candidate in the Economics for the Anthropocene Project (E4A) in the Department of Natural Resource Sciences at McGill University, Montreal. She works on incorporating ethics for a mutually enhancing human-Earth relationship at different levels of water governance in Canada: from governance structures to creating community tools that help shape policy, projects, and curriculum.

Jen Gobby is a PhD candidate at McGill University, Montreal. In her research, she works closely with social movements in Canada to *think together* about how to bring about radical transformation towards justice and sustainability. She also organizes with Climate Justice Montreal.

Teresa Auntora Gomes is a Master's in Environmental Studies candidate at York University, Toronto, and recent graduate from the University of Toronto where she studied Critical Development Studies, Public Policy & Human Geography. Currently, she is a Legislative Policy Researcher at Environment and Climate Change Canada. As a first generation Bangladeshi-Canadian woman, she hopes to explore the politics of disaster governance, environmental displacement, legal & environmental history, critical geopolitics and environmental health in the workplace through her academic career. Through her love for community, she is the founder of an educational initiative for women in Bangladesh, "Education & Equity for Woman in Nilphamari Bangladesh." She is also a filmmaker and her work, "Footprints of Environmental Justice" was screened at the 2017 Toronto Reel Asian International Film Festival.

Lindsay Gray is a two-spirited Anishnaabe from the Aamjiwnaang First Nation, located in the middle of Canada's Chemical Valley. Faced with injustice and environmental racism, their organization Aamjiwnaang and Sarnia Against Pipelines hosts environmental events such as the Aamjiwnaang Water Gathering and Toxic Tour.

Daniel Horen Greenford is a PhD student at Concordia University in Montreal, studying under the supervision of climate scientist Damon Matthews in the Department of Geography, Planning and Environment. Born and raised in Montreal, he is actively engaged in organizing efforts to decarbonize Canada.

Gabriel Yahya Haage is currently pursuing a PhD in the Department of Natural Resource Sciences (Neotropical concentration) at McGill University, Montreal. His research interests include freshwater systems and understanding water demands in the ecological, social, and economic spheres. His current research focuses on lakes and watersheds, particularly the Bayano watershed in Panama.

Kelly Hamshaw is a PhD student at the Rubenstein School of Environment and Natural Resources at the University of Vermont (UVM). Working with Vermont's mobile home park communities during Tropical Storm Irene recovery efforts in 2011, both as a researcher and as a volunteer with the Central Vermont Long Term Recovery Committee, she has witnessed first-hand the impacts of climate change on rural communities. She is a Lecturer in the UVM Department of Community Development and Applied Economics, where she engages students in a variety of service-learning projects.

Claire-Hélène Heese-Boutin is pursuing a Master's in Environmental Studies at York University, Toronto. She holds a BA in Environmental Studies and Caribbean Studies from the University of Toronto. Over the last three years, she has been working in personal financial services with a focus on responsible investment management and household financial planning.

Michaela Hynie is a social and cultural psychologist in the Department of Psychology and the Centre for Refugee Studies at York University, Toronto, and past president of the Canadian Association for Refugee and Forced Migration Studies. Dr. Hynie conducts both qualitative and quantitative community-based research with a focus on situations of social conflict and forced displacement, and the development and evaluation of interventions that can strengthen social and institutional relationships to improve health and well-being in different cultural, political, and physical environments. Her work in Canada, India, Nepal, Rwanda, and South Africa has been funded by the *Social Sciences and Humanities Research Council*, *Grand Challenges Canada*, and the *Canadian Institutes for Health Research*. She is currently leading *Syria.LTH*, a 5-year longitudinal study on Syrian refugee integration in Canada.

Alia Karim is a PhD candidate in the Faculty of Environmental Studies at York University, Toronto. Her latest research focuses on the Canadian labour movement's engagement with Indigenous workers and their communities in the Southern Ontario region, in the hopes of building common understanding and solidarity between Indigenous and settler peoples. Her interests include labour–community coalitions, community organizing, Indigenous and settler relations, and eco-socialism. She is also a community organizer in the Fight for $15 and Fairness campaign, which has demanded a $15/hour minimum wage and stronger labour laws in Ontario.

Beth Lorimer is the Ecological Justice Program coordinator at KAIROS Canada. KAIROS unites ten Christian churches and religious organizations to deliberate on issues of common concern, advocate for social justice, and join with people of faith and goodwill on action for social transformation. Beth has also worked on community-based watershed awareness projects in South Africa and Kenya.

Caitlin Bradley Morgan is a doctoral student of Food Systems at the University of Vermont. Her previous education includes a Master's in Food Systems, also from UVM, and a BS in Food Literacy from the University of California at Berkeley. In her younger years, Caitlin was involved in an affordable, cooperative housing non-profit, both as a member and as an elected leader, and later worked in community nutrition education and youth empowerment. She believes that food is a crucial component of environmental and social justice.

Prateep Nayak is Associate Professor in the School of Environment, Enterprise and Development at the University of Waterloo, Ontario. He has an academic background in political science, environmental studies, and international development, and holds a PhD in Natural Resources and Environmental Management from the University of Manitoba. He engages in interdisciplinary scholarship with an active interest in combining social and ecological perspectives. Prateep's research focuses on the understanding of complex human-environment connections (or disconnections) in coastal-marine systems with attention to social-ecological change, its drivers, their influence and possible ways to deal with them. His main research interests include coastal commons, environmental change, migration and governance, resilience, social-ecological regime shifts, environmental justice, and political ecology. Prateep is a past SSHRC Banting Fellow, Trudeau Scholar, a Harvard Giorgio Ruffolo Fellow in Sustainability Science and a recipient of Canada's Governor General's Academic Gold Medal.

Patricia E. Perkins is Professor in the Faculty of Environmental Studies at York University, Toronto, where she teaches ecological economics, community economic development, climate justice, and critical interdisciplinary research design. She has worked with civil society and university-based partners and young researchers in Canada, Brazil, and several African countries on community-based research regarding priorities in resource and water management, environmental education, and gender issues in relation to climate change. Her research centres around feminist ecological economics, participatory watershed and commons governance, and climate justice. She is currently serving as a Lead Author for the chapter on "Demand, services, and social aspects of mitigation" in the Intergovernmental Panel on Climate Change's 6th Assessment Report, to be published in 2021.

Alicia Richins is a recent graduate of York University's Master in Environmental Studies, Planning Concentration, in Toronto, and a member of the Economics for the Anthropocene research collective. As a dual citizen of Canada and Trinidad & Tobago, her graduate research sought to apply the theoretical framework of ecological economics to the practice of planning in the Global South, or international development. Alicia is now pursuing a career that combines ecological sustainability with international development, to contribute to the work of adaptation and resilience in areas most vulnerable to environmental shocks.

Aaron Saad completed his PhD in Environmental Studies at York University, Toronto, in 2017, and wrote a dissertation focusing on climate justice. During his time at York, he was one of the leaders of the Fossil Free York divestment campaign. He is currently a columnist with Ricochet Media where he writes on climate change politics and is a professor at Humber College, Toronto.

Lynda H. Schneekloth is on the Board of the Western New York Environmental Alliance, serves as Advocacy Chair, and has been involved in its Youth & Climate Justice Initiative since its initiation. She works with partners in the space of climate justice, to help build a movement toward a just transition that is inclusive of diverse interests. She is Professor Emerita of the School of Architecture and Planning at the University at Buffalo, New York, where she has focused her research on *placemaking*, seeking to engage people and institutions in the work of protecting and healing their communities and the earth while joining together in making beloved places.

Martin Sers is a PhD candidate in the Faculty of Environmental Studies at York University, Toronto. He is developing mathematical models of the energy transition as well as working on systems dynamics and energy in ecological economics modelling.

Barbara Sniderman is a parent, educator, and activist. Her work focuses on pedagogy for innovative inter-generational change. The aim of her work is to use inquiry-based, problem-based, arts or creative-based learning, often in outdoor settings, to enable young people to think and live in systems. Her approach is to teach mental, physical, emotional, and ecological health by fostering a love for preserving the local and not-so-local systems we have in our midst.

Nowrin Tabassum is a PhD Candidate in the Department of Political Science at McMaster University in Hamilton, Ontario. Her dissertation focuses on climate change, policy, and migration in Bangladesh.

Rebekah A. Williams is a community organizer and trainer from Western New York, founder of Food for the Spirit (an organization committed to racial healing towards ecological justice and equitable food systems), and a campaign organizer with the Massachusetts Avenue Project (MAP), working to bring the Good Food Purchasing Program to Buffalo Public Schools. She has more than 20 years' experience with youth leadership, social/racial justice, environment, food, and arts organizing and activism in Buffalo. She holds a Bachelor's degree in Social Structure, Theory and Change from SUNY Empire State College, and has completed training with Training for Change in Philadelphia, PA, Movement Generation in Oakland, CA, the Buffalo Montessori Teacher Education Program, and North American Students of Cooperation in Chicago, IL.

LAND ACKNOWLEDGEMENT[1]

The Great Lakes Watershed, the focus of this book, has been the home of Indigenous peoples for thousands of years. They include Anishinaabe peoples (the Ojibwe/Chippewa, Algonquin, Abenaki, Odawa, Potawatomi, and Mississauga, among others), the Cree, the Métis, the Wyandot or Huron, and other Haudenosaunee/Iroquois peoples (the Seneca, Cayuga, Onondaga, Oneida, Tuscarora, and Kahien'kehá:ka, also called Mohawk), and those called the Neutrals. Many treaties cover the territories discussed in this book; we who live here are all Treaty people. The Dish with One Spoon Wampum Belt Covenant is an agreement between the Haudenosaunee Confederacy and the Anishinaabe Three Fires Confederacy to peaceably share and care for the lands around the Great Lakes. Many place names and waterway names record the First Nations' reverence and care for the land and waters. Ontario means "large, beautiful lake" in the now-dormant Wendat language; Niagara is from an Iroquoian word for "the strait"; Quebec is from the Mikmaq word meaning "strait" or "narrows"; Tkaronto means "where there are trees in the water"; Canada is from the Iroquoian word "kaná:ta" meaning "town" or "land."

The ongoing leadership of Indigenous elders and activists regarding how to live sustainably in these territories is an immense gift. Non-Indigenous residents of the watershed, and of Turtle Island and the world, continue to enjoy a wide range of privileges today because of colonialism, while Indigenous residents

1 Land Acknowledgements have become normal practice at most public meetings in Canada, especially those related to environment, equity, social justice, and education, and in publications. They "insert an awareness of Indigenous presence and land rights in everyday life ... [and] can be a subtle way to recognize the history of colonialism and a need for change in settler colonial societies" (https://native-land.ca/territory-acknowledgement/). The proof that we take land acknowledgements seriously lies in our actions.

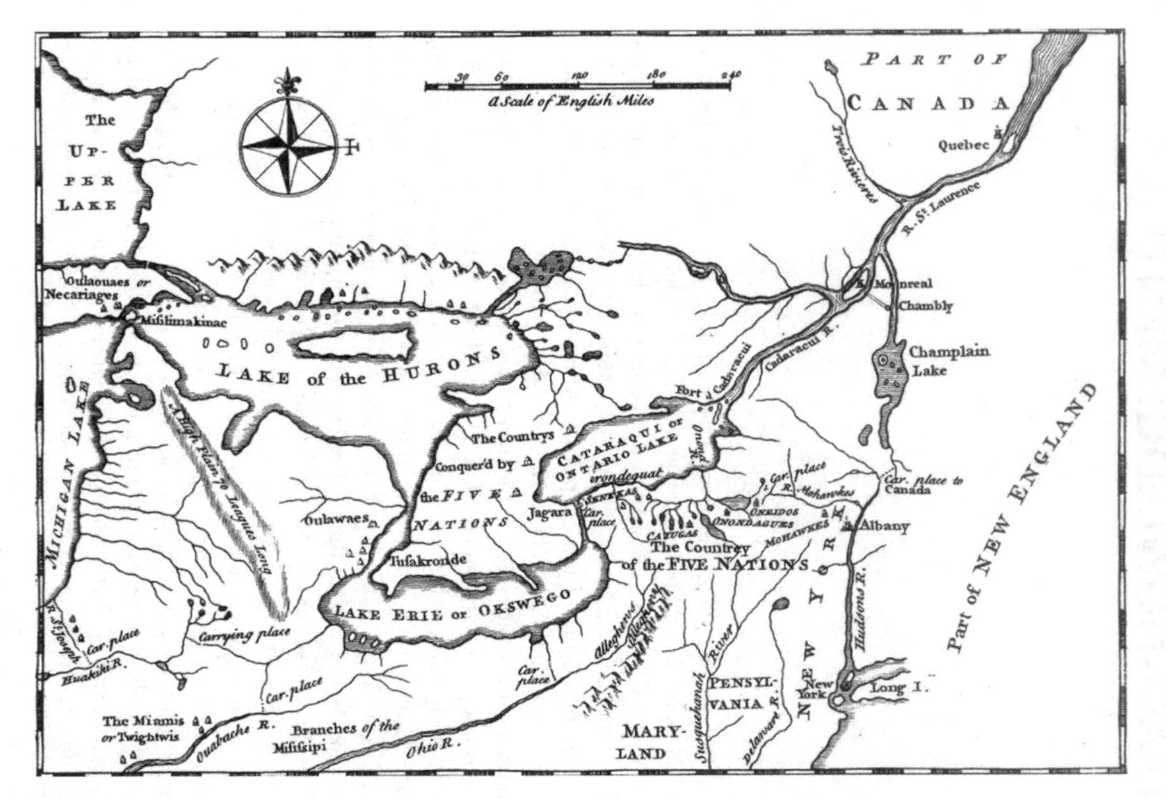

FIGURE 0.1 A 1755 map of the Great Lakes. This early map shows how important Indigenous peoples and place-names were in the early years of European colonization. Carrying places, sometimes indicated as "car place," were where canoes could be portaged from one waterway to another.

continue to suffer its impacts. It is our responsibility to keep learning about this and striving to disrupt and dismantle colonialism as an essential part of our work for climate justice.[2]

2 See the Tribal and First Nations Great Lakes Water Accord, 2004 (http://www.nofnec.ca/PDF/2012/Tribal-and-First-Nations-Great-Lakes-Water-Accord.pdf); W. Singel and M.L.M. Fletcher, "Indian treaties and the survival of the Great Lakes," 2006, Michigan State Law Review 1285, https://digitalcommons.law.msu.edu/cgi/viewcontent.cgi?article=1159&context=facpubs; and https://commons.wikimedia.org/wiki/File:Anishinaabe-Anishinini_Distribution_Map.svg

1

INTRODUCTION

Climate justice, the Great Lakes, and the Earth

Patricia E. Perkins (with Martin Sers)

Climate change is the major crisis of our time—perhaps the most daunting challenge in human history. Between 20,000 years ago and 2016, human activity increased the levels of carbon dioxide in the Earth's atmosphere from 240 to more than 400 parts per million (ppm), causing the global average temperature to rise more than 1°C. If emissions continue at the present rates, atmospheric CO_2 will reach 500 ppm before 2070, raising temperatures 3°C or more, which will cause rapid sea level rise, mass migrations, species extinctions, extreme weather, and threatened food supplies (Jones 2017; see Box 1.1).

Addressing climate change is especially difficult because it worsens existing inequities among people. Some geographic locations are hit harder than others by extreme weather events or rising sea levels. Those living in poverty, or who are marginalized because of factors such as their gender, race, or ethnicity, have fewer resources at their disposal, fewer options, and less resilience to meet catastrophes caused by the changing climate. These may include disease epidemics, food and housing shortages, infrastructure destruction, fires, floods, and droughts, forced migration and political violence.

The vulnerable include other species too. Humans have caused mass species extinctions, with an estimated 30 to 50% of all animal and plant species possibly threatened with extinction by mid-century (Center for Biological Diversity 2017).

BOX 1.1: THE SCOPE AND SCALE OF CLIMATE CHANGE

By Martin Sers

Previous shifts in Earth's climate, which include those caused by asteroid impacts, have led to mass extinctions of 50 to 90% of life on Earth, sometimes over millions of years. The speed at which human-produced CO_2 emissions are rising, in a geological blink of time, means that we humans are now in a race against a closing window of time. The burden and obligation of avoiding the worst aspects of climate change fall on those humans who are alive now, in the first century of this millennium.

In the next several decades, the choices made by societies may largely determine the fate of posterity for humans and many other species. The reality that the Earth-system possesses dynamics that, once set in motion, are largely unstoppable on human time scales, implies that we have essentially one chance at mitigating our impacts.

As such, the choices of just a few generations will echo into geological time and potentially determine the fate of much of life's commonwealth. Climate justice and just action imply a special and enormous burden on current generations; the obligation to posterity implies that we, in the small amount of time remaining, find some way to avoid the darkest aspects of Earth-system change.

The inherent difficulties associated with determining what constitutes just action, especially when considering the problem of scale (temporal and physical) and the concerns of non-human life, imply that any single conventional justice framework may be inadequate. Just as the impacts of climate change are unequally distributed, so too is the responsibility, obligation, and ability to make meaningful change.

Excerpted from Sers, M. (2017), "Climate justice and the closing window for meaningful change," unpublished working paper, Economics for the Anthropocene. https://docs.google.com/viewer?a=v&pid=sites&srcid=ZGVmYXVsdGRvbWFpbnxtYXJ0aW5yb2JlcnRzZXJzfGd4OjRlYTMyM2JjYzRhOWE4NDE

The term "climate justice" expresses the visionary *hope* that humanity as a whole will be able not only to stop climate change, but to do so fairly (see Box 1.2). The most vulnerable deserve special support and protection; their consumption of fossil fuels is low so their responsibility for the problem is tiny in relation to richer, wealthier consumers. This is as true within rich societies as it is between wealthy countries and poorer ones, although the global injustices are orders of magnitude larger. But climate-related injustices exist at all scales, from the local to the global, and this includes inequities in power within the socio-political systems which have created climate change in the first place.

Thus, the climate justice *research and policy perspective* means acknowledging that these inequities exist (rather than ignoring them or assuming them away), and actively investigating their causes, specific impacts, and the links between addressing them and resolving climate change itself. For example, Indigenous peoples' opposition to pipelines and extraction calls attention to not only the violence and unjust impacts of colonialism on Indigenous peoples, but also the ecologically and socially sustainable Indigenous governance systems which colonialism attempted to crush. By centring ecological sustainability, reciprocity, and resilience, Indigenous governance shows that humans need not inevitably wreck the Earth (Whyte 2017, 2018a, 2018b; Trosper 2009). This imparts particular gravity, respect, and importance to Indigenous leadership and activism to protect land and waters, working for multi-level climate justice.

"Climate justice" also names a *movement* of people worldwide who insist that equity must be front and centre in all climate-related activism and policy. As a species, human welfare is interconnected: from communicable diseases to food production and distribution in times of famine to migration options, local ecological knowledge and skills transmission, people's destinies are interlinked. Our humanity and our survival depend on whether and how the most vulnerable are able to survive and thrive.

BOX 1.2: WHAT IS CLIMATE JUSTICE?

"Climate justice" is defined by activists and academics in various ways. Here is a sampling:

> A group of global non-governmental organizations developed the Bali Principles of Climate Justice in 2002, which remains a comprehensive climate justice rationale and program.
>
> *(Corpwatch 2002)*

"Climate justice includes a focus on the root causes of climate change and making the systemic changes that are therefore required, a commitment to address the disproportionate burden of the climate crisis on the poor and marginalized, a demand for participatory democracy in changing these systems which require dismantling the fossil fuel corporate power structure, and a commitment to reparations and thus a fair distribution of the world's wealth. Some articulate climate justice more loosely as the intersection of environmentalism and social justice, drawing on the intersectionality analysis developed by feminists and critical race theorists to understand the interlocked workings of race and gender" (Hall 2014).

"A complicated array of equity and social justice issues besets efforts to fashion global responses to climate change. They include specific questions as to who bears the responsibility for the legacy of accumulated greenhouse gas emissions, and whether such emissions were essential for livelihood support or resulted primarily from the growing affluence of populations. Climate justice also involves issues of how climate change is associated with other broad inequalities in wealth and well-being, dissociations between those who will benefit from and those who will bear the burdens and damage associated with climate change, procedural justice issues as to how decisions have been made in structuring international approaches to assess scientific issues and creating the institutions of the global climate change regime to address the problems, and how equity issues and adaptation strategies may interact" (Dow, Kasperson and Bohn 2006:79). "Climate justice thus includes elements of both distributive and procedural justice" (Adger et al. 2006:2–4 and 263–264).

> "Climate justice is a moral and political framework that 1) identifies the various moral concerns that are either causing, caused by, or otherwise raised by climate change and 2) organizes and manifests in responses to climate change, attempting to address those injustices."
>
> *(Saad 2017:13)*
>
> *(See also Jafry 2019:3.)*

This book explores examples of how people are working together to advance climate justice—starting close to home, and with global awareness. The approaches, skills, and projects which are effective at the local scale bear many similarities to those that work at broader scales. So local-level activism helps people learn what to do and how to do it, and it also brings them together with others who have similar commitments and passions, building the climate justice movement—the movement for stopping climate change as soon as possible, and doing it fairly.

This is not just a bottom-up process in local communities. Diaspora networking, where recent migrants use their language skills and roots in other places to exchange information, funds, and connections to strengthen global movements, is an important though under-studied mechanism for climate justice and international redistribution (Mohamoud, Kaloga and Kreft 2014; see also Goldring and Krishnamurti 2007). Policy activism is important, where people with different degrees of political access learn from each other, build powerful movements, and hold leaders to account. Online organizing via social media, long-distance internet, whistle-blowing, art, and film are key to many global campaigns related to climate justice. The social media communications technologies of fossil-fuelled societies have made it possible for groups to self-organize and demand political action, with unprecedented global implications. An example is the dynamic leadership of Indigenous peoples (especially women) for water and land protection, in opposition to fossil fuel extraction, pipelines, violence against women, environmental destruction, and colonialism (Women's Earth Alliance and Native Youth Sexual Health Network 2016; Whyte 2014).

Youth activists, too, have special credibility, motivation, and impact; wherever in the world they live, their options and future prospects are already being shaped by the accelerating climate crisis. As 16-year-old Swedish activist Greta Thunberg, who in August 2018 began to demonstrate in front of the Swedish parliament every Friday, has stated to world leaders, "I don't want you to be hopeful. I want you to panic. I want you to feel the fear I feel every day.

And then I want you to act" (Watts 2019:n.p.). During the Global Climate Strike for the Future on Friday, March 15, 2019, an estimated 1.6 million students in more than 120 countries participated in more than 2,200 local demonstrations to call attention to global inaction on climate change (Haynes 2019:n.p.).

In some places, democratic governance allows people to organize for climate justice within the context of existing political systems. In other places or for other people, the state is repressive, hardly exists, or its legitimacy is questionable, and people find a way to organize their own responses to climate catastrophes. Social capital, community ties, and grassroots institutions for managing common resources such as land, food production, collective work and service-sharing, and watersheds—protecting their sustainability and preventing their privatization to benefit a few—are a central way that people have always maintained their livelihoods. Climate justice activism transcends market, state, and commons frameworks. Effective methods for addressing climate impacts can be seen as "unfolding along a continuum" from local, context-specific actions through participatory community programs to broad-scale government and international interventions (Singer 2019:230).

In this book, each chapter tells a story about how people are facing a small piece of the climate justice challenge. The chapters are written by activists, students, teachers, youth and community organizers, and environmentalists who are working for climate justice in some way, at some scale from the local to the global, and/or spanning the world. Many of the chapter authors met each other through the Economics for the Anthropocene project, an ecological economics research and education partnership linking York University in Toronto, McGill University in Montreal, and the University of Vermont in Burlington (on the shores of Lake Champlain, sometimes called the "fifth Great Lake," which flows into the Richelieu River, a tributary of the St. Lawrence). The authors are all Great Lakes watershed residents.

We have chosen the lower Great Lakes watershed as the geographic focus and organizing principle for this book for several reasons. Since climate change causes extreme weather events and fluctuations, most of which are largely experienced through precipitation or drought, watersheds are more important than ever before: dams, agriculture, industry, energy, forestry, food and water supplies, pollution, and transportation all affect people upstream and downstream in interconnected ways. Watersheds link people and their responses to climate change. Climate justice, with its focus on inequities in climate impacts, is perhaps easiest to understand in reference to how people within the same watershed may be affected very differently by the very same weather events depending on their economic status, geographic position, country, gender, ethnicity, whether they are Indigenous, and other markers of difference and relative options—which can interact to create heightened "intersectional" impacts as well (Perkins 2019).

Second, the authors all live in this watershed, and we believe in starting close to home! The injustices of climate change affect us here in the Great Lakes watershed, and many people here are working to reduce those inequities.

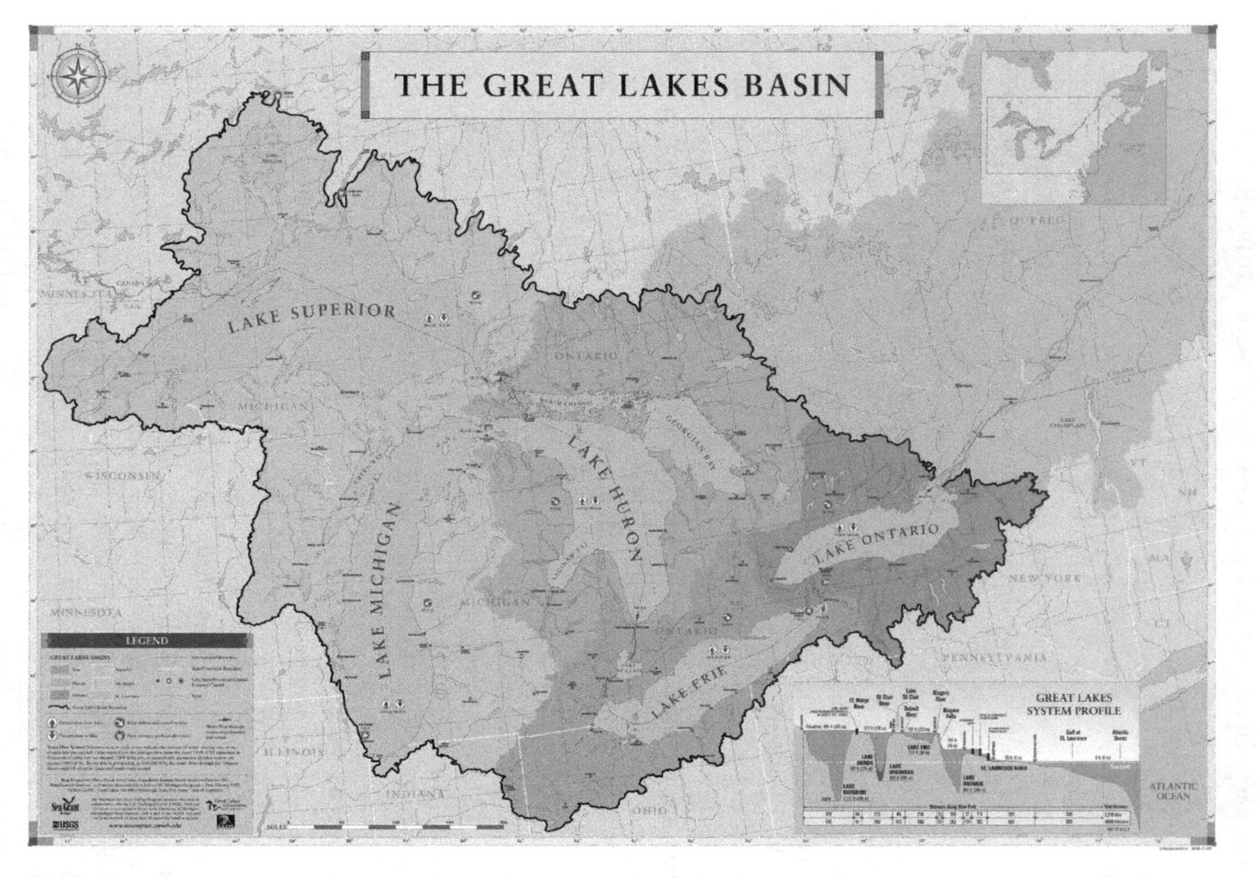

FIGURE 1.1 Great Lakes map. This map is one of the few to show Lake Champlain in Vermont as a tributary of the St. Lawrence River, and to centre the whole watershed rather than emphasizing the US–Canada border. It also shows the lakes' depth. Map credit: Michigan Sea Grant.

The Great Lakes watershed is international (spanning Canada and the United States), large, diverse, and dynamic, so there are many stories to tell which we hope will strike a relevant chord for people in other places too.

The Great Lakes and St. Lawrence watershed is home to more than 30 million people, as well as an estimated 179 native species of fish, 75 native mammals, and a rich biodiversity of birds, reptiles, amphibians, insects, plants, and trees (Freedman 2018). More than 30% of the Canadian population, and about 10% of the US population, lives in the watershed, which produces 25% of Canada's and 7% of US agricultural goods. The Great Lakes store 84% of North America's surface fresh water—21% of the world's supply (US Environmental Protection Agency n.d.). Parts of eight US states and two Canadian provinces are included in the watershed, which is the largest source of surface freshwater in the world (Great Lakes and St. Lawrence Cities Initiative n.d.).

The International Joint Commission (IJC) was formed in 1909 to help mediate conflicts between the United States and Canada related to their Great Lakes and other boundary waters. These mainly involved shipping, dams, and water diversions at first, but later environmental problems arose, such as urban, industrial, and agricultural pollution, algal blooms, lead contamination of drinking water, plastic microbeads, habitat degradation, extinct species such as the passenger pigeon, endangered species, and invasive non-native species such as zebra mussels, Asian carp, and purple loosestrife (Freedman 2018). An IJC poll conducted in 2018 found that an overwhelming majority—88%—of watershed residents feel it is essential to protect the Great Lakes, and most would pay to do so (IJC 2018).

Climate change is already affecting the Great Lakes watershed. Average annual temperatures have increased by 1.1°C in the region, and by 2050 they are projected to rise another 1°C–3°C. The frost-free season increased by 9 days between 1958 and 2012. Annual average ice cover on the lakes declined by 71% between 1973 and 2010. By 2050, in a typical year there will be little significant ice cover even on Lake Superior. Total annual precipitation has increased by about 11% since 1900; in the future, precipitation is expected to fluctuate more widely. Severe storms have increased in frequency and intensity, and this trend is likely to continue. This increases the risk of erosion, agricultural runoff, algal blooms, sewage overflow, transportation problems, flood damage, public health challenges, varying lake water levels, and stress on plants, animals, and ecosystems (Great Lakes Integrated Science Assessments 2014).

These kinds of climate change impacts are unfairly distributed. Economic inequalities throughout the region, high unemployment and poverty rates, gutted housing markets in some cities where industries have closed down, depleted tax bases, and collapsing infrastructure mean that climate resilience is shaky in many parts of the Great Lakes region. While this is a relatively water-rich region, not everyone has access to clean water, which is important both to address current challenges and for future prosperity (Roller 2017).

Actions taken to protect the Great Lakes, and to advance equity, in response to these threats include the creation of the Great Lakes Commons (See Chapter 17 in this book), which was called for in a 2011 report by Maude Barlow and the Council of Canadians (Barlow 2011) and carried out through ongoing initiatives including arts-based public education; collaborative development of the Great Lakes Commons Charter Declaration, with extensive consultation with Indigenous water defenders; and involvement of many organizations and individuals, representing "an unusual and promising alliance of people from across Nations, geography, ancestry and traditions" (Great Lakes Commons n.d.).

The Canadian Environmental Law Association's "Healthy Great Lakes" program is coordinating networks of individuals and organizations to shape and make use of laws protecting the Great Lakes and St. Lawrence basin, including citizen engagement, equitable distribution of government action across the watershed to address the needs of disadvantaged communities, and action-oriented engagement with First Nations and Métis communities (Canadian Environmental Law Association 2017).

Universities and citizen activists are partnering to develop public education, research and policy strategies, such as at the "Untrouble the Waters Summit" at the University of Illinois Freshwater Lab in 2015 (Havrelock 2017), which highlighted that marginalized communities with higher non-white and low-income populations are more likely to suffer from poor water quality (e.g. in Flint, Michigan), deteriorating infrastructure (Chicago, Detroit), poor enforcement of water quality regulations, and less access to policy processes. A climate justice project based at the University of Michigan is analysing social-economic indicators in relation to water and environmental hazards in the Huron River watershed, and developing a way to measure climate justice threats using an index that combines environmental hazards, flooding hazards, and social vulnerability indicators (Mohai et al. 2017). The Global Footprint Network has begun working with York University in Toronto to produce its National Footprint Accounts, which show each country's overall demand on nature and progress toward meeting the Sustainable Development Goals (Global Footprint Network 2018).

Certainly more research is needed to document the climate injustices in the Great Lakes watershed so they can be better understood and tackled. As global warming advances, these injustices are almost certain to increase, since poverty reduces people's climate resilience. Moreover, the "ecological footprint" of consumption by people who live in the Great Lakes watershed extends far beyond the physical boundaries of the watershed—much of the food and other goods we consume is imported from other places, so the related greenhouse gas emissions occur elsewhere—but we are still responsible. Even the emissions which take place within the watershed (from the cars we drive, buildings we heat in the winter and cool in the summer, fertilizer we spread and industrial production in our workplaces) affect the entire Earth's atmosphere. The relative affluence of North Americans who live around the Great Lakes entails a global responsibility for climate justice, both individually and collectively. Awareness of this fact, and what we can do about it, is spreading—as witnessed by the chapters in this book.

These stories also contribute to process research on what works: how climate injustices can successfully be reduced, both from the top-down via policy, and from the bottom-up via grassroots action. For both improved policy and better grassroots understanding, ever more reflection, education, and participatory action research are needed (Bradbury 2019).

The chapters in this book are grouped into sections which roughly correspond to the scale at which the actions they describe are undertaken. The first section, "Fairness in public policies," includes chapters about ways to advance climate justice institutionally, by working within unions, local communities, city governments, and via local, regional, and national government policy advocacy. In Chapter 2, Alia Karim discusses trade union initiatives for a Just Transition to low-carbon economies, along with models for bringing social equity and respect for Indigenous land rights into the climate justice struggle. How the transition beyond fossil fuels can contribute to social redistribution is taken up by Daniel Horen Greenford in Chapter 3, and in Chapter 4, Douglas Baxter outlines community initiatives at the local level to reduce energy poverty through community energy planning. The City of Toronto's strategy to address climate change is the focus of Chapter 5, by Monica Krista De Vera. Chapters 6, 7, and 8 discuss different perspectives on refugee policy and its effectiveness: the "right to remain" by Meagan Dellavilla; how diaspora scholars shed light on refugee options and global policies in the case of Bangladesh, by Nowrin Tabassum; and climate justice lessons from Canada's recent experiences with refugee resettlement, by Michaela Hynie, Prateep Kumar Nayak, Teresa Auntora Gomes, and Ifrah Abdillah. In Chapter 9, Alicia Richins explains the concept of climate debt, and how it summarizes countries' and individuals' relative responsibility for the climate crisis. She also overviews how global financial flows, investment, taxes, and the Green New Deal fit into the climate justice picture.

The second section, "Personal action and local activism," includes chapters describing climate justice actions taken at the personal scale, or that branch out from individual initiatives. In Chapter 10, Aaron Saad overviews the issue of fossil fuel divestment. Food's relation to climate justice is the topic of Chapters 11 and 12, which take up insights for personal food choices (Caitlin Bradley Morgan), and how to address problems with global food markets (Sam Bliss). The final three chapters in this section discuss ways of building climate resilience while advancing social equity at the community level. In Chapter 13, Stephen Clare shows how social capital helps the vulnerable in climate-related crises. How Vermonters built and made use of those community connections is Kelly Hamshaw and JoEllen Calderara's focus in Chapter 14, and in Chapter 15 Laura Gilbert and Claire-Hélène Heese-Boutin outline flooding vulnerabilities and how to address them in Toronto.

The book's final section, "Education, consciousness-raising, and collective visions," brings together chapters which lay out diverse ways to teach, share information, and otherwise contribute to the cultural shift that is making everyone less disempowered regarding climate change and more ready to just pitch in and do something about it. In Chapter 16, Lindsay Gray describes the Toxic Tours at Aamjiwnaang First Nation that introduce many visitors to the "sacrifice zone" around Chemical Valley in Ontario, highlighting both the high costs of fossil fuel-based economies for Indigenous people and their climate justice leadership. Great Lakes Commons, discussed by Paul Baines in Chapter 17, is an organization that sees the Great Lakes and their Indigenous caretakers as inspiration for new forms of organizing to equitably protect the waters and the land. In Chapter 19, Beth Lorimer outlines another watershed-based format for workshops to consider social justice along with environmental and climate action. Progressive educational ideas to foster youth leadership and avoid squelching children's passions in schools are the topic of Chapters 18 and 22—suggestions for climate justice teachers (Barbara Sniderman), and global models for child-friendly schools (Gabriel Yahya Haage and Natália Britto dos Santos). Two more chapters outline special climate justice initiatives in Montreal (Chapter 20, by Jen Gobby) and the Youth Climate Justice Fellows program in Buffalo (Chapter 21, by Lynda H. Schneekloth, Rebekah A. Williams, and Emily Dyett). Finally, in Chapter 23, Alison Adams shows how art inspires climate justice awareness and action, changing our culture by showing us the future and what we can do about it.

The Action Glossary lists and explains strategies and climate justice-related approaches which are mentioned throughout the book. It is meant to serve as both a quick reference on these methods and concepts, and as a cross-link to the chapters where each is discussed in more detail.

I would like to thank all the chapter authors; it has been a joy and a wonderful learning experience to assemble this collection of work from so many dynamic youth activists, community organizers, and creative climate justice leaders. In particular, I would like to thank Nowrin Tabassum, who provided crucially important assistance at a key point with the bibliography, references, and index.

We share these stories in the hope that they will inspire, motivate, and assist others to join the fight for climate justice. Throughout all the chapters runs our shared goal: advancing equity for all in the face of climate change impacts, in the Great Lakes watershed and throughout the Earth.

References

Adger, N. et al. (2006). *Fairness in Adaptation to Climate Change*. MIT Press.

Barlow, M. (2011). "Our Great Lakes Commons: A people's plan to protect the Great Lakes Forever." https://canadians.org/sites/default/files/publications/GreatLakes%20Commons%20report%20-%20final-Mar2011.pdf

Bradbury, H. (2019). "Introducing the AR special issue: ART and climate transformation." *Action Research*, 17(1), March, n.p. https://actionresearchplus.com/introducing-the-ar-special-issue-art-and-climate-transformation/?mc_cid=b13be1bce9&mc_eid=e502303f60 (Accessed 23 March 2019).

Canadian Environmental Law Association (CELA) (2017). "Healthy Great Lakes." www.cela.ca/healthy-great-lakes

Center for Biological Diversity (2017). "The extinction crisis." www.biologicaldiversity.org/programs/biodiversity/elements_of_biodiversity/extinction_crisis/

Corpwatch (2002). "Bali principles of climate justice." https://corpwatch.org/article/bali-principles-climate-justice (Accessed 27 March 2019).

Dow, K., Kasperson, R.E., and Bohn, M. (2006). "Exploring the social justice implications of adaptation and vulnerability." In N. Adger et al. (eds.), *Fairness in Adaptation to Climate Change*. Cambridge, MA: MIT Press, pp. 79–96.

Freedman, E. (2018). "New research tackles Great Lakes regional problems." Great Lakes Echo, February 2. http://greatlakesecho.org/2018/02/02/new-research-tackles-great-lakes-regional-problems/

Global Footprint Network (2018). "Global Footprint Network 2018." www.footprintnetwork.org/2018/04/17/york-university-footprint-un-sustainable-development-goals/ (Accessed 23 March 2019).

Goldring, L. and Krishnamurti, S. (eds.) (2007). *Organizing the Transnational: Labour, Politics, and Social Change*. Vancouver/Toronto: UBC Press.

Great Lakes and St. Lawrence Cities Initiative (n.d.). "About the Great Lakes and St. Lawrence Cities Initiative." https://glslcities.org/about/about-the-great-lakes-and-st-lawrence-region/ (Accessed 21 June 2019).

Great Lakes Commons (n.d.). "Who we are." www.greatlakescommons.org/who-we-are/

Great Lakes Integrated Science Assessments (2014). "Climate change in the Great Lakes region." http://glisa.umich.edu/media/files/GLISA_climate_change_summary.pdf

Hall, R. (2014). "What is climate justice?" Peaceful Uprising, February 4. www.peacefuluprising.org/tag/rebecca-hall

Havrelock, R. (2017). "Great Lakes shared future considered at 'Untrouble the Waters' summit." Great Lakes Connection, July 10. http://ijc.org/greatlakesconnection/en/2017/07/great-lakes-shared-future-considered-untrouble-waters-summit/

Haynes, S. (2019). "'It's literally our future.' Here's what youth climate strikers around the world are planning next." *Time*, March 20. http://time.com/5554775/youth-school-climate-change-strike-action/ (Accessed 23 March 2019).

International Joint Commission (IJC) (2018). "Overwhelming public support for protection of Great Lakes, says new poll." July 10. www.cbc.ca/news/canada/windsor/great-lakes-protection-poll-1.4741665

Jafry, T. (2019). *Routledge Handbook of Climate Justice*. London/New York, NY: Earthscan/Routledge.

Jones, N. (2017). "How the world passed a carbon threshold and why it matters." Yale Environment 360, January 26. https://e360.yale.edu/features/how-the-world-passed-a-carbon-threshold-400ppm-and-why-it-matters

Mohai, P., Cheng, C., Kalcic, M., and Esselman, R. (2017). "Exploring empirical evidence for climate justice in the Huron River watershed." http://graham.umich.edu/activity/28619

Mohamoud, A., A. Kaloga, and S. Kreft (2014). *Climate Change, Development and Migration: An African Diaspora Perspective*. Bonn/Berlin: Germanwatch. www.germanwatch.org/sites/germanwatch.org/files/publication/9112.pdf (Accessed 23 March 2019).

Perkins, P.E. (2019). "Climate justice, gender, and intersectionality." In T. Jafry (ed.), *Routledge Handbook of Climate Justice*. New York/London: Routledge, pp. 349–358.

Roller, Z. (2017). "An equitable water future: Opportunities for the Great Lakes region." U.S. Water Alliance. http://uswateralliance.org/sites/uswateralliance.org/files/publications/uswa_greatlakes_021318_FINAL_RGB.PDF

Saad, A. (2017). "Climate justice: Its meanings, struggles and prospects under liberal democracy and capitalism." PhD dissertation, York University, p. 13. https://yorkspace.library.yorku.ca/xmlui/handle/10315/34301 (Accessed 23 March 2019).

Sers, M. (2017). "Climate justice and the closing window for meaningful change." https://docs.google.com/viewer?a=v&pid=sites&srcid=ZGVmYXVsdGRvbWFpbnxtYXJ0aW5yb2JlcnRzZXJzfGd4OjRlYTMyM2JjYzRhOWE4NDE

Singer, M. (2019). *Climate Change and Social Inequality: The Health and Social Costs of Global Warming*. London/New York, NY: Routledge/Earthscan.

Trosper, R. (2009). *Resilience, reciprocity, and ecological economics: Northwest Coast sustainability*. New York / London: Routledge.

U.S. Environmental Protection Agency (EPA) (n.d.). "Great Lakes facts and figures." www.epa.gov/greatlakes/great-lakes-facts-and-figures

Watts, J. (2019). "Greta Thunberg, schoolgirl climate change warrior: 'Some people can let things go. I can't.'" *The Guardian*, March 11. www.theguardian.com/world/2019/mar/11/greta-thunberg-schoolgirl-climate-change-warrior-some-people-can-let-things-go-i-cant (Accessed 23 March 2019).

Whyte, K.P. (2014). "Indigenous women, climate change impacts, and collective action." *Hypatia*, 29(3), pp. 599–616.

Whyte, K.P. (2017). "Indigenous climate change studies: Indigenizing futures, decolonizing the anthropocene." *English Language Notes*, 55(1–2), pp. 153–162.

Whyte, K.P. (2018a). "Way beyond the lifeboat: An indigenous allegory of climate justice." In D. Munshi, K. Bhavnani, J. Foran, and P. Kurian (eds.), *Climate Futures: Reimagining Global Climate Justice*. Berkeley, CA: University of California Press.

Whyte, K.P. (2018b). "Sovereignty, justice and indigenous peoples: An essay on settler colonialism and collective continuance." In A. Barnhill, T. Doggett, and A. Egan (eds.), *Oxford Handbook of Food Ethics*. Oxford, UK: Oxford University Press, pp. 345–366.

PART I
Fairness in public policies

One way to advance climate justice is to ensure that equity considerations are included in public policies—especially those related to energy, urban form, transportation, taxation, finance, employment, food, infrastructure, education, social services, and immigration—since all these policy areas (and more) involve causes and effects of climate change. For example, reducing government subsidies for the fossil fuel industry (or increasing the stringency and application of environmental regulations governing pipelines and oil/gas extraction) have the effect of raising oil and gas prices for consumers, which can make those fuels less competitive and hasten the energy transition away from fossil fuels. These price effects, however, matter more to poorer people who have to spend a larger proportion of their income on fuels, so the impacts on low-income people should be mitigated *as part of the energy transition policies*. Another example is that reducing urban sprawl lessens the need for water pumping, road construction, and private cars, allowing road and infrastructure taxes to be reduced for everyone while also reducing individual expenditures on automobiles. These policies, however, may be inequitable if they disproportionately affect the jobs, family expenditures, or economic position of lower-income people who have less flexibility than wealthier people to adjust, deal with, and buffer those changes.

Climate change, and the urgency of the needed energy transition, heighten the importance of integrating equity into public policy.

The political sustainability of climate-related policies, as well as their effectiveness and fairness, depends on taking an equity perspective. At the local level, it is possible to gauge options, consult more effectively, and design policies more fairly than at broader, more populous, and complex levels of governance (state/province, regional, national, and international). Citizens with experience in how to apply a justice perspective in local policy processes can help people see how similar processes also play out in other local jurisdictions across the world—and how the impacts of policies at broader scales can have perhaps unintended unjust outcomes. Climate policy engagement can thus provide a hands-on workshop in climate justice.

The chapters in this first section of the book discuss climate justice in relation to government policies of various kinds—at jurisdictional levels from the local to the global.

2

CARBON CUTS, NOT JOB CUTS

Toward a Just Transition in Canada

Alia Karim

For unionized workers accustomed to hard-fought struggles over jobs and wages, climate justice often means trying to protect their economic position in relation to bosses and large corporations, through the turmoil caused by climate change. The rapidity of the needed energy transition away from fossil fuels can leave both labour unions and corporations struggling to define their positions and protect their interests. These struggles also elucidate a range of equity-related challenges brought about by climate change, which are related to climate justice questions more broadly.

In 2016, the largest federation of trade unions in the United States, the American Federation of Labor (AFL-CIO), endorsed the construction of the $3.8 billion Dakota Access Pipeline (DAPL) as the world watched unconscionable acts of militarized brutality unfold against the Standing Rock Sioux Tribe. Immediately after the Tribe and their allies established an encampment in opposition to DAPL, two divergent perspectives emerged concerning its construction. The Tribe and their allies opposed the pipeline in order to protect the Missouri river reservoir and their sacred burial grounds; meanwhile, Energy Transfer L.P., a multibillion-dollar firm, pushed for construction of DAPL as a way to allegedly create 4,500 "high-quality, family-supporting" jobs (Trumka 2016, para 2). In a shocking decision that defied the Tribe and the climate justice movement's urgency to transition away from oil extraction, AFL-CIO President Richard Trumka denounced the Tribe's encampment, arguing that "attacking individual construction projects is neither effective nor fair to the workers involved" (ibid., para 3). To Trumka, the Tribe's resistance of the DAPL was an illegal occupation and a threat to good jobs. His argument resonated with other union leaders, who also repeated the sentiment that opposing pipeline construction meant that they are against providing jobs for working-class people. Laborers' International Union of North America (LIUNA) General President, Terry O'Sullivan, wrote to his members: "Some of our so-called brothers and sisters in the trade union movement have abandoned solidarity with the working class and are instead throwing in with environmentalists who have co-opted the tribes in their effort to fight pipelines" (Gruenburg 2016, para 12). He warned that his union "will not forget the *reprehensible actions*" by the encampment (ibid., emphasis added). In effect, Trumka and supporting unions advanced the false dichotomy of "jobs versus the environment" that pits workers in carbon-based industries against the climate justice movement which has increasingly called for the defence of Indigenous self-determination.

But not all AFLCIO members complied with Trumka's endorsement of the DAPL and showed urgency to challenge this common trope. Trade unions representing nurses, bus drivers, communications workers, and electricians opposed DAPL on the basis that it would infringe on Indigenous peoples' sovereignty over their territories, risk public health and safety, and further exacerbate climate change. Indigenous trade unionists spoke out, such as Melissa Stoner of the Navajo Nation and member of the American Federation of Teachers Local 1474 of the AFL-CIO. Stoner suggested that oil and pipeline workers are "doubly" vulnerable since their jobs rely on volatile, carbon-based industries, but they rely on these jobs to support their livelihoods (Anderson 2016). While she empathized with workers looking for good jobs, she also argued how important water protection is to the Navajo Nation. "When we pray, we pray with water. We bless ourselves with water … Water is healing, and has to be respected," she explained (ibid.). Furthermore, by protecting water and land, she insisted that the Standing Rock Sioux Tribe did not intend to hold jobs hostage or to blame pipeline workers: she remarked that,

> The water protectors are not attacking the workers. This isn't the Wild West … There are highly educated people in the camp that are having the same conversation we are having right now: that the construction workers aren't the bad guys.
>
> *(ibid.)*

Stoner was not the only AFL-CIO member to express support for the Standing Rock Sioux Tribe and the need to transition from carbon-intensive industries. Nearly 16,000 people signed a petition to Trumka in solidarity with the Tribe. As self-described trade unionists and social justice activists, they declared, "Workers' rights are inseparable from Indigenous rights" (Urgent call 2016). In addition, they made a clear connection between the shared interests of Indigenous peoples and the labour movement, as the petition read:

> DAPL continues more than 500 years of settler-colonialism, dispossession, and genocide against Indigenous people in the Americas, who are defending the Earth's vital resources against the *same corporate greed, state violence,* and *repression that violate workers' rights on a daily basis.*
>
> *(ibid., emphasis added)*

The multi-layered conflicts between union leaders, workers, climate justice activists, and Indigenous nations remind us that the intersection of the need for good jobs, climate justice, and Indigenous sovereignty is not easy to navigate. In recent years, the North American labour movement has shown a burgeoning interest in climate justice demands—at best, pushing for a "green" New Deal to address climate change and economic inequality. Yet, many unions still have a hard time parting with the carbon-intensive industries that their members currently rely on for their jobs. This raises an important question: How can we collectively push for *both* high-quality jobs based on renewable energy and social equity goals like the right to Indigenous self-determination?

High-carbon industries and expendable workers

> I'm a heavy equipment mechanic who has worked for several years on the front line of oil and gas carbon extraction. I've seen the impact that resource extraction has on the communities where the carbon deposits are, and I've been a part of the chaos that the up-and-down nature of the oil business can bring to families.
>
> *Clayton Strang, Alberta, Canada*
> *(Iron & Earth 2016, p. 28)*

The North American labour movement has come to rely on carbon-intensive jobs in the emergent colonial-capitalist economy. As mercantile trade expanded and evolved into large, carbon-intensive industries using coal-based steam power, extraction, mining, shipping, and railway construction demanded more and more workers—mostly European immigrants but also a significant number of Indigenous and Asian workers—to keep up with rapid industrial growth. The need for workers in industrial workplaces in World War I and the New-Deal era in the 1940s and 1950s and mobilization of the labour movement enabled trade unions to grow in number. These industrial workers formed the "backbone" of the North American labour movement. Workers resisted overwork, low pay, and extremely unsafe and exploitative working conditions. In Canada, some of the largest and most influential union movements were held by industrial workers in large, carbon-based industries. Notable strikes included The Ford Windsor strike (1945) that saw 10,000 United Auto Workers walk out of the Ford Motor Company factories in Windsor, Ontario to demand a new contract; 5,000 workers who struck for better health conditions and wages in Québec asbestos mines (1949); and the 1978 miners' strike featuring over 2,400 workers at Inco's Sudbury, Ontario mine operations that broke out in response to massive job cuts in their industry (Bradbury 1985; Cruikshank and Kealey 1987).

Carbon-intensive industries, like all capitalist firms, put downward pressure on workers' wages in order to extract as much surplus value (i.e. profit) as possible. But carbon-intensive industries are uniquely positioned as their production is also highly dependent on fluctuations in oil prices affected by the vagaries of international reserves and geopolitics, domestic government subsidies, and bilateral and multilateral trade agreements (Stanford 2008; Weis et al. 2016). The global oil price crash in 2014, along with massive wildfires on Canada's West Coast and extreme weather events like flooding in the Prairies, made Canadian climate justice activists question our reliance on such volatile energy industries, especially to sustain working-class jobs and livelihoods (Klein and Lewis 2015). Nearly 43,000 jobs were lost in mining, forestry, fishing, quarrying, and oil and gas during the oil price downturn between December 2014 to April 2016 (Johnson 2016). Alberta's oil energy sector, reliant on the Fort McMurray Tar Sands—the world's largest megaproject—scaled back production due to low oil prices and climate catastrophe when wildfires forced the temporary closures of production facilities in spring 2016 (Lewis and McCarthy 2016). The losses in these industries rippled across the province as Alberta's overall unemployment rate climbed to 8.6% in July 2016—its highest mark in nearly 22 years (ibid.). Even the fossil fuel-dependent auto industry in Southern Ontario, once a stronghold for good jobs, dramatically downsized and laid off close to 30,000 positions from 2001 to 2015. In 2018, corporate giant General Motors decided to close all assembly operations in Oshawa and move manufacturing to the Global South (Ferreras 2017). The downsizing of these industries has an economic "domino" effect on small businesses who service them, and force workers to move to new industries (often into low-wage service industries) or go into early retirement (Cooling et al. 2015). The extreme resulting economic instability has debilitating impacts on workers and their families, often resulting in drug and alcohol addiction, increased domestic violence, and collapsed housing markets and community activities (ibid.). Entire communities have been ravaged and wiped out by job layoffs in carbon-intensive industries, and many are questioning the ability of the free-market and existing governmental climate policy to meet climate targets and simultaneously generate high-quality jobs with long-term security.

Despite economic setbacks and extreme weather patterns, however, oil corporations, private banks, and investors are intensifying extraction by pressuring governments to approve the expansion of fossil fuel extraction. For instance, the Canadian federal government has aggressively pushed for the expansion of the Alberta Tar Sands bitumen production to foreign markets, even despite its lower returns on investment compared to areas with cheaper production costs (Weis et al. 2016). The Conservative Party significantly invested in subsidies, lobbying, and advertising of the Tar Sands, spending $40 million in 2013 to promote the oil and gas sector (Cheadle 2013). Industrial expansion is lucrative to workers who believe that project agreements will mean a "boom" in good jobs. Canadian Natural Resources Minister Jim Carr estimated the Keystone XL pipeline will provide about 32,000 jobs in Alberta (Gillies 2017). These promises lure workers toward right-wing populist movements, evident in the 2016 US Presidential election when Donald Trump made a protectionist, "America-first" pledge to end the long-term slump in the American coal industry, which he deemed to be a "war on coal" (Wolfgang 2017). His pledge promised to increase domestic mining and manufacturing jobs and oppose free-trade agreements like the North American Free Trade Agreement (NAFTA) for the benefit of American workers above all others. Trump received mass support from coal miners, many of whom said they voted for him based on his promise to generate more industrial jobs (Zito 2017). The election shows that without a vision for climate justice that explicitly addresses the issue of employment, working-class people—particularly those entering the job market—will see climate action as a threat and they will refuse to join the climate justice movement. So how can workers push back to generate low-carbon jobs, engage in meaningful relationships with Indigenous nations and equity-seeking groups, and work toward climate justice?

The vision of a Just Transition

As the North American economy transitions to renewable energy, job losses in the carbon-intensive energy sector will be more severe. Anticipating mass layoffs in these industries, workers in the global labour movement have endorsed an equity-based vision put forth by the Just Transition—a set of economic and policy reforms which recognizes that the rapid transformation from a fossil fuel energy system to one built on renewable energy is imperative but the workers directly affected by this energy transformation must be retrained in new industries and everyone must have a secure livelihood as a part of that process. The Just Transition came into being in the early 1990s when Tony Mazzocchi, Secretary-Treasurer of the Oil, Chemical, and Atomic Workers (OCAW) union in the United States, advocated for financial support and education for workers displaced by environmental protection policies (Mazzocchi 1993). He saw OCAW workers in a similar situation to US veterans who were supported in their transition from military to civilian life. Mazzocchi also recognized how workers interact with nature through their labour and he knew that pollution starts in the workplace and then moves into the community and the natural environment. Therefore, workers and their communities directly affected by this energy transformation need a transition with safe and healthy workplace environments, good jobs and quality of life as a part of that process. He wrote in 1993:

> Paying people to make the transition from one kind of economy to another is not welfare. Those who work with toxic materials on a daily basis … in order to provide the world with the energy and the materials it needs deserve a helping hand to make a new start in life.
>
> (*ibid.* 40)

Mazzocchi pushed for OCAW workers to receive full government support for the education and training required to transition from carbon-intensive industries to renewable energy industries. Soon after, Canadian union activist, Brian Kohler, pushed unions to face the question of climate change and employment. In 1998, he wrote, “The real choice is not jobs or environment.

It is both or neither" and made the first mention of the Just Transition concept in a union newsletter (Kohler 1998). OCAW and the Communications Energy and Paperworkers Union of Canada endorsed Just Transition in 1997, which sparked discussions about how to "reconcile efforts to provide workers with decent jobs and the need to protect the environment" through the Canadian labour movement (ILO 2018). Meanwhile, groups like the Just Transition Alliance in California, a coalition of labour and climate justice groups founded in 1997, resisted air pollution and impacts of mining on land and water use (Newell and Mulvaney 2013). They began to come together explicitly under the call for a Just Transition for workers. In 2001, the Service Employees International Union, the largest union in the United States consisting of healthcare workers, issued an official energy policy that included a call for Just Transition (ibid.).

International Trade Union Confederation (ITUC) policy advisor, Anabella Rosemberg (2010), explained that high-quality jobs do not have to be sacrificed when achieving climate justice goals. "Job losses are not an automatic consequence of climate policies, but the consequence of a lack of investment, social policies and anticipation," she argued (ibid. 134). In accordance with the growing number of debates and activities on climate change, and increased public attention to climate change worldwide, trade unionists began to participate in UN climate discussions. The ITUC released a strong statement to make the link between employment and climate change for the 2007 COP 13 conference:

> The effects on the economy—including on employment—will be catastrophic if ambitious measures are not taken to reduce GHG emissions. While employment protection has often been used by certain developed country governments as a reason for not engaging in GHG emissions reductions, emerging evidence indicates that climate change mitigation has positive net employment effects … Trade unions are aware that certain sectors will suffer from efforts aimed at mitigating climate change. Sectors linked to fossil fuel energy and other energy intensive sectors will be profoundly transformed by emissions reduction policies.
>
> *(ITUC 2007, p. 4)*

ITUC also brought the Just Transition to the global stage by campaigning for its principles to be included in the *2015 Paris Climate Agreement*. It stated that its parties would, "[Take] into account the imperatives of a just transition of the workforce and the creation of decent work and quality jobs in accordance with nationally defined development priorities" (UN 2015). ITUC (2015) responded by stating that the language is a good step to build on, but it does not go far enough since it lacks a defined path of transition for workers. "The race to stabilize the climate has begun but tragically, too many governments still lack ambition for the survival of their people," argued Sharan Burrow, the ITUC General Secretary. Since then, Just Transition policies have been widely proposed in unions and workers' organizations around the globe, although their guidelines have evolved in different jurisdictions. They tend to feature the following recommendations (ILO 2018; Mertins-Kirkwood 2017; Newell and Mulvaney 2013; Rosemberg 2010):

- Major government investment via long-term policies to create new, high-quality "green" jobs in wind power, solar power, tidal power, and geothermal industries
- Social services, including social insurance and universal access to healthcare, renewable energy, water, housing, and sanitation
- Apprenticeships, job placement, and retraining for older workers to transition to "green" industries, and income support programs
- Participatory and democratic decision-making in communities and workplaces
- Climate change adaptation and mitigation strategies specified by region or jurisdiction

Just Transition effectively dispels the "jobs versus environment" trope that good jobs must come at the expense of climate targets. Job gains could be massive in emerging renewable energy industries. For example, the Canadian Centre for Policy Alternatives (CCPA) projects as many as 20,000 potential jobs linked to carbon stewardship in British Columbia's forestry sector (Parfitt 2011).[1] CCPA researcher Ben Parfitt says a central challenge for the industry is that it will have less raw material to work with in future years because of a looming timber supply crisis linked to the mountain pine beetle outbreak and weak reforestation efforts, putting rural communities at risk for their jobs and livelihoods. Parfitt calls for economic diversification for job growth, such as greater secondary forest products manufacturing and maximizing the use of forest industry wood waste in a range of bio-products. Government policies are trying to play catch up but largely miss the need to re-skill workers. Provincial policies like *British Columbia's Climate Leadership Plan* (2016) promises to create up to 66,000 jobs over 10 years, although they do not propose significant jobs education and retraining.[2] In the same year, scholars Robert Pollin and Brian Callaci (2016) developed a Just Transition framework for US workers in domestic fossil fuel production. Looking at employment data of fossil fuel production contraction and accounting for costs of retraining and relocation for laid-off workers,

1 Parfitt (2011) identifies areas for job growth including 2,630 manufacturing jobs turning raw logs that are currently exported into higher-value forest products; 2,400 jobs per year converting usable logs left behind at logging sites (and often burned) into forest products instead; 10,000-plus jobs over time in the production of secondary forest products; and 5,200 seasonal tree-planting and tree nursery jobs, funded by an annual $100 million public investment in reforestation.

2 More than 40,000 of those jobs are in the transportation sector, and a further 20,000 are in agriculture and forestry and other jobs in construction jobs related to new public transit infrastructure.

they estimated that 3 million high-quality jobs could be created over a 20-year period through new investments and basic regional economic development and industrial diversification (ibid.). Ambitious proposals for a Just Transition have inspired unions and community organizations to argue that there is opportunity to transform the economy while creating high-quality jobs. However, such a massive economic transformation brings up necessary questions like, who is going to invest in renewable energy industries, who is the transition going to affect, and what are the processes in negotiating the transition?

A new paradigm of industrial production led by workers?

Just Transition policies often target governments to end fossil fuel subsidies, raise income tax rates on corporations, and increase resource royalties and "progressive" carbon taxes. Some proponents call for a "green" New Deal reminiscent of US President Theodore Roosevelt's creation of federal programs, public work projects, financial reforms, and regulations enacted in response to the 1930's Great Depression. Critics in favour of this approach demand that governments provide stimulus funds for low-carbon, "climate jobs" and the expansion of renewable energy industries, but some scholars also call for "socially responsible" private investment (Albo and Yap 2016). The *Green New Deal* resolution introduced by US Representative Alexandria Ocasio-Cortez in February 2019 greatly raised national public discussion of the need for public investment in climate jobs, but also met backlash from Republican-aligned think tanks and politicians who argued that it would be too expensive (Natter 2019). While it is entirely possible for governments to implement measures of Ocasio-Cortez's *Green New Deal* and similar proposals for a Just Transition, some critics warn that the willingness of state institutions to oversee the transition is unlikely in today's context where government provide massive subsidies for fossil fuel industries and seek to appease private firms who have direct control over energy production and distribution (Carter 2016). Massive institutional coordination is needed but it is not clear how to implement this without challenging the power of private firms who continue to lobby for subsidies and focus on greater and greater profits for their shareholders.

Scholars Peter Newell and Dustin Mulvaney (2013) argue that there is a broader political economy of climate justice that needs to be addressed. They say that the capitalist growth paradigm will always result in exploitative labour practices and the relentless extraction of the natural environment. Efforts to improve worker's rights and wages and ensure ecologically sustainable production, as the Just Transition proposes, will increase the costs of production which clash with the capitalist goal to maximize profits, at the expense of workers' livelihoods. For a Just Transition working toward the political goal of eco-socialism, it's necessary for decisions on production and distribution of energy, and the division of labour, to be made by workers themselves. Eco-socialists who agree with the Just Transition call for collective bargaining to propose carbon-mitigation strategies and job transition plans, a diversified economy based on public ownership of essential social services and worker-owned cooperatives, and democratic participatory processes of socially planned renewable energy production and supply—together with socialist goals like free universal healthcare, education, transit and childcare, and shortening the work day (Albo and Yap 2016).

Already there are groups of workers coming together to think about workers' self-management and retraining programs. Iron & Earth, a group of oil industry workers, developed the *Workers' Climate Plan* (2016) to survey workers and make recommendations to build up the national renewable energy manufacturing sector and their own self-operated Solar Skills Training Program to support upskilling of 1,000 Alberta tradespeople as renewable energy design and installation professionals. These actions by oil workers are not entirely surprising considering the history of workers' push for health and safety. In 1973, for example, 4,000 Shell Oil refinery workers and OCAW members in California, Washington, and Louisiana went on a successful 5-month strike amidst the speedups to fuel the Vietnam War, cutbacks in maintenance, and direct chemical exposure (Weisberg 1973). Their strike was met by mass community actions in solidarity with the workers, such as the creation of Shell Strike Support Groups, work stoppages by longshoremen in San Francisco who refused to unload Shell cargo, cancellations of Shell contracts by city councils, and boycotts of Shell gasoline and products by New York taxi drivers. Around the United States, the slogan "their fight is our fight" spread. The Sierra Club, the largest environmental organization at the time, endorsed the strike (ibid.). The nationwide boycott cut sharply into Shell's sales and the coming together of union and community groups showed the potential to collectively act against low wages and unsafe work environments, inevitably linked to the degradation of workers' bodies and the natural environment. More recently, in 2015, more than 5,200 United Steelworkers of America (USW) (2015) members struck at 11 refineries to protest "dangerous and often deadly" working conditions and the use of temporary contract workers rather than hiring and retraining USW-represented permanent employees. The main issue for workers is that they do not have decision-making power in production, but their ability to strike and coordinate solidarity actions shows huge potential for a "bottom-up" movement where workers lead climate justice actions (ibid.). What will be a major challenge for the future, however, is to build enough grassroots support and momentum for an entirely new paradigm of production, distribution, and livelihoods founded on equitable workers' movements and low-carbon energy production.

Union-community movements in Toronto

The endorsement of Just Transition is gaining momentum in the Canadian labour movement as of late. The Canadian Labour Congress (2015) and the Green Economy Network put forth a proposal to reduce Canada's fossil fuel emissions by 35% over 10 years. Their One Million Climate Jobs campaign launched in

2015, much like the equivalent in the U.K. and South Africa both launched in 2010, demands the federal government to invest in clean renewable energy, build retrofits, and expand public transit and high-speed rail. They detail how spending a mere 5% of Canada's federal budget could create a million jobs and cut emissions by 25% in the next decade. In Toronto, the Good Jobs for All (GJFA) coalition was founded in 2008 with the explicit goal of using Just Transition goals while simultaneously addressing economic and racial inequalities in the job market. Their founding meeting at the Toronto Metro Convention Centre had nearly 1,000 trade unionists, environmentalists, and members of racialized communities (Egan 2016). As long-time unionist Carolyn Egan (2016) described, "They came together recognizing that we were faced with an economic crisis, an environmental crisis and a crisis of equity in our society, and that we needed to build the unity necessary to overcome all three."

Toronto's diverse population is highly polarized based on extreme economic inequality and social inequities in the labour market for youth, new immigrants, Indigenous and racialized peoples. Scholars Cheryl Teelucksingh and Laura Zeglen (2016) argue that the city faces a widening gap between its increasingly racialized population and the job recruitment challenges for workers who have been historically excluded from skilled, unionized jobs. GJFA bases their activism on this social equity goal and they use the Just Transition to strategically target employment for these marginalized groups. In 2016, GJFA and the Toronto Community Benefits Network demanded and won a minimum of 10% of work hours to be done by workers from low-income, racialized neighbourhoods on all future mass transit builds (Egan 2016). They worked in coordination with the Toronto and York Region Labour Council who released their *Greenprint for Greater Toronto*, which outlines reforms on workplace improvements around energy use and waste, enhancing solar and geothermal energy, district cooling and heating, and employment equity (Cartwright 2016). Additionally, in early 2017, GJFA successfully lobbied against privatization of municipal garbage with CUPE Local 416, and they launched a campaign for Toronto hydro to use the subsidies in Ontario's *Green Energy Act* (2009) to expand renewable energy generation, attract "green" manufacturers to Toronto, and create new jobs for racialized workers (Rider and Powell 2017).

GJFA's strategy for change is to build a union-community coalition and lead political and social reforms, such as winning community benefits agreements, that work toward a new paradigm of production, distribution, and livelihoods based on climate justice demands and a worker-led vision. Since the formation of the GJFA, activists in the Toronto labour movement have made links between the struggles for high-quality jobs, social equity, and climate justice. Egan commented on GJFA's future:

> The coalition is continuing to work with racialized communities prioritizing their needs in the fight for climate jobs for all and is pressing city council to adequately fund *Transform TO*, a climate action plan for the City

> of Toronto featuring building retrofits, more mass transit, increased use of sustainable energy sources and waste diversion (from landfills).
>
> *(Egan C. 2017, personal communication, Dec 17)*

Other unionists have expressed bold visions for a Just Transition based on *The Leap Manifesto* (2015), a proposal by Indigenous and non-Indigenous climate justice activists for renewable energy and "energy democracy" under democratic community control, with explicit recognition of Indigenous peoples' sovereignty. The 2017 Canadian Labour Congress convention featured a meeting with about 65 delegates including members of the International Brotherhood of Electrical Workers from both Alberta and Ontario, Fort McMurray oil workers, General Motors autoworkers from Oshawa, and labour councils, postal workers, steelworkers and others, which they called Workers for the Leap (Egan 2017).

Through my experiences organizing in the Fight for $15 and Fairness, a campaign launched in 2015 to demand a $15 an hour minimum wage and better labour protections in Ontario, I have witnessed precarious workers challenge the "jobs versus environment" trope.[3] In our Fight for $15 and Fairness community meetings, organizers have made clear links with the climate justice movement by discussing how the majority of low-wage jobs are already situated in low-carbon industries. The problem is that these precarious jobs sacrifice workers' quality of life. So, while there is a need for growth in unionized, high-quality jobs in renewable energy industries, there is also a need to dramatically improve working conditions for the most precarious in the workforce. Pam Frache, Provincial Coordinator of the Fight for $15 and Fairness explains:

> By linking the climate justice movement with the Fight for $15 and Fairness, we can see that all workers across the carbon spectrum have a role in building a future that is sustainable for the planet and people: workers in high-carbon industries can demand a just transition to jobs that don't pollute their workplace and the environment, and low-paid workers in low-carbon jobs can demand better wages and conditions.
>
> *(Frache P. 2019, personal communication, Mar 5)*

The Fight for $15 and Fairness established a climate justice caucus in 2018 and produced a leaflet called *Climate Justice and the Fight for $15 and Fairness*, which views the campaign demands "through a climate lens" (Lannon V. 2019, personal communication, Mar 5). For example, the leaflet explains how the right to organize into unions gives workers the confidence to challenge employers to

3 In November 2017, the Fight for $15 and Fairness coalition won an historic increase to Ontario's minimum wage to $14.00 an hour on January 1, 2018, and an additional increase to $15.00 on January 1, 2019, 10 paid emergency leave days, and better scheduling in *Bill 148, Fair Workplaces, Better Jobs Act*, 2017 (Government of Ontario 2017).

fight for workplaces that are environmentally safe and sustainable. The caucus believes that pushing for basic labour reforms like increasing the minimum wage could provide an incentive for job growth *outside* of carbon-intensive industries like the oil and gas sector, so that workers have a viable choice to transition from high-carbon to low-carbon industries. The fight for an increase in the minimum wage simultaneously fulfils pressing social equity goals, such as the need for basic workplace protections for low-wage, non-unionized workers—the majority of whom are racialized women, immigrants, Indigenous, and who have injuries and disabilities. These workers are more likely to experience racism, sexism, and Islamophobia. But by winning wage increases in the Fight for $15 and Fairness campaign, low-wage workers have built the confidence to fight for more.

Climate justice activist Naomi Klein observed the connection between low wage, precarious work, and the carbon-based economy in her speech to Canada's largest private-sector union, Unifor, at their founding convention in 2013:

> It's not just boilermakers, pipefitters, construction workers and assembly line workers who get new jobs and purpose in this great transition. There are big parts of our economy that are already low-carbon. They're the parts facing the most disrespect, demeaning attacks and cuts. They happen to be jobs dominated by women, new Canadians, and people of colour. And they're also the sectors we need to expand massively: the care-givers, educators, sanitation workers, and other service sector workers ... Turning low-paying low-carbon jobs into higher-paying jobs is itself a climate solution and should be recognized as such.
>
> *(Klein 2013, para 79–82)*

Klein calls for a Just Transition that addresses the complex goal of generating new high-quality, low-carbon jobs that workers in polluting industries can transition into. However, the goal of this transition is not only to create these new jobs as she argues that the government should invest in the "caring economy"—the vast majority of service sector jobs that are *already situated* in low-carbon industries and not directly advancing climate change. Some examples of jobs in the "caring economy" include educators, healthcare workers (doctors, nurses, paramedics, etc.), careworkers, sanitation workers, government and administrative workers, and other service sector workers. She argues that instead of imposing austerity in the public sector who hosts these jobs, the government needs to make mass reinvestments to fundamentally expand the public sector and ensure that all workers must have access to a high quality of life. Of course, as we know in Canada and elsewhere, the struggle for decent wages and job security has been a long haul for the most marginalized, such as migrant farmworkers and temporary foreign workers—it is an uphill battle to fight for the most basic workplace protections. Klein and Frache insist that these workers must be included in the vision for a more holistic Just Transition.

In Toronto, local labour-community coalitions are trying to implement Klein's theoretical framework in their campaigns. For example, in 2018, organizers of *The Leap Manifesto* and community organizations, such as the Toronto Environmental Alliance, the BlueGreen Alliance, and BlueDot, attended the Fight for $15 and Fairness climate caucus meetings to amplify the climate justice demands within the campaign (Lannon V. 2019, personal communication, Mar 5). As a result of this growing coordination between these groups, and the current media interest in a "green" New Deal, they are seeing a significant growth in interest in their caucus, and more broadly in community discussions for a Just Transition (ibid.). The cases of GFJA and the Fight for $15 and Fairness offer some ways to foster the building blocks of labour-community coalitions which could form the basis of larger climate justice movements that could explicitly address the intersections in social equity and pertinent work and labour issues. However, it should be noted that the ability of these groups to reach a broad scope of workers is very limited. The GFJA started off with widespread interest in meetings but they have not sustained the number of people who participated in their founding meeting. GFJA meetings primarily consist of a few union leaders and community members, without the direction and mass participation of rank-and-file members to guide their collective vision and work. Ontario's Fight for $15 and Fairness has a much larger base of community organizers, anti-poverty activists, unionized and non-unionized workers, and students, but it is a constant struggle to sustain an explicit focus in low-wage labour *and* climate justice, as the connection between them is not immediately apparent to the public. In addition, one major challenge that the broader Canadian labour movement faces is its moments of internal fragmentation and divisions as they struggle to respond to restructuring workplaces and declining union density. These crises are incredibly disorienting for their members and often distract from union engagement in the climate justice movement.

The Just Transition and Indigenous people's right to self-determination

Any vision of a Just Transition must address ongoing colonization of Indigenous peoples that has resulted in their large-scale dispossession, gendered violence, and deplorable living conditions on First Nations reserves. As the struggle over the Dakota Access Pipeline reminds us, industrial and carbon-intensive megaprojects disproportionately affect the livelihoods of Indigenous peoples who live near them. A few of these megaprojects include: Sarnia's "Chemical Valley" that has exposed Aamjiwnaang First Nation to deadly carcinogens from nearby oil refineries; the Alberta Tar Sands extraction in Lubicon Cree territory which has resulted in oil spills and massive ecological destruction on their lands; and hydroelectric projects on Indigenous peoples' territories in Manitoba and Québec that have not only created ecological disasters like flooding, deforestation, contaminated water, and displaced game, but also exacerbated deep social conflicts, such as sexual abuse imposed on Indigenous women by nearby non-Indigenous hydro workers (Luginaah et al. 2010; Wera and Martin 2008). In Canada, the history of the relationships between Indigenous peoples and carbon-intensive industries is one fraught with conflict and violence against Indigenous peoples due to the need for these industries to gain control over land. Indigenous peoples have been forced to endure long, up-hill legal battles to fight rich corporations and the Canadian state in order to regain jurisdiction over their territories. Movements for Indigenous sovereignty gained momentum in the late 1960s and 1970s in response to large-scale resource development in Canada's North, which triggered legislative and political protest by Indigenous-led organizations, such as the National Indian Brotherhood (Coulthard 2014). In the 1973 Calder case, the Nisga'a people in British Columbia sought a declaration that their title had never been extinguished. Later, the Inuvialuit, Gwich'in, and Dene spoke out against the building of the Mackenzie Valley pipeline on their lands, which put forth radical land claims and demands for Indigenous peoples' participation in resource management and self-government (ibid.). Since then, Indigenous nations have staged protests, filed land claims, and won court challenges to gain recognition of their rights to land, such as the 1997 Delgamuukw decision, in which the Supreme Court observed that Aboriginal title constitutes an ancestral right protected by the Constitution Act. Pressure on the Canadian government to protect Indigenous peoples' right to self-determination made Cree Member of Parliament, Romeo Saganash, introduce a private members' bill to introduce the United Nations Declaration on the Rights of Indigenous Peoples (UNDRIP) in 2016 (Galloway 2017). The Canadian government has expressed endorsement of UNDRIP but it has still not been passed, and the state has been heavily criticized by Indigenous nations for not adhering to free, prior, and informed consent, and rights to self-determination before proceeding with industrial megaprojects (Brake 2019). Land claims remain a contentious issue in Canadian politics but they must be a core part of the Just Transition in order to redress the historical exploitation of Indigenous peoples' lands, as well as remaining conflicts of ongoing dispossession and its resulting violence against Indigenous women, and deep racial tensions between Indigenous and non-Indigenous peoples.

Another significant tension in resource development on Indigenous lands is the fact that First Nations do not tend to reap economic profits via resource royalties from industrial megaprojects on their lands. Isolated, rural First Nations like Attawapiskat First Nation in Northern Ontario are exploited by extractive corporations who have gained record profits from resource development but have not negotiated fair impact benefit agreements with First Nations band councils.[4] In late 2011 during the height of the Idle No More movement in Canada, Attawapiskat's complaints came to light against De Beers, the world's

4 Impact benefit agreements are the primary mechanism used to secure jobs and economic benefits from resource extraction on Indigenous territories (Mills 2011).

largest diamond company, who established the Victor diamond mine in 2007 (Celli 2015). De Beers allegedly pulled $2.5 billion worth in diamonds since opening and paid meagre resource royalties to the Government of Ontario, to which Attawapiskat First Nation received nothing. Attawapiskat First Nation Chief, Theresa Spence, went on a hunger strike in Ottawa to bring attention to the community's housing and suicide crises from the neglect by De Beers and the provincial and federal governments. Even before the Victor diamond mine began extraction, it had been revealed that De Beers disposed sewage sludge into the community's lift station, causing a backup that brought toxic damage to houses in Attawapiskat (APTN 2011). Shockingly, during and following Spence's hunger strike, she was consistently blamed by mainstream media for adopting bad governance practices within her band council and accused of failing to negotiate in good faith with De Beers (Quesnel 2017). Extractive corporations are still keen on expanding industrial production to try to lure local, unemployed Indigenous youth by claiming that it will bring them economic prosperity. For example, Canadian-based oil and gas corporation Husky Energy (2019) invested over $1 million in education and skills training for First Nations, and Enbridge (2014) created educational training programs for First Nations located near their projects, such as Whitefish Lake First Nation. While these jobs may benefit some First Nations populations, this does not change the top-down dominance that extractive corporations have over Indigenous peoples. It is uncertain whether First Nations will be afforded fairer impact benefit agreements for industrial projects on their lands in the near future. But it is crucial that a vision for a Just Transition entails the elimination of colonial-capitalist exploitation of Indigenous peoples—they must have the right to freely determine their political status as autonomous nations, and be able to pursue their economic, social, and cultural development.

Despite the plunder against Indigenous peoples for access to their territories, Indigenous climate justice activists are taking practical steps toward enacting a Just Transition. Some First Nations are making small but impactful strides by starting up community-run projects to create more "green" jobs and reduce reliance on fossil fuels. For instance, the Piitapan Solar Project, an 80-panel solar power station in Lubicon Cree territory, was established in 2015 in response to decades of detrimental ecological impacts from the Alberta Tar Sands, which included a massive oil spill in Lubicon territory (Lubicon Solar 2019). As organizer Melina Laboucan-Massimo (2015) explained about the project: "Even in the heart of the tar sands we can build a different kind of economy with clean energy and green jobs without compromising our families and communities." The Piitapan Solar Project falls into the vision of Indigenous activists who argue that the rejection of partnerships with corporate employers and the Canadian state is the only way to ensure Indigenous peoples' self-determining political autonomy. Lubicon Solar, who operates the solar project, aims to empower other Indigenous activists asserting their right to self-determination by donating solar panels to help them generate renewable energy. In 2018, they donated solar panels to the Tiny House Warriors from the Secwepemc Nation, which powered tiny houses along the Kinder Morgan Trans Mountain pipeline route that runs through their territory, ultimately aiming to resist oil and gas extraction that would threaten Indigenous lands and livelihoods (Laboucan-Massimo 2018).

In addition to the courageous work by Indigenous climate justice activists, non-Indigenous trade unionists have made steps necessary for a Just Transition, including addressing racism against Indigenous peoples, negotiating employment equity plans, and supporting Indigenous-led struggles for self-determination. In 2012, the Public Service Alliance of Canada (PSAC) developed an Indigenous peoples' awareness component within its anti-harassment programs, and likewise, the Canadian Union of Public Employees in Saskatchewan developed a similar course with the Saskatchewan Federation of Labour, entitled *Unionism on Turtle Island* (Mills and McCreary 2012). The course "helps to dispel the historical basis of non-Indigenous racism towards Aboriginal people by emphasizing the history of colonialism and links between this history and present-day Aboriginal experiences" and brings to light the features of settler-colonialization in Canada to dispel claims of unfair privilege for Indigenous peoples (ibid., p. 122). Later that year, the Canadian Union of Postal Workers (CUPW) endorsed the Idle No More movement, followed by several other unions who expressed concern with the neglect of Indigenous peoples by the Canadian state. CUPW National President Denis Lemelin wrote in solidarity with Idle No More:

> We recognize the racist and genocidal history of Canada and that the attempts to assimilate and silence Indigenous voices have been rife with failure and abuse. The ongoing theft of Indigenous lands, the refusal to honour agreements made in the British Crown reveal a sadly dishonest and indefensible relationship.
>
> *(CUPW 2012, p. 225)*

Some Indigenous nations and trade unions have come together in united opposition to the Canadian state and rich employers, claiming that their needs for healthy livelihoods are not being addressed—for Indigenous nations, they point to the failure of the Canadian government to enact the right to Indigenous self-determination and improve living conditions for First Nations, whereas trade unionsy demand the Canadian government and employers provide unionized, high-quality jobs. However, their interests have overlapped in the past. For example, in 1975, 20 organizations representing First Nations, trade unions, and community groups adopted a resolution to focus on local control over resource development decision-making in Northwest British Columbia (Mills 2011). The secretary of the Terrace-Kitimat Labour Council argued that industrial development without treaties was akin to working without a contract. More recently, in 2014, several public sector unions and Unifor joined First Nations and environmental groups in opposition to the Northern Gateway pipeline, showing that they are interested in departing from carbon-intensive industries even though their members may benefit from future job opportunities (Cox and O'Keefe

2014). As a member of Unifor Local 950 (2013) claimed, "I personally believe that there is also a growing awareness within the labour movement that 'there are no jobs on a dead planet,' and that we have to radically change our economies and reduce our dependence on fossil fuels." A couple of years later, several trade unionists showed support for Indigenous struggles by fundraising money for the Chippewas of the Thames and Clyde River First Nations in their 2016 Supreme Court challenge to the National Energy Board to allow Alberta Tar Sands and fracking companies to use their land without their consent. Union and student groups also protested the Muskrat Falls hydroelectric megaproject (Stoodley 2016). Furthermore, unions have put their resources into recent campaigns to demand that the Canadian government address the deplorable living conditions on First Nations reserves. On National Aboriginal Day 2016, PSAC launched its Thirsty for Justice campaign, in partnership with the community of Grassy Narrows, to fight for safe public drinking water on reserves (First Nations Drum 2018). Yet, despite this apparent shift in the attitudes of some non-Indigenous trade unionists to move away from fossil fuel production and to defend the rights of Indigenous peoples, scholar Suzanne Mills argues that there are still significant challenges within the Canadian labour movement. Non-Indigenous peoples often believe that concerns about Indigenous land claims and its related conflicts remain separate from, and less important than, bread-and-butter workers' struggles for higher wages and improved benefits. If non-Indigenous workers push for shared struggles with Indigenous nations and overcome these "traditional" beliefs, then Indigenous movements and the Canadian labour movement could be a strong force to push for a vision of Just Transition. Unions will have to address the complex terrain between Indigenous peoples' distinct relationships to land, self-determination, and title to lands and resources, and decisions regarding resource management that include governments and impact benefit agreements.

References

Albo, G. and Yap, L. (2016). "From the tar sands to 'green' jobs? Work and ecological justice." In T. Weis, T. Black, S. D'Arcy, and J. K. Russell (eds.), *A Line in the Tar Sands: Struggles for Environmental Justice*. Oakland, CA: PM Press, pp. 297–309.

Anderson, B. (2016). "Labor leaders support the Dakota Access Pipeline—But this native union member doesn't." http://yesmagazine.org/planet/big-labor-supports-the-dakota-access-pipeline-but-this-native-union-member-doesnt-20161018 (Accessed 21 March 2019).

APTN National News (2011). "De Beers Decision to Dump Sewage into Attawapiskat Played Role in Current Housing Crisis." https://aptnnews.ca/2011/12/13/de-beers-decision-to-dump-sewage-into-attawapiskat-played-role-in-current-housing-crisis/ (Accessed 23 March 2019).

Bradbury, J. H. (1985). "International Movements and Crises in Resource-Oriented Companies: The Case of Inco in the Nickel Sector." *Economic Geography*, 6(2), pp. 129–143.

Brake, J. (2019). "On international day to end racial discrimination, advocates urge swift passage of UNDRIP bill in senate." https://aptnnews.ca/2019/03/21/on-international-day-to-end-racial-discrimination-advocates-urge-swift-passage-of-undrip-bill-in-senate/ (Accessed 22 March 2019).

British Columbia's Climate Leadership Plan (2016). "Province of British Columbia." www2.gov.bc.ca/assets/gov/environment/climate-change/action/clp/-clp_booklet_web.pdf (Accessed 19 March 2019).

Canadian Labour Congress (CLC) (2015). "Making the shift to a green economy." http://canadianlabour.ca/issues-research/making-shift-green-economy (Accessed 20 March 2019).

Canadian Union of Postal Workers (CUPW) (2012). "Letter to Chief Theresa Spence." In The Kino-nda-niimi Collective (eds.), *The Winter We Danced: Voices from the Past, the Future, and the Idle No More Movement*. Winnipeg, MB: ARP Books, pp. 225–226.

Carter, A. V. (2016). "Petro-Capitalism and the Tar Sands." In T. Weis, T. Black, S. D'Arcy, and J. K.Russell (eds.), *A Line in the Tar Sands: Struggles for Environmental Justice*. Oakland, CA: PM Press, pp. 25–35.

Cartwright, J. (2016). "Greenprint for greater Toronto: Working together for climate action, adapting Canadian Work." www.adaptingcanadianwork.ca/digital-library/greenprint-for-greater-toronto-working-together-for-climate-action/ (Accessed 21 March 2019).

Celli, R. (2015). "Diamond royalties a closely guarded secret in Ontario." www.cbc.ca/news/business/diamond-royalties-a-closely-guarded-secret-in-ontario-1.3062006 (Accessed 23 March 2019).

Cheadle, B. (2013). "Ottawa spending $40-million to pitch Canada's natural resources." http://theglobeandmail.com/news/politics/ottawa-spending-40-million-to-pitch-canadas-natural- resources/article15641360/ (Accessed 18 March 2019).

Cooling, K., Lee, M., Daub, S., and Singer, J. (2015). "Just Transition: Creating A green social contract for B.C.'s resource workers." www.policyalternatives.ca/-sites/default/files/uploads/publications/BC%20Office/2015/01/ccpa-bc_JustTransition_web.pdf (Accessed 19 March 2019).

Coulthard, G. S. (2014). *Red Skin, White Masks: Rejecting the Colonial Politics of Recognition*. Minneapolis, MN: University of Minnesota Press.

Cox, E. and O'Keefe, D. (2014). "Kinder Morgan is elephant in room for B.C. federation of labour." http://ricochet.media/en/221/kinder-morgan-is-elephant-in-room-for-bc-federation-of-labour (Accessed 20 March 2019).

Cruikshank, D. and Kealey, G. (1987). "Strikes in Canada, 1891–1950: I. Analysis." *Labour/Le Travail*, 20, pp. 85–145.

Egan, C. (2016). "Toronto community benefits network wins green jobs." http://socialist.ca/node/3367 (Accessed 21 March 2019).

Egan, C. (2017). "Workers for the Leap: A great step forward." http://socialist.ca/node/3221 (Accessed 21 March 2019).

Enbridge. (2014). "Pipeline training program a 'win-win' for Aboriginal community." www.enbridge.com/stories/2014/september/whitefish-lake-pipeline-training (Accessed 18 March 2019).

Ferreras, J. (2017). "GM jobs are just the tip of what Canada has lost to Mexico." http://globalnews.ca/news/3211196/gm-jobs-are-just-the-tip-of-what-canada-has-lost-to-mexico/ (Accessed 17 March 2019).

First Nations Drum (2018). "PSAC continues its Thirsty for Justice Campaign to address water crisis in First Nations communities." www.firstnationsdrum.com/-2018/03/

psac-continues-its-thirsty-for-justice-campaign-to-address-water-crisis-in-first-nations-communities/ (Accessed 22 March 2019).

Galloway, G. (2017). "Ottawa drops objections to UN resolution on Indigenous consent." http://theglobeandmail.com/news/politics/ottawa-drops-objections-to-un-resolution-on-indigenous-consent/article34802902/ (Accessed 20 March 2019).

Gillies, R. (2017). "Despite 'good news' of Keystone XL approval, Canada needs oil customers beyond US: Jim Carr." http://cbc.ca/news/canada/manitoba/-keystone-xl-pipeline-jim-carr-1.4041222 (Accessed 21 March 2019).

Government of Ontario (2017). "Fair Workplaces, Better Jobs Act, 2017, S.O. 2017, c. 22 – Bill 148." www.ontario.ca/laws/statute/S17022 (Accessed 21 March 2019).

Gruenburg, M. (2016). "Unions still disagree on Dakota Access Pipeline." http://peoplesworld.org/article/unions-still-disagree-on-dakota-access-pipeline (Accessed 19 March 2019).

Husky Energy (2019). "Indigenous & community relations." https://huskyenergy.com/responsibility/indigenous-relations/ (Accessed 20 March 2019).

International Labour Organization (ILO) (2018). "Just Transition towards environmentally sustainable economies and societies for all." Available at: www.ilo.org/wcmsp5/-groups/public/---ed_dialogue/---actrav/documents/publication/wcms_647648.pdf (Accessed 18 March 2019).

International Trade Union Confederation (ITUC) (2007). "Trade Union Statement to COP 13 United Nations Framework Convention on Climate Change." www.ituc-csi.org/IMG/pdf/COP13_Statement.pdf (Accessed 18 March 2019).

International Trade Union Confederation (ITUC) (2015). "ITUC response to Paris Climate Summit conclusions." www.ituc-csi.org/ituc-response-to-paris-climate (Accessed 18 March 2019).

Iron & Earth. (2016). "Workers' climate plan report: A blueprint for sustainable jobs and energy climate plan." http://workersclimateplan.ca (Accessed 17 March 2019).

Johnson, T. (2016). "Just how many jobs have been cut in the oilpatch?" http://cbc.ca/news/canada/calgary/oil-patch-layoffs-how-many-1.3665250 (Accessed 21 March 2019).

Klein, N. (2013). "Why unions need to join the climate fight." http://naomiklein.org/articles/2013/09/why-unions-need-join-climate-fight (Accessed 19 March 2019).

Klein, N. and Lewis, A. (2015). "Naomi Klein and Avi Lewis: How the climate crisis can change Canada for the Better." *Toronto Star.* http://thestar.com/opinion/-commentary/2015/09/15/naomi-klein-and-avi-lewis-how-to-avoid-a- climate-catastrophe.html (Accessed 19 March 2019).

Kohler, B. (1998). "Just transition—A labour view of sustainable development." *CEP Journal*, vol. 6, no. 2.

Laboucan-Massimo, M. (2015). "What a Just Transition needs to look like." http://greenpeace.org/canada/en/blog/Blogentry/what-a-just-transition-needs-to-look-like/blog/53109/ (Accessed 20 March 2019).

Laboucan-Massimo, M. (2018). "Is there anything these Secwepemc solar powered tiny houses can't do?" https://davidsuzuki.org/story/can-solar-powered-tiny-houses-stop-a-pipeline/ (Accessed 20 March 2019).

Lewis, J. and McCarthy, S. (2016). "Major oil sands producers scale back output, shut down as wildfire rages." http://theglobeandmail.com/report-on-business/industry-news/energy-and-resources/oil-sands-firms-cut-production-as-fort-mcmurray-wildfire-rages/article29850997/ (Accessed 21 March 2019).

Lubicon Solar (2019). "About: Solar in the Tar Sands." www.lubicon-solar.ca/#about (Accessed 21 March 2019).

Luginaah, I., Smith, K., and Lockridge, A. (2010). "Surrounded by chemical valley and 'living in a bubble': The case of the Aamjiwnaang First Nation, Ontario." *Journal of Environmental Planning and Management*, 53(3), pp. 353–370.

Mazzocchi, T. (1993). "A superfund for workers." *Earth Island Journal*, 9, pp. 40–41.

Mertins-Kirkwood, H. (2017). "Evaluating government plans and actions to reduce GHG emissions in Canada: Just Transition policies." *Adapting Canadian Work and Workplaces Project*, Working Paper # ACW–103.

Mills, S. (2011). 'Beyond the blue and green: The need to consider aboriginal peoples' relationships to resource development in labour-environment campaigns,' *Labor Studies Journal*, 36(1), pp. 104–121.

Mills, S. and McCreary, T. (2012). "Social unionism, partnership and conflict: Union engagement with Aboriginal peoples in Canada." In S. Ross and L. Savage (eds.), *Rethinking the Politics of Labour in Canada*. Halifax, UK: Fernwood Publishing, pp. 116–131.

Natter, A. (2019). "Alexandria Ocasio-Cortez's Green New Deal could cost $93 trillion, group says." www.bloomberg.com/news/articles/2019-02-25/group-sees-ocasio-cortez-s-green-new-deal-costing-93-trillion (Accessed 21 March 2019).

Newell, P., and Mulvaney, D. (2013). 'The political economy of the 'just transition,'' *The Geographic Journal*, 179(2), pp. 132–140.

Parfitt, B. (2011). "Making the case for a carbon focus and green jobs in BC's forest industry." http://policyalternatives.ca/-newsroom/news-releases/bc-missing-out-greener-more-sustainable-jobs-forest-industry-report-finds (Accessed 18 March 2019).

Pollin, R. and Callaci, B. (2016). "The economics of Just Transition: A framework for supporting fossil fuel-dependent workers and communities in the United States." http://peri.umass.edu/economists/robert-pollin/item/762-the-economics-of-just-transition-a-framework-for-supporting-fossil-fuel-dependent-workers-and-communities-in-the-united-states (Accessed 19 March 2019).

Quesnel, J. (2017). "Attawapiskat could have more prosperity, but not by getting local diamond mines to shut down". https://business.financialpost.com/opinion/attawapiskats-refusal-to-support-a-diamond-mine-misses-an-opportunity-for-better-prosperity (Accessed 23 March 2019).

Rider, D. and Powell, B. (2017). "Tory trashes his own plan to privatize waste collection." www.thestar.com/news/gta/2017/01/31/tory-trashes-his-own-plan-to-privatize-waste- collection.html (Accessed 19 March 2019).

Rosemberg, A. (2010). "Building a Just Transition: The linkages between climate change and employment." *International Journal of Labour Research*, 2(2), pp. 125–161.

Stanford, J. (2008). "Staples, deindustrialization, and foreign investment: Canada's economic journey back to the future." *Studies in Political Economy*, 82, pp. 7–34.

Stoodley, A. (2016). "Hundreds rally against austerity in downtown St. John's." http://cbc.ca/news/canada/newfoundland-labrador/rally-march-antiausterity-budget-1.3807022 (Accessed 20 March 2019).

Teelucksingh, C. and Zeglen, L. (2016). "Building Toronto: Achieving social inclusion in Toronto's emerging green economy." https://metcalffoundation.com/-stories/publications/building-toronto/ (Accessed 21 March 2019).

The Leap Manifesto. (2015). "The Leap." www.leapmanifesto.org/wp-content/uploads/Leaplet-digital-NEW.pdf (Accessed 20 March 2019).

Trumka, R. (2016). "Dakota Access Pipeline provides high-quality jobs." www.aflcio.org/press/releases/dakota-access-pipeline-provides-high-quality-jobs (Accessed 21 March 2019).

Unifor Local 950 (2013). "A case against Tar Sands pipelines." www.unifor950.com/case-against-tar-sands-pipelines/ (Accessed 21 March 2019).

United Nations Framework Convention on Climate Change (2015). "United Nations." http://unfccc.int/files/essential_background/convention/application/pdf/-english_paris_agreement.pdf (Accessed 20 March 2019).

United Steelworkers of America (2015). "National oil bargaining talks break down: USW calls for work stoppage at nine oil refineries, plants." Available at: www.usw.org/news/media-center/releases/2015/national-oil-bargaining-talks-break-down-usw-calls-for-work-stoppage-at-nine-oil-refineries-plants (Accessed 17 March 2019).

"Urgent call on the AFL-CIO: Reverse support for the Dakota Access Pipeline" (2016). *Change.org*. www.change.org/p/urgent-call-on-the-afl-cio-reverse-support-for-the-dakota-access-pipeline (Accessed 21 March 2019).

Weis, T., Tobin, B., D'Arcy, S., and Russell, J. K. (2016). "Introduction: Drawing a line in the Tar Sands." In T. Weis, T. Black, S. D'Arcy, and J. K. Russell (eds.), *A Line in the Tar Sands: Struggles for Environmental Justice*. Oakland, CA: PM Press, pp. 1–20.

Weisberg, B. (1973). "Our lives are at stake: Workers fight for health and safety: Shell strike." www.marxists.org/history/usa/workers/ocaw/-1973/shell-strike.pdf (Accessed 17 March 2019).

Wera, R. and Martin, T. (2008). "The way to modern treaties: A review of hydro projects and agreements in Manitoba and Québec." In T. Martin and S. M. Hoffman (eds.), *Power Struggles: Hydro Development and First Nations in Manitoba and Québec*. Winnipeg, MB: University of Manitoba Press, pp. 55–74.

Wolfgang, B. (2017) "White House Declares 'the war on coal is over' as Trump begins unraveling Obama's climate agenda." http://washingtontimes.com/news/-2017/mar/28/white-house-declares-war-coal-over/ (Accessed 20 March 2019).

Zito, S. (2017) "Don't be so quick to dismiss Trump's coal mining initiative." http://nypost.com/2017/06/17/dont-be-so-quick-to-dismiss-trumps-coal-mining-initiative/ (Accessed 20 March 2019).

3

WHY ENDING OIL AND GAS PRODUCTION IN CANADA IS ESSENTIAL TO A JUST TRANSITION BOTH AT HOME AND ABROAD

Daniel Horen Greenford

Introduction: A brief history of extraction in Canada

Canada's economy and national identity were founded upon resource extraction. What began as a network of fur trading posts quickly became a European colony designed to concentrate and export enormous amounts of natural resources. Today, Canada's economy and polity are largely unchanged, but the country has transitioned to extracting mostly oil, gas, metals, minerals, timber, and crops.[1] Also similar to its founding model, many of these resources are taken from the farthest reaches of the North and channelled southward to industrial and urban centres, with the bulk of this bounty being sent to Canada's former, arguably present, and possible future parent states.[2] To put it mildly, Canada does not have a strong history of justice in relation to extraction.

Although the majority of Canada's economy today is found in the "service sector," services are often auxiliary to the extractive industry. Approximately 8% of the economy is based on extraction of non-renewable resources from mining, quarrying, oil and gas extraction, and about 2% comes from agriculture, forestry, fishing, and hunting.[3] These industries form both the historical base of the country's economy and the basis for many of Canada's services—from industries directly implicated in supporting extraction like real estate, food services, and public infrastructure that serves the regions at the extractive frontier, to industries like the banking and financial services that invest and manage much of the capital that's instrumental for extraction. It is also noteworthy that even though oil and gas production only make up about 6% of GDP,[4] half of which is from the tar sands, the Canadian dollar fluctuates in step with oil prices.[5] Suffice it to say—you can't talk about the Canadian economy without talking about oil.

It is crucial to examine the parts of the economy that both drive and rely on the extractive sector, the most important of which is the financial sector. Sitting far upstream of extractive industries, many do not see the financial sector to be part of the extractive sector, but the reality is that much of their business is contingent on resource extraction, and in turn is the engine driving these industries. Globally, 33 international banks have invested $1.9 trillion in fossil fuels since the signing of the Paris Agreement,[6] and financing is on the rise each year even though the scientific and policy community have made it resoundingly clear that we have only 12 years to undertake a rapid drawdown of emissions needed to avoid catastrophic climate change.[7]

After decades of calling for increased investments in oil and gas production, the International Energy Agency (IEA) is finally indicating that investments are in excess of what is needed to match expected demand. During a summit in Ottawa in February of 2019, IEA executive director Fatih Birol publicly urged Canada to scale back its planned expansion of the tar sands (Meyer 2019). Barring the occasional public appeal, the IEA calls for investments in oil and gas extraction that are trillions in excess of what would be needed to meet the Paris Agreement, and would result in production that would exhaust 1.5°C and 2°C carbon budgets by 2023 and 2040, respectively (Muttitt et al. 2018). Whether the IEA will revise its models to match an energy future compatible with Paris remains to be seen.

There is presently a lot of discussion about the "Just Transition," signifying the end of fossil fuels and a shift towards a society that is both sustainably powered and equitable. There are many reasons for why any energy transition humanity attempts should (and even must) be achieved in an equitable manner, for example: extreme and unjustifiable inequality is immoral,[8] wealth inequality drives consumption and likewise emissions (e.g. "keeping up with the Joneses"), and more egalitarian societies are more environmentally sustainable. Indeed, the case for why the energy transition must be a Just Transition is compelling, rather than excess work for an already intractable challenge as many in the mainstream climate policy world believe. The most intuitive and succinct argument is found in the words of Martin Luther King, Jr.: "You can't reach good ends through evil means, because the means represent the seed and the end represents the tree."[9] Unfortunately, this chapter does not allow for a longer digression. We will now focus on what the Just Transition means for everyday Canadians as well as the world's most marginalized groups. The Just Transition implies the end of both fossil fuel consumption (demand-side) and production (supply-side). We will explore the climate justice ramifications of supply-side measures to constrain fossil fuel production in Canada.

In countries like Canada—where a significant amount of individuals' and the government's income depends on oil and gas extraction—the conversation invariably ends up at jobs and revenue. The discussion boils down to short-term vs. medium-term arguments. It is true that transitioning away from fossil fuel extractivism will unavoidably affect the lives of workers, but not necessarily in a negative way. Government revenue and spending will have to be fundamentally restructured, but it is neither impractical nor undesirable to do so. Propping up a failing industry like that of oil production in the Alberta tar sands may keep many well-paying jobs from disappearing—but for how long, and at what cost to our economy and the earth's climate?

Exiting fossil fuel production now may be politically challenging, but the alternative is unacceptably worse. Delaying the transition out of oil and gas production to more economically and environmentally sustainable industries will further contribute to the suffering of communities most vulnerable to the impacts of climate change, and oil and gas workers will still ultimately be forced out of their jobs as global oil demand wanes. Many leading scientists and economists have called for an immediate start to the Just Transition, beginning with a moratorium on new oil production in the tar sands.[10] Phasing out oil and gas production in Canada is a prerequisite for safeguarding the economy, the environment, and the climate.

Canadians have already witnessed the results of doubling down on the oil and gas sector—a decision made during the Harper administration (2006–2015) and respected by the Trudeau administration. Unrestricted oil and gas production has created an enormous amount of wealth in the short term for many workers, but is also responsible for the precarity and distress of these same people. And now that the world is finally attempting to embark on an energy transition (not necessarily a just one), it will mean even more hardship for this sunset industry. Canadian oil resources are amongst the most marginal in the world, and therefore will be among the first to be abandoned as most reserves become obsolete (McGlade and Ekins 2015), rendering jobs obsolete too. Pair this reality with the rise of automation in the oil and gas sector, and workers stand to bear the brunt of the economic sacrifice soon to be foisted upon the industry. The decision here is whether Canada will prepare adequately for this eventuality or not.

For these economic reasons alone, the case for a voluntary transition away from oil production towards more sustainable livelihoods should be compelling. In an ecological society, environmental sustainability should be synonymous with economic sustainability. The Just Transition in Canada—particularly for oil workers in the tar sands—presents an unrivalled opportunity to demonstrate what this might look like in practice. It also can be achieved relatively soon, and must be, if we are to avoid warming in excess of 1.5°C or even 2°C above preindustrial times. By even conservative estimates, we have already extracted almost all the burnable oil from the tar sands.[11] So what are some practical steps we can take to get to an equitable post-extraction economy in Canada?

Canada's dependence on oil extraction

The Canadian economy, as just discussed, is one founded on extractivism, with this legacy continuing in many ways today. Oil and gas production does contribute substantially to Canada's economy, but perhaps less than many Canadians (especially Albertans) would surmise from discourse opining the economic indispensability of the sector. As we noted earlier, while 6% of Canada's GDP comes from oil and gas, only 3% of that is from oil produced from tar sands. Services that support oil and gas extraction comprise an additional 1% of GDP. So together, at 7%, oil and gas with auxiliary services make up a substantial base, tying with construction and finance and insurance, only slightly less foundational than the real estate (13%) and manufacturing (10%) sectors.

The oil economy runs much deeper than its geological reserves. In Canada, other sectors such as auxiliary services, banking and financial industries, and even public spending on infrastructure as varied as roads and social services like schools and hospitals, are all built on top of, and to serve, the extractive frontier.

Indirect economic impacts are touted by the industry as reasons to support oil and gas extraction in Canada. Industry reports that oil and gas production employs 420,000 people in direct and indirect capacities, and creates $44.1 billion in value added each year (KPMG-SECOR 2013). Though significant, the majority of these cascading impacts—like manufacturing of machinery used in excavation and refining, or services like geological expertise—are confined to the provinces with the bulk of extraction. Oil and gas production amount to 11% and 5% of Alberta and Saskatchewan's GDPs, respectively—Alberta roughly double that of the national average but still nowhere close to a majority share of its prairie economy. Note that these contributions to provincial output are on par with the national average; the prairie economies are still less diverse than eastern provinces, but far more than the petrostates of the Persian Gulf and South America. Other provinces all benefit far less—indirect activities accounting for less than 1% of their GDPs. Two-thirds of this economic impact is from investment in new production, while the remaining third is from existing operations. This means that as soon as demand for Canadian oil and gas starts to decline, the economic impact of the sector will also drop precipitously, cut down overnight to less than a third of its current size.

These jobs and wealth have become an integral part of the Canadian economy—most critically for the prairie oil-producing provinces—but this does not mean they must, or even can, continue to be. Nor does this imply that the industry's rise to prominence was due to either chance or merit. A series of deliberate decisions made by industry and politicians over the last half-century have resulted in an oil economy and attendant cultural identity in the prairies (e.g. Edmonton's hockey team was founded in 1972 as the "Alberta Oilers"). That said, oil production is still a relatively new enterprise which has a bleak future for both its volatility as well as its local environmental and global climate impacts.

Research on transitioning to a post-carbon energy economy demonstrates that more long-lasting, well-paying jobs would be created in the switch. Canada can create 375,000 permanent jobs operating renewable energy infrastructure (Jacobson et al. 2017),[12] meaning most direct and indirect jobs would be replaced by those in or related to the domestic renewable energy sector. The remainder of jobs could also come from renewable energy, for export. Just as Canada is an energy exporter now in the form of oil and gas, there's no reason the country can't continue being one in a post-carbon energy system. Given that renewables create more employment per unit of power, Canada can employ even more people than presently work in the oil and gas industry if it prioritizes a strong renewables sector.

This still says nothing about the part of the economy that is harder to isolate but just as dependent on extraction as the extractive industry itself. Consider the financial markets in Canada. It is impossible to invest broadly in Canada's economy without owning oil and gas in some way, or owning stock in industries that drive extraction. The Toronto Stock Exchange (TSX) is home to over 1,500 companies with a combined value of nearly $3 trillion. Extractive industries (oil and gas, mining, and auxiliary services to both sectors) comprise a quarter of the listed value on the TSX. Of this oil and gas, and pipelines, make up 8% and 7%, respectively. Financial services and Exchange Traded Funds make up 29% and 6% of the market, respectively. Exactly how much of this 35% market share is invested in oil and gas extraction in Canada is uncertain; however, the bulk of financing for these industries does come from Canadian financial institutions. The Royal Bank of Canada is the country's largest stakeholder, with over $100 billion invested in fossil fuels, and is the biggest funder of the tar sands.[13] Other Canadian banks are also heavily implicated in domestic extraction.

Most of the resources we extract leave our country without any value being added. As mentioned earlier, this "rip and ship" economic strategy harks back to the old days. Take oil, for example—Canada is a net exporter of petroleum, producing about 5% of the world supply, while consuming 2.5% of it, with virtually all exports going to our southern neighbour, the United States. This doesn't mean we use what we produce and ship the rest off: 79% of oil production is for export, and we import 63% of what we consume. Why is this the case? First, the North American Free Trade Agreement (NAFTA) contained a proportionality clause that forces Canada to export oil and gas at above 74% and 52% of its production, respectively, even if it experiences supply shortages—a problematic condition that is a bad deal for Canadians and the environment. This clause has been scrapped in the new trade deal which may replace NAFTA in the coming months. However, most of the oil extracted in Canada will continue to be exported to the United States until we've built up our domestic refining capacity, since this is more economical given the abundance of capacity for upgrading and refining heavy oil that already exists in the United States.

Canada could become energy independent while having more control over its energy transition if it refined its own oil for domestic consumption. Green Party leader Elizabeth May has proposed that we refine and use our own oil domestically. This approach is not pro-oil, but a pragmatic move for the Canadian economy. If we went one (big) step further, the smart thing would be to nationalize the oil industry, buying (or expropriating) the remaining tar sands operations so Canadians could own and operate them during the managed decline of the industry. We would receive all revenue from extraction, be able to protect workers, and employ more people by adding more value to the product while also having more control over supply, positioning the country to ease off production quickly but reliably to best mitigate the economic repercussions of this structural economic shift from fossil fuels to post-carbon industry. To ensure a Just Transition at home, this should be paired with worker retraining or "upskilling" programs, employment insurance, and retrofitting the manufacturing sector to build components for photovoltaics, wind power, and electric vehicles. Much of this worker retraining and facility retooling would be relatively modest, since so much of the basic manufacturing infrastructure already exists. And all this could be paid for with subsidies now going to support oil and gas production (see "How subsidies perpetuate colonialism and hurt the poor").

In sum, Canada's economy is very much tied up in oil and gas extraction, and most of this product isn't even for domestic consumption. But what about the sector's overall contribution to revenue? When you look at the rents from oil production—"rents" being the profit made on a resource before taxes or royalties—the story of Canada's dependence on oil revenue is even less compelling. For comparison, countries who produce their oil at lower costs derive more wealth—Kuwait, Iraq, and Saudi Arabia derive 44%, 42%, and 26% of their country's GDP from oil rents, while Norway, the UK, and Canada's rents only make up 3.8%, 0.29%, and 0.25% of their GDP.[14] For Canada specifically, as a very rough estimate, at a corporate tax rate of 15%, Canada would collect $750 million from oil rents when GDP is $2 trillion. If Canada taxed oil rents at 100%, as Norway has done to build its sovereign wealth fund that is responsible for supporting its premier welfare state, it would instead collect $7 billion per year. This revenue would be very useful for our own welfare state as well as funding the transition.

The case for fossil fuel producer responsibility

There are many reasons why fossil fuel producers should shoulder much—if not most—of the responsibility for climate change. These reasons can be broadly split into three categories: the moral, the political, and the economic. As one may expect, these dimensions overlap considerably.

First, there is a difference between "retrospective" and "prospective" responsibility. Retrospective responsibility looks back at what has already come to pass, and judges historical actions, or "fault." Prospective responsibility looks ahead at what has yet to occur to assess the onus to act, often referred to simply as "responsibility." Both can be connoted in the positive or the negative, however, most people tend to emphasize fault when looking backwards, and responsibility in the affirmative when looking to the future (as in "you have a responsibility to act").

When we speak about climate change specifically, historic fault and responsibility to act are two sides of the same coin. Those who have emitted the most are most at fault for existing levels of climate change as well as most capable of doing something to address it, since they have reaped the economic and political benefits from fossil-fueled industrialization—which implies their responsibility to address it as well (Gardiner 2010, 2015; Hayward 2012; Caney 2010; Garvey 2008).

Additionally, as we know, those who have contributed least to the problem will suffer the most from climate change since they are amongst the poorest nations and therefore ill-equipped to adapt, as well as often the most geographically vulnerable to the impacts of climate change (Althor et al. 2016). A recent report by Philip Alston, the U.N. special rapporteur on extreme poverty and human rights, warns that the world is heading towards "climate apartheid," showing that although accounting for only 10% of global emissions, the world's poorest half of the population are set to incur at least 75% of the costs of climate change.[15] Conversely, those who have contributed most to the problem stand to suffer the least and are much better positioned to adapt to whatever impacts their nations may face. This reverse correlation between negative behaviour and negative consequences sets the harsh tone of injustice of the climate crisis.

When it comes to the entities that have and continue to produce these fuels, it is appropriate to take aim at the fossil fuel industry (Frumhoff et al. 2015). The case for their historic culpability is compelling. Perhaps most damning is how major fossil fuel corporations have actively misled the public and policymakers for decades, with recent evidence showing that they even suppressed internal research that confirmed the findings of the scientific community (Supran and Oreskes 2017), while instead spending trillions on new extraction, and billions lobbying government to quash climate policy (Brulle 2018) and disseminating misinformation to cast doubt in the public mind (Oreskes and Conway 2010).

The fossil fuel industry is also the originator of the vast majority of historic emissions—63% of historical carbon dioxide and methane emissions can be traced back to 90 fossil fuel and cement producers, a third of these emissions being from investor-owned producers and two-thirds from state-owned producers (Heede 2013). Finally, these businesses also had more than enough opportunity to change their behaviour and restructure their business models to produce energy in a sustainable, climate-safe way. Given that these corporations a) knew the effects of fossil fuel combustion on the climate and b) hid their own research and actively misled the public to delay climate action as much as possible so that they could continue profiting from a dangerous business, it follows that these corporations are responsible for much (if not all) of these historic emissions.

Whether they consent or not, Canadians are also implicated in oil and gas production. Billions of public dollars are spent every year supporting the industry (see "How subsidies perpetuate colonialism and hurt the poor"), and the federal government's purchase of the Trans Mountain Expansion (TMX) pipeline project (announced in May of 2018) has made the public's stake in the industry increasingly visible.

The TMX purchase has also helped to bring the climate impact of the oil and gas industry in Canada to centre stage. By increasing supply by 590,000 barrels a day, the emissions embodied in the oil shipped by this expanded pipeline would be approximately 6 billion tonnes (Gt) of carbon dioxide equivalents (CO_2e)

over its 40-year lifetime.[16] When accounting for oil displaced by this increase in supply, global consumption would rise 0.4 barrels for every barrel of added supply, resulting in a net increase of 2.5 Gt CO_2e globally.[17] The upstream emissions alone would be 0.75 Gt CO_2e of this increase, meaning the majority of the pipeline's climate impact would be due to emissions from combustion of this oil occurring *outside our borders*. For comparison, Canada's annual emissions are on par with what would be the proposed project's cumulative upstream emissions (750 million tonnes (Mt) CO_2e), and the United States' annual emissions are on the order of the full 6 Gt CO_2e which would be embodied in oil transported by the pipeline over its lifetime, if built.

How subsidies perpetuate colonialism and hurt the poor (or why the Just Transition is about international equity as much as it is about protecting workers)

Fossil fuel subsidies present one of the most obvious barriers to the energy transition. Globally, countries continue to spend anywhere from $373 billion to nearly $500 billion per year of public money to support the production of coal, oil, and gas production (OECD 2018). While these subsidies have declined over the last decade, this has been due to declining support within developing countries, while developed nations have plateaued. The vast majority of these subsidies come from G20 nations, reported to be $444 billion in 2015 (Bast et al. 2015). Public money makes fuels viable that otherwise would not be lucrative enough to produce. In the United States, nearly half of oil is made profitable by subsidies (Erickson et al. 2017). This oil is equivalent to 6 Gt CO_2—on par with the annual emissions of the United States itself. Phasing out these subsidies could make the difference between continued inaction and a rapid comprehensive transition to renewable or other carbon-free energy sources.

This year marks the tenth anniversary of the G20 commitment to phase out "inefficient" fossil fuel subsidies. Despite renewed vows, Canada has made limited progress and remains the largest provider of fiscal support to oil and gas production in the G7 relative to its economic size.[18] This raises the question of whether there is such a thing as an "efficient" subsidy, since any subsidy to fossil fuel production still incentivizes the production of something which must be curtailed[19].

Canada's latest budget reaffirms its commitment to phasing out inefficient subsidies: "Canada will continue to review measures that could be considered inefficient fossil fuel subsidies with a view to reforming them as necessary."[20] Such subsidies are defined in the budget as those that "encourage wasteful consumption, impede investment in clean energy sources and undermine efforts to fight the threat of climate change." Given that most subsidies in Canada support production of oil from bitumen reserves, and virtually all of such oil is in excess of what is allowed under a 1.5°C to 2°C carbon budget, it follows that all such subsidies are inefficient by the above definition.

In order to better understand the pernicious effects of oil subsidies, it is worth exploring other examples of subsidization. Agricultural subsidies are considered by many to be the most pernicious form of financial intervention. Undoubtedly, there are many positive impacts of subsidizing food production, and financial support can be used to protect both farmers and the people they feed. But what happens when subsidized products from wealthy nations enter markets of those that can't compete with deflated prices?

Subsidies to food production can cause harm both at home and abroad. Take American corn, for example. A highly subsidized crop, it is a major ingredient in processed foods, much of it in the form of high-fructose corn syrup. This artificially lowers the prices of junk foods, making them more accessible to the poorest in America. It is therefore no coincidence that these cheap and highly addictive foods inflict a disproportionate segment of America's poor with ailments like obesity and heart disease. The analogy here is clear — by making something dangerous cheaper, you incentivise the consumption of something that harms poor and vulnerable people. The damage inflicted by cheap oil is perhaps more difficult to directly attribute, but no less real.

The harmful impact of subsidies on food (and energy) sovereignty and security are discussed more often. There are many countries that rely on domestic production for both domestic consumption and export revenues. (This is true of both food and oil.) Most developing countries cannot afford to subsidize their own industries, at least not on the level of developed countries. Exports of staple crops with artificially lower prices undermine food security and sovereignty of the global poor. For example, when cheap (or free) American rice floods markets in the Global South, it undermines the ability for these nations to produce their own food since there are already cheaper alternatives available in adequate abundance. For countries that rely on revenue from domestic consumption and exports of surplus food, cheap food imports undermine their food sovereignty and destabilize their economy. Similar to cases where agricultural subsidies of one nation undermine the food sovereignty of another, oil subsidies in wealthy nations can also undermine the energy sovereignty or economic stability of poorer ones. Every barrel Canada exports of its crude competes directly with other oil from countries, and often, these countries depend on the revenue more.

This kind of trade domination becomes a form of colonialism. As wealthy nations grow their market share, they do so to the detriment of the nations they undercut with their subsidized product. Poorer nations then become increasingly dependent on the rich for economic security. This dynamic is accompanied and exacerbated by predatory lending to embattled countries. Usurious interest rates and stipulations to deregulate economies make weakened economies more vulnerable to the interests of foreign capital. Resources, labour, and wealth continue to flow upward from Global South to Global North, from poor to rich, in keeping with our "reverse Robin Hood" history of colonialism.

Also similar to agricultural subsidies, countries that support fossil fuel production make it more cheaply available, encouraging the poor to choose it over "healthier" options, though in this case the results are global. If developed nations intend to lower global emissions, they must help provide energy solutions to abate emissions, not abet their growth. Of course, this will mean forgoing rents on what are seen to be otherwise viable reserves.

Canada at present continues to subsidize the oil and gas industry with billions of public dollars. Pinpointing the exact figure is difficult, as much of the subsidies come in the form of tax and royalty exemptions that increase with production. Recent estimates show that between federal and provincial support, Canadians could spend up to *$5.7 billion subsidizing oil and gas this year*, rising from a baseline of $3.3 billion (Touchette and Gass 2018)[21] with the addition of $0.8 billion new subsidies from the Alberta government (Environmental Defence and Gass 2019) and $1.6 billion in additional financial support from the federal government.[22] This brings Alberta's contribution to roughly $2 billion each year, and doubles the federal government's existing support of oil and gas production.

This figure still doesn't include other tacit subsidies that industry makes use of. For oil and gas, these come in many forms; for example, as environmental protection, clean up and reclamation efforts, or disaster response programs. The Oceans Protection Plan came at the cost of $1.5 billion (repeating every 5 years),[23] and was in direct response to concerns that the Kinder Morgan Trans Mountain Expansion would greatly increase the risk of spills from tanker traffic. It promises to protect Vancouver's harbour and other ports where oil is shipped by a tanker, despite the fact that cleaning up waterborne bitumen is not feasible (National Academies of Sciences 2016). There is also $2.1 billion in financial assurances set aside for any land-based spills from the pipeline.

Finally, oil and gas transport infrastructure itself, like pipelines and rail, may also qualify as substantial subsidies. The highest profile public intervention to date has been the federal government's purchase of the Trans Mountain Expansion Project, which will cost anywhere from $4.5 billion to $9 billion[24], but the government asserts this measure is temporary, and that the project will be sold back to a private operator.[25] Until then, taxpayers are now the official owners of this controversial energy project. If the project were ultimately returned to the private sector, but at a loss, the amount of this loss would constitute a subsidy to the oil industry. Less conspicuously, the Alberta government has recently leased $3.7 billion worth of new rail services to increase oil transport by rail by 100,000 barrels a day.[26] This is money the public will never recoup, but the government promises that revenues from the oil this rail brings to the Gulf of Mexico will make the venture profitable. The same cannot be said about subsidies unilaterally, and Canadians should be asking how government revenues compare to the subsidies paid out for the industry as a whole. Other public contributions include all the other infrastructural accoutrements supplied to develop the north such as roads, schools, and hospitals at the extractive frontier. After tallying these projects up, Canadian taxpayers could be providing around $12 billion more to bolster this ailing industry. The federal and Albertan governments have shown they will spare no expense when it comes to saving oil and gas production in

Canada—but when Canadians ask where the money for the energy transition is, it's nowhere to be found.

We have already discussed the general equity implications of these kinds of subsidies. What are the global ramifications of Canada's oil and gas subsidies? Many nations rely more heavily on oil revenues than Canada does. Consider any developing country whose economy is implicated in fossil fuel production to the point where it would cease to function without it (sometimes called a "petrostate"). Nigeria and Venezuela are two prominent examples. Compare Canada to these African and Latin American oil producers. In Nigeria, oil revenue accounts for 40% of GDP and 80% of government earnings, and therefore the country is far more dependent on oil production than Canada is. Venezuela produces extra-heavy oil similar in carbon content and breakeven price to that originating in the Western Canadian Sedimentary Basin (sometimes slightly more efficiently in both regards than Canada). Venezuela also relies heavily on oil exports to power its economy, with nearly 17% of its GDP coming from oil exports.

When these countries are excluded from global markets, their citizens suffer. This is particularly acute today for Venezuela, whose government supports its social welfare state with export revenue. The country now finds itself under heavy economic sanctions imposed by the United States that restrict its ability to sell crude oil or use the proceeds from past sales. The toll on the country's economy when trade and revenue are cut off is severe, and hurts everyday Venezuelans. Canada could help diplomatically by condemning U.S. sanctions, and materially by limiting exports of competing crude oil. This could be achieved in part by using more of our own product domestically, which as discussed earlier has been and continues to be a missed opportunity for the Canadian economy. Phasing down production would leave more room for Venezuelan oil to be sold abroad during the global energy transition.

Oil production is not a zero-sum game, but new supply does displace some existing production. When wealthy nations subsidize their oil, and then increase global supplies when exporting it to other markets, they do so to the detriment of countries that sell oil of similar quality and price, since they rarely can compete with this subsidized product. Further, in a world where global demand wanes as countries decarbonize their energy systems in an effort to curtail global emissions, this contraction squeezes producers even more, and competition between "have" and "have not" countries will only intensify. Equity should be a part of decisions on who should be allowed to extract what little of fossil fuel reserves remain, under the Paris Agreement (Piggot et al. 2017; Lazarus et al. 2015; Erickson et al. 2018; Kartha et al. 2016; Kartha et al. 2018).

Exiting the fossil fuel economy (and why Canada must lead)

The task of moving Canada beyond oil is certainly a daunting challenge, but the country is well positioned to make the transition. Canada is one of the wealthiest countries on earth, and has one of the highest levels of post-secondary education too. With financial prosperity and a highly educated population, Canada is in a position to choose its future. Compare this to countries far more entrapped in the fossil fuel economy.

As much as it may be desirable to be able to act locally (here, for example, as citizens of the Great Lakes bioregion), climate change and climate justice demand a global perspective and collective action that targets problems regardless of location. Canada would be meeting its (paltry and wholly inadequate) GHG emissions targets if it weren't for the tar sands (Saxifrage 2018). Oil and gas production is the largest contributor to Canadian GHG emissions, and oil production from tar sands remains the fastest-growing single source of emissions in the country. If Alberta were a country, its per capita emissions, at about 70 tonnes of carbon dioxide equivalents (t/CO_2e), would be the highest in the world, much higher than even any small oil-producing state like Kuwait (55 t/CO_2e) or Qatar (37 t/CO_2e). Environment and Climate Change Canada now predicts that Canada will miss its 2030 target by 103Mt CO_2e, with oil and gas emissions now representing 32% of national emissions by 2030 (Environment and Climate Change Canada 2019). This gap has continued to grow from 66Mt and 44Mt in models from 2018 and 2016, respectively (Weber 2018). And this says nothing of the emissions we export, embodied in every barrel sold to other countries. Emissions from combusting our exported oil *are equal to our territorial emissions* (Lee 2017).

Developing more fossil fuel reserves—especially "reserves" of marginal value and high emissions like Canada's vast bitumen stores—is simply not an option. Canada's contribution to climate action must be to abandon future extraction and immediately begin phasing out oil and gas production. Those now employed in the sector could be protected with wage insurance, and retrained to work in the burgeoning industries of a more humane and ecological world. This would take modest effort and cost roughly what we spend presently subsidizing the oil and gas industry (IISD 2019). When tens of billions continue to be spent propping up an outdated industry, it has become clear that the only real impediment to a Just Transition in Canada is a lack of vision and will to make it so. In the 20th century, the world mobilized entire economics to fight global wars; it shouldn't take any longer to restructure our economy now during peacetime to combat a danger on par with the greatest dangers to our collective existence.

The 2019 Canadian federal budget did earmark money for beginning the Just Transition for those employed in coal power generation. Federal investments of $35 million over 5 years will fund worker transition centres "that will offer skills development initiatives and economic and community diversification activities in western and eastern Canada" (Government of Canada 2019:88). The program also sets aside $150 million for an infrastructure fund "to support priority projects and economic diversification in impacted communities" (ibid.). The federal government also plans to work with affected workers to protect wages and pensions.

This program is a tremendous symbolic achievement, as it opens the door to a real discussion (in an election year for both Alberta and Canada) about measures specifically needed to transition Canada out of fossil fuel production and consumption. Why haven't any funds been directed towards beginning a just transition out of the oil and gas production sector? Instead of following this logical course, and exercising strong leadership for the future of the country, the federal and Albertan governments have instead doubled down on their industry support. Billions, not millions, will be needed to transition and diversify Canada's economy. Canadians must demand that government at all levels shift priorities and divert oil and gas subsidies to the Just Transition effort.

Prime Minister Justin Trudeau echoed the sentiment of his predecessor Stephen Harper when he said, "No country would find 173 billion barrels of oil in the ground and leave them there" (CBC 2017)—but that is exactly what Canada must do in order to contribute to a Just Transition both at home and abroad.

Notes

1. If you count industrial agriculture as an extractive industry, which in many cases turns soil into a non-renewable resource by degrading it irreversibly.
2. Here I allude to the United States being our present parent state—a role transferred informally along with the trade flows. As our largest trading partner who is far less dependent on our resources than we are from its wealth, it holds disproportionate economic and political power over us. The United States also is our geographic neighbor, with whom we share thousands of kilometres of border. This says nothing of the military superiority, which imbues them with additional geopolitical and diplomatic power. Instead of being codified in colonial ties of Canada and the British Empire, it is more subtly embedded in trade agreements between our nations. Future 'parent states' may continue to include the United States or perhaps other rising global powers.
3. Statistics Canada. Table 36-10-0434-03 Gross domestic product (GDP) at basic prices, by industry, annual average (x 1,000,000) URL: www150.statcan.gc.ca/t1/tbl1/en/tv.action?pid=3610043403, Retrieved March 25, 2019
4. Statistics Canada. Table 36-10-0434-01 Gross domestic product (GDP) at basic prices, by industry, monthly (x 1,000,000) URL: www150.statcan.gc.ca/t1/tbl1/en/tv.action?pid=3610043401, Retrieved March 25, 2019
5. Since it must be traded in proportion to our exports, and our dollar is a commodity subject to the same laws of supply and demand as our oil is.
6. See the latest report on fossil fuel financing, "Banking on Climate Change: Fossil Fuel Finance Report Card 2019," available at: https://www.banktrack.org/article/banking_on_climate_change_fossil_fuel_finance_report_card_2019#inform=1, Retrieved March 25, 2019
7. Greenhouse gas emissions must be cut in half by 2030 and in full by 2050 if we are to have a reasonable chance of limiting warming to 1.5°C. See the IPCC Special Report on Global Warming of 1.5°C for more details, available at https://www.ipcc.ch/sr15/
8. Most political philosophers and individuals generally agree that extreme inequality is untenable, and therefore unjust. Concessions made for inequality usually are justified if they improve the wellbeing of the poor (Rawlsian theory), while others argue that there are no such real world examples and all inequality is unjustifiable (Marxist theory, e.g. Gerald Cohen). Libertarians would argue that self-ownership is a virtue

in itself that justifies inequality, regardless of the outcome or impacts on the world's poorest and most vulnerable (Nozickian theory). It is not the place of this chapter to elaborate further, but I will assert that, at least, most Western societies are more Rawlsian than Nozickian, and claim not to tolerate egregious levels of inequality.

9 The full quote by Dr. Martin Luther King, Jr., taken from the 1967 "A Christmas Sermon on Peace," is worth including here: "So, if you're seeking to develop a just society, they say, the important thing is to get there, and the means are really unimportant; any means will do so long as they get you there. They may be violent, they may be untruthful means; they may even be unjust means to a just end. There have been those who have argued this throughout history. But we will never have peace in the world until men everywhere recognize that ends are not cut off from means, because the means represent the ideal in the making, and the end in process, and ultimately you can't reach good ends through evil means, because the means represent the seed and the end represents the tree."

10 See an open letter from June 2015 calling for an immediate moratorium on tar sands development signed by over 100 prominent scientists, including 2 Fellows of the Royal Society of Canada, 22 Members of the U.S. National Academy, 5 Recipients of the Order of Canada, and a Nobel Prize winner. Available at www.oilsandsmoratorium.org/

11 Author's calculation based on estimates of oil remaining from Canadian bitumen reserves, (McGlade & Ekins 2015) subtracting production since the study's publication. McGlade and Ekins (2015) found that 85% of bitumen was unburnable under market optimal global decarbonization in line with 2°C (50%, IEA model), leaving 7.5 billion barrels of burnable reserves. Since the study, approximately 6.5 billion barrels have been extracted. With 1 billion remaining, at 2017 production levels of approximately 0.75 billion barrels per year, this would be exhausted in 1.3 years. I round up to 2 years since production presently is in decline. This figure is not intended to be precise, but to demonstrate that continued extraction slated by the Alberta and federal governments is unequivocally incompatible with climate ambitions enshrined in the Paris Agreement. Keep in mind that this 2°C allowance is already extremely conservative for the uppermost end of what has been agreed upon as maximum acceptable global warming levels in Paris. See also Muttit et al. (2016).

12 See data from the Solutions Project by Jacobson and colleagues found in the "all countries" spreadsheet, available at http://web.stanford.edu/group/efmh/jacobson/Articles/I/WWS-50-USState-plans.html

13 See the latest report on fossil fuel financing: Kirsch, A., Opena Disterhoft, J., Marr, G., Aitken, G., Hamlett, C., Louvel, Y., et al. (2019). Banking on Climate Change (Fossil Fuel Finance Report Card 2019). Rainforest Action Network. Retrieved from https://www.ran.org/wp-content/uploads/2019/03/Banking_on_Climate_Change_2019_vFINAL1.pdf

14 Data on oil rents provided by the World Bank. Retrieved from: https://data.worldbank.org/indicator/NY.GDP.PETR.RT.ZS?year_high_desc=true

15 Find the report and press release here https://www.ohchr.org/EN/NewsEvents/Pages/DisplayNews.aspx?NewsID=24735&LangID=E; Retrieved on July 1, 2019

16 A barrel of tar sands oil contains approximately 714 kg CO_2e. Author's calculation using total GHG emissions content of a barrel of oil produced from bituminous sands in the Western Canadian Sedimentary Basin, using weighted average of production from the region, using estimates by the Oil Climate Index. See more information at https://oci.carnegieendowment.org/

17 Author's calculation using elasticity estimates from Erickson (2018), with methodology from Erickson and Lazarus (2014).

18 See the response to the 2019 federal budget from Canadian civil society at https://climateactionnetwork.ca/2019/03/19/climate-action-network-reacts-to-the-2019-federal-budget/

19 One clever pun circulating in the supply-side climate policy community is emissions *abetment* (c.f. emissions abatement)
20 Found in the latest Canadian federal budget "2019 budget: Fulfilling Canada's G20 Commitment," p. 90. Retrieved from: www.budget.gc.ca/2019/docs/plan/toc-tdm-en.html
21 See work by the International Institute for Sustainable Development and partner organizations like the Global Subsidies Initiative. A good starting point to understanding subsidies in Canada is at their FAQ page: www.iisd.org/faq/unpacking-canadas-fossil-fuel-subsidies/
22 See press statement from Government of Canada: www.canada.ca/en/natural-resources-canada/news/2018/12/government-of-canada-announces-support-for-workers-in-canadas-oil-and-gas-sector.html.
23 See the Government of Canada's announcement here: https://pm.gc.ca/eng/news/2016/11/07/prime-minister-canada-announces-national-oceans-protection-plan. Also see the response from civil society: https://climateactionnetwork.ca/2019/03/19/budget-2019-reaffirms-canadas-commitment-to-remove-subsidies-to-oil-gas-yet-offers-new-handouts-to-the-sector/
24 See for example Allan, R. (May 29, 2019). Kinder Morgan bailout to cost north of $15 billion. *National Observer.*retrieved from: https://www.nationalobserver.com/2018/05/29/analysis/kinder-morgan-bailout-cost-north-15-billion and Nikiforuk, A. (May 29, 2019). Canada's Dirty $20-Billion Pipeline Bailout. *The Tyee.* Retrieved from https://thetyee.ca/Opinion/2018/05/29/Canada-Dirty-Pipeline-Bailout/and Erickson and Lazarus 2013.
25 See coverage of the announcement of the federal government's purchase of the Trans Mountain Expansion Project here: www.cbc.ca/news/politics/liberals-trans-mountain-pipeline-kinder-morgan-1.4681911
26 See coverage of the announcement of the Alberta government rail lease here: https://business.financialpost.com/commodities/alberta-to-spend-3-7b-to-get-oil-to-prized-gulf-coast-market-by-rail

References

Allan, R. (May 29, 2019). Kinder Morgan bailout to cost north of $15 billion. National Observer.retrieved from: https://www.nationalobserver.com/2018/05/29/analysis/kinder-morgan-bailout-cost-north-15-billion

Althor, G., Watson, J.E.M., and Fuller, R.A. (2016). "Global mismatch between greenhouse gas emissions and the burden of climate change." *Scientific Reports*, 6, p. 20281.

Barnett, R.C., Bhattacharya, J., & Bunzel, H. (2009). "Choosing to keep up with the Joneses and income inequality." *Economic Theory*, 45(3), pp. 469–496. http://doi.org/10.1007/s00199-009-0494-5

Bast, E. et al. (2015). "Empty promises: G20 subsidies to oil, gas and coal production." Overseas Development Institute, Oil Change International. www.odi.org/sites/odi.org.uk/files/odi-assets/publications-opinion-files/9957.pdf.

Brulle, R.J. (2018). "The climate lobby: A sectoral analysis of lobbying spending on climate change in the USA, 2000 to 2016." *Climatic Change*, 149(3–4), pp. 289–303.

CBC–Canadian Broadcasting Company. (2017). "Trudeau: 'No country would find 173 billion barrels of oil in the ground and leave them there'." March 10. www.cbc.ca/news/world/trudeau-no-country-would-find-173-billion-barrels-of-oil-in-the-ground-and-leave-them-there-1.4019321.

Caney, S. (2010). "Climate change and the duties of the advantaged." *Critical Review of International Social and Political Philosophy*, 13(1), pp. 203–228.

Christen, M., & Morgan, R.M. (2005). "Keeping Up With the Joneses: Analyzing the Effect of Income Inequality on Consumer Borrowing." *Quantitative Marketing and Economics*, 3(2), pp. 145–173. http://doi.org/10.1007/s11129-005-0351-1

Environment and Climate Change Canada. (2019) "Canada's greenhouse gas and air pollutant emissions projections 2018." http://publications.gc.ca/collections/collection_2018/eccc/En1-78-2018-eng.pdf.

Environmental Defence and Gass, P. (2019). "Doubling down with taxpayer dollars: Fossil fuel subsidies from the Alberta government." Environmental Defence, International Institute for Sustainable Development. https://d36rd3gki5z3d3.cloudfront.net/wp-content/uploads/2019/02/EDC_IISD_AlbertaFFSReportFINAL.pdf?x84918.

Erickson, P. (2018). *Confronting Carbon Lock-in: Canada's Oil Sands.* Stockholm: Stockholm Environment Institute.

Erickson, P. and Lazarus, M. (2013). *Discussion Brief: Assessing the Greenhouse Gas Emissions Impact of New Fossil Fuel Infrastructure,* Stockholm Environment Institute. July 3, 2019. www.sei-international.org/mediamanager/documents/Publications/SEI-DB-2013-Assessing-GHGs-fossil-fuel-infrastructure.pdf.

Erickson, P. and Lazarus, M. (2014). "Impact of the Keystone XL pipeline on global oil markets and greenhouse gas emissions." *Nature Climate Change*, 4(9), pp. 778–781.

Erickson, P. et al. (2017). "Effect of subsidies to fossil fuel companies on United States crude oil production." *Nature Energy*, 2(11), pp. 891–898.

Erickson, P., Lazarus, M., and Piggot, G. (2018). "Limiting fossil fuel production as the next big step in climate policy." *Nature Climate Change*, 8(12), pp. 1037–1043.

Frumhoff, P.C., Heede, R., and Oreskes, N. (2015). "The climate responsibilities of industrial carbon producers." *Climatic Change*, 132(2), pp. 157–171.

Gardiner, S.M. (2010). "Ethics and climate change: An introduction." *Wiley Interdisciplinary Reviews: Climate Change*, 1(1), pp. 54–66.

Gardiner, S.M. (2015). "Ethics and global climate change." *Ethics*, 114(3), pp. 555–600.

Garvey, J. (2008). "Right and wrong." In *The Ethics of Climate Change*. Continuum. Edited by James Garvey, Jeremy Stangroom, pp. 33–49.

Government of Canada (2019). "2019 budget: A Just Transition for Canadian coal power workers and communities." July 3, 2019. www.budget.gc.ca/2019/docs/plan/toc-tdm-en.html (Accessed 27 March 2019).

Hayward, T. (2012). "Climate change and ethics." *Nature Climate Change*, 2(12), pp. 843–848.

Heede, R. (2013). "Tracing anthropogenic carbon dioxide and methane emissions to fossil fuel and cement producers, 1854–2010." *Climatic Change*, 122(1–2), pp. 229–241.

IISD–International Institute for Sustainable Development (2019). Global subsidies initiative. www.iisd.org/faq/unpacking-canadas-fossil-fuel-subsidies/ (Accessed 27 March 2019).

Jacobson, M.Z. et al. (2017). "100% clean and renewable wind, water, and sunlight all-sector energy roadmaps for 139 countries of the world." *Joule*, 1(1), pp. 108–121.

Kapeller, J., & Schütz, B. (2014). "Conspicuous Consumption, Inequality and Debt: The Nature of Consumption-driven Profit-led Regimes." *Metroeconomica*, 66(1), pp. 51–70. http://doi.org/10.1111/meca.12061

Kartha, S., Lazarus, M., and Tempest, K. (2016). *Fossil Fuel Production in a 2°C World: The Equity Implications of a Diminishing Carbon Budget*. Stockholm: Stockholm Environment Institute.

Kartha, S. et al. (2018). "Whose carbon is burnable? Equity considerations in the allocation of a 'right to extract.'" *Climatic Change*, 150(1–2), pp. 117–129.

KPMG-SECOR. (2013). *Economic Impacts of Western Canada's Oil Industry*. Fédération des Chambres de Commerce du Québec.

Lazarus, M., Erickson, P. and Tempest, K. (2015). *Supply-Side Climate Policy: The Road Less Taken*. Stockholm Environment Institute.

Lee, M. (2017) "Extracted Carbon: Re-examining Canada's contribution to climate change through fossil fuel exports." Parkland Institute. www.parklandinstitute.ca/extracted_carbon.

McGlade, C. and Ekins, P. (2015). "The geographical distribution of fossil fuels unused when limiting global warming to 2 [deg]C." *Nature*, 517(7533), pp. 187–190.

Meyer, C. (2019). "U.S. and Canadian oil production pushing planet's climate goals out of reach, says IEA." *National Observer*. July 3, 2019. www.nationalobserver.com/2019/03/08/news/us-and-canadian-oil-production-pushing-planets-climate-goals-out-reach-says-iea.

Motesharrei, S., Rivas, J., & Kalnay, E. (2014). "Human and nature dynamics (HANDY): Modeling inequality and use of resources in the collapse or sustainability of societies." *Ecological Economics*, 101, pp. 90–102. http://doi.org/10.1016/j.ecolecon.2014.02.014

Muttitt, G. et al. (2016). *The Sky's Limit: Why the Paris Climate Goals Require a Managed Decline of Fossil Fuel Production*, C. Rees, ed., Oil Change International.

Muttitt, G., Scott, A. and Buckley, T. (2018). *Off Track*, C. Rees, ed., Oil Change International. http://priceofoil.org/content/uploads/2018/04/Off-Track-IEA-climate-change1.pdf.

National Academies of Sciences. (2016). *Spills of Diluted Bitumen from Pipelines: A Comparative Study of Environmental Fate, Effects, and Response*. Washington, DC: The National Academies Press.

OECD. (2018). *OECD Companion to the Inventory of Support Measures for Fossil Fuels 2018*. OECD.

Ordabayeva, N., & Chandon, P. (2011). "Getting ahead of the Joneses: When equality increases conspicuous consumption among bottom-tier consumers." *Journal of Consumer Research*, 38(1), pp. 27–41. http://doi.org/10.1086/658165

Oreskes, N. and Conway, E.M. (2010). *Merchants of Doubt: How a Handful of Scientists Obscured the Truth on Issues from Tobacco Smoke to Global Warming*. New York, NY: Bloomsbury Press.

Piggot, G. et al. (2017). *Addressing Fossil Fuel Production Under the UNFCCC: Paris and Beyond*. Stockholm: Stockholm Environment Institute.

Saxifrage, B. (2018). "There's some good news about Canada's 2020 climate target." *National Observer*. July 3, 2019. www.nationalobserver.com/2018/03/06/analysis/theres-some-good-news-about-canadas-2020-climate-target.

Supran, G. and Oreskes, N. (2017). "Assessing ExxonMobil's climate change communications (1977–2014)." *Environmental Research Letters*, 12(8), p. 084019.

Touchette, Y. and Gass, P. (2018). "Public cash for oil and gas: Mapping federal fiscal support for fossil fuels." International Institute for Sustainable Development. July 3, 2019. www.iisd.org/sites/default/files/publications/public-cash-oil-gas-en.pdf.

Ulph, D. (2014). "Keeping up with the Joneses: Who loses out?" *Economics Letters*, 125(3), pp. 400–403. http://doi.org/10.1016/j.econlet.2014.10.029

Weber, B. (2018). "Report suggests growing gap between greenhouse gas projections and promises." *National Post*. https://nationalpost.com/pmn/news-pmn/canada-news-pmn/report-suggests-growing-gap-between-greenhouse-gas-projections-and-promises.

4

SHOULD THE POOR PAY MORE? COMMUNITY ENERGY PLANNING AND ENERGY POVERTY IN ONTARIO

Douglas Baxter

Community energy planning (CEP) is a strategy to help offset the inequitable implications of low carbon transition policies. This chapter discusses the potential of community energy planning in Ontario.

In recent years, several regions across Canada have engaged in initiatives to reduce their contribution to greenhouse gas (GHG) emissions. To the credit of the regional and provincial institutions who undertook such initiatives, the potential for active engagement in GHG reduction to alleviate or prevent several of the negative social, economic, and environmental implications of climate change has been frequently highlighted. However, less frequently discussed is the potential for GHG reduction initiatives to have undesirable consequences—specifically for at-risk demographic groups. In Ontario, Canada, the impact of low-carbon transition policies on low-income households illustrates how GHG reduction initiatives can sometimes inadvertently propagate social injustices.

As part of an initiative to improve the health of its residents and reduce its contribution to global GHG emissions, in 2003 Ontario committed to completely phasing out coal as a method of electricity generation in the province (Government of Ontario 2017). Instead, Ontario became increasingly reliant on less carbon-intensive forms of energy production—nuclear, wind, solar, natural gas, and hydro (Government of Ontario 2017). However, while good for the environment, the transition to less carbon intensive energy production has been directly linked to dramatic increases in electricity prices within the province (McKitrick and Adams 2014).

When the Ontario government opted for a cleaner energy supply mix, the Independent Energy System Operator—the government agency responsible for managing Ontario's electricity market—signed long-term contracts with various producers to increase generation from renewable sources (Independent Energy System Operator 2018). These contracts guarantee producers fixed rates for the renewable energy they generate, and effectively ensure that they will actively support the development of green energy infrastructure (Independent Energy System Operator 2018).

In Ontario, any costs associated with the increased green energy generation are then passed on to consumers via their electricity bills, in what is referred to as the "global adjustment" (Independent Energy System Operator 2019). Since the commitment to phase out coal, and the subsequent development and implementation of cleaner energy initiatives, electricity prices in Ontario have almost doubled during off-peak times (3.5¢ per kWh in 2006 to 6.5¢ per kWh in 2017) and increased almost 30% during peak usage (10.5¢ per kWh in 2006 to 13.2¢ per kWh in 2017) (Ontario Energy Board 2017). Much of this increase can be attributed to rising global adjustment fees, which have increased rapidly and consistently since 2008 (Independent Energy System Operator 2018). It should also be noted that the large increase in the cost of electricity during off-peak usage relative to peak usage (3.5¢:10.5¢ versus 6.5¢:13.2¢) has also significantly reduced the potential monetary savings for consumers who avidly monitor their time of use to keep costs down.

Community energy planning

The increased financial strain on all Ontarians driven by clean energy transitions is almost undeniably felt most acutely by low-income families and communities. In Canada, energy poverty is defined as any situation where a household must spend more than 10% of its household budget on energy expenditure—lighting, heating, etc. (Green et al. 2016). A 2016 report by the Fraser Institute highlighted that as of 2013 the number of Ontario households facing energy poverty hovers around 7.5% (just under 390,000 households) (Green et al. 2016). As energy prices continue to rise, it is likely that the percentage of households threatened by energy poverty will increase proportionally. Thus, mechanisms that can help reduce the vulnerability of communities to climate injustice in the form of energy inaccessibility, and aid in a more equitable transition to a low-carbon economy, should be explored.

With Ontario's electricity rates unlikely to decrease, several Ontario municipalities have begun developing community energy plans as a means of better, and more equitably, managing energy use. A community energy plan is best defined as a composition of policy, infrastructure, stakeholder, and utility goals intended to reduce fossil fuel dependence and energy intensity while promoting the equitable distribution of energy resources (Laszlo et al. 2016). They function to optimize energy allocation and where possible reduce demand within specific communities. In theory, a properly constructed community energy plan should represent a two-pronged approach to tackling issues associated with both climate change and a low-carbon energy transition. On one hand, it directly attempts to address one of the fundamental causes of climate change—consumption of non-renewable resources—and on the other, it aims to facilitate inexpensive energy delivery to all residents in the target community (Laszlo et al. 2016).

Community energy plans represent an innovative approach to energy management. Rather than relying on top-down mechanisms to govern how energy is produced and distributed in each region, strong community energy plans are created through collaborative planning between local governments and community stakeholders (St. Denis and Parker 2009). This ensures that the needs of the community are met while still contributing to a reduction in carbon emissions. Community energy planning focuses primarily on the conservation of energy, or "demand-side management," which refers to the total amount of energy required by the community to function optimally (Laszlo et al. 2016). Community energy plans may be used to address a wide range of energy-related issues, from building heating and cooling to the location of transportation infrastructure, and even the education of residents and business owners on more energy efficient practices (Laszlo et al. 2016). Placing the focus on reducing consumption rather than supply allows for significantly more control with regards to various aspects of energy management. This in turn allows for easier strategic decision-making by municipalities, who are better able to identify community interests. The potential for a community energy plan to reduce energy dependence and household spending on energy expenditure—particularly in Ontario—can be seen when examining the community energy plan of Toronto's Mount Dennis neighbourhood.

Mount Dennis community energy plan: Case study

The Mount Dennis area of Toronto is a relatively large mixed-use commercial and residential neighbourhood. Its residents are predominantly low income, and many are tenants in one of several multi-unit buildings in the area. A 2011 income survey revealed the median income per household to be under $30,000—significantly lower than the national average (City of Toronto 2016). Additionally, a recent evaluation of living conditions in the area also revealed that many of the residential buildings are improperly insulated (City of Toronto 2016). This leads to excessive heating and cooling costs for the residents and increases the vulnerability of community members to abnormal extreme weather events like heat and cold waves. As a result, the residents in the Mount Dennis area represent a demographic that is extremely sensitive to increasing electricity prices.

The Mount Dennis community energy plan was originally drafted in 2016 by the City of Toronto. The first draft was created as a collaborative effort between the City of Toronto's environment and energy division, the provincial transit agency (Metrolinx), the regional electricity distributor (Toronto Hydro), and the community residents who were able to both make and review propositions (City of Toronto 2016). It highlights several areas of interest regarding energy demand, consumption behaviour, and carbon emissions.

In Mount Dennis, poorly designed residential infrastructure was highlighted as one of the main contributing factors to high energy consumption by residents.

The plan notes that apartment buildings account for 70% of total energy use in the region, and approximately 41% of total emissions (City of Toronto 2016). Recognizing this, the CEP places a focus on strategies for future infrastructure development, building retrofits, and backup power generation for residents and commercial buildings. These retrofits are intended to reduce greenhouse emissions while simultaneously reducing the costs associated with heating and cooling for residents. The CEP then takes this a bit further, proposing requirements for new residential building construction in the region to ensure that new residents will enjoy the same energy savings as homeowners involved in the retrofit program. The CEP also highlights several other community initiatives, such as the inclusion of renewable technology in the form of solar photovoltaics on building roofs to further reduce energy costs (City of Toronto 2016). Ultimately, the municipality concluded that energy savings for residents could exceed 22% when all strategies in the plan are fully implemented (City of Toronto 2016)—significant savings, especially for low-income residents who pay a higher proportion of their income for housing and utilities than wealthier people do.

Conclusion

Efforts to address climate change can also threaten to create social injustices as well. Increasing rates of energy poverty in Ontario represent just one example of this. By reducing the scale of adaptation and mitigation strategies to the community level, the likelihood for proposed initiatives to succeed is significantly higher (Lazslo et al. 2016). The first draft of a community energy plan for Toronto's Mount Dennis area is a relatively clear example of how community energy planning can be used as a tool to address energy poverty. Without having to tackle the more complex issues affecting energy prices, the plan still provides several adaptation strategies to reduce the community's vulnerability to continued increases in electricity prices.

References

City of Toronto (2016). "Mount Dennis community energy plan—Draft." www.toronto.ca/legdocs/mmis/2016/pe/bgrd/backgroundfile95624.pdf (Accessed 18 June 2017).

Government of Ontario (2017). "Climate Change Mitigation and Low-carbon Economy Act." www.ontario.ca/laws/statute/16c07 (Accessed 18 June 2017).

Green, K. P., Jackson, T., Herzog, I., and Palazios, M. (2016). "Energy costs and Canadian household: How much are we spending." www.fraserinstitute.org/studies/energy-costs-and-canadian-households-how-much-are-we-spending (Accessed 18 June 2017).

Independent Energy System Operator (2018). "A progress report on contracted electricity supply, third quarter 2018." www.ieso.ca/Power-Data/Supply-Overview/Transmission-Connected-Generation (Accessed 15 June 2019).

Independent Energy System Operator (2019). "Price overview—Global adjustment." www.ieso.ca/en/Power-Data/Price-Overview/Global-Adjustment (Accessed 15 June 2017).

Laszlo, R., Gilmour, B., Marchionda, S., Drapeau, S., and Lee, M. (2016). "Community energy planning in Ontario: A competitive advantage for your community." www.questcanada.org/maps/community-energy-planning-on-comp-adv (Accessed 18 June 2017).

McKitrick, R., Adams, T. (2014). "Ontario's soaring electricity prices and how to get them down." www.fraserinstitute.org/research/what-goes-up-ontarios-soaring-electricity-prices-and-how-get-them-down (Accessed 16 June 2017).

Ontario Energy Board (2017). Historical electricity rate. www.oeb.ca/rates-and-your-bill/electricity-rates/historical-electricity-rates (Accessed 18 June 2017).

St. Denis, G. and Parker, P. (2009). "Community energy planning in Canada: The role of renewable energy." *Renewable and Sustainable Energy Reviews*, 13(8), pp. 2088–2095.

5

VULNERABLE COMMUNITIES AND MUNICIPAL CLIMATE CHANGE POLICY IN TORONTO

Monica Krista de Vera

Introduction

This chapter provides a snapshot of municipal climate initiatives and policies in Toronto—the largest city in Canada and the fourth largest city in North America—as an example of the possibilities and challenges inherent in addressing climate change equitably at the local level, where inequities abound (City of Toronto 2013). Toronto approved an ambitious climate change policy in 2017 and has been named one of the most resilient cities in the world, but when we look closely into the politics and policy, we see uneven adaptation (Grosvenor Group 2014). Those who are marginalized—including Indigenous, racialized, differently abled people, women, non-conforming gendered, and those living in poverty and homelessness—experience the negative effects of climate change much more than the rest of the population. This chapter is about efforts to ensure climate justice at the local level in a relatively prosperous global city.

A former world leader

In 2000, the city's first Environmental Task Force released a report entitled *Plan for an Environmentally Sustainable Toronto* (City of Toronto 2000). It provided Toronto's foundation for climate change policy and outlined a vision the city still strives for today—walkable communities, clean water, a green energy transition and economy, and an affordable transit system (City of Toronto 2000). Former Mayor David Miller (2003–2010), as chair of C40, a network of cities dedicated to fighting climate change, advocated for cities' climate action at the 2009 Copenhagen climate change conference (City of Toronto 2019b). Toronto implemented the first major organics bin initiative into its solid waste management system in 2002, and in 2003 a city by-law banned most chemical pesticide use (Toronto Environmental Alliance 2016a, p. 14; Toronto Environmental Alliance 2009). Toronto once had visions of a public-transit oriented city where all its constituents were connected by light rail, but this project was largely abandoned by former Mayor Rob Ford in 2010 (CBC News 2010). Grand visions of a sustainable and equitable Toronto still exist, and some of the groundwork is already in place, but residents and organizations have to keep working as a community to achieve a greener, liveable, and more adaptive city.

Toronto's vulnerable communities

As in other cities, vulnerable residents in Toronto disproportionately suffer the impacts of climate change, such as difficulty keeping cool during times of extreme heat, or warm in extreme cold, and lack financial resources to recover property when damages are incurred by extreme weather events, including snowstorms and flooding. A number of reports document these vulnerabilities. In the 2008 report, *Ahead of the Storm: Preparing Toronto for Climate Change*, the city identified low-income seniors, homeless people, and those with illness and disabilities as examples of vulnerable people (City of Toronto 2008, p. 14).

Despite awareness of the disproportionate effects of climate change on vulnerable populations, research into which communities are specifically affected has not been undertaken. According to the Toronto Community Housing Corporation's (TCHC) 2015 Resident Survey, most residents were age 50 or older, and primarily female (Toronto Community Housing 2015, p. 93). Two in five people also self-identified as someone living with or caring for a person with a disability (39% of survey respondents) (Toronto Community Housing 2015, p. 93). The demographic results from the survey also gave insight into who was affected during severe weather events in Toronto Community Housing complexes. During the 2013 ice storm, which affected 300,000 people across the city, one Toronto Community Housing complex went 2 weeks without heat—a prime example of how extreme weather can affect the disadvantaged (Toronto Environmental Alliance n.d.; Casey 2014). Due to the power outage, the boiler had shut down (Alamenciak 2014). Heating pipes had also burst due to air bubbles in the heating lines (Alamenciak 2014).

The City of Toronto's (2010) *Poverty Profile* identified key communities where low income was prevalent. These communities included Regent Park (64%), Oakridge (45%), Thorncliffe Park (43%), Flemingdon Park (40%), Kensington-Chinatown (38%), South Parkdale (38%), and Black Creek (36%) (City of Toronto 2010, p. 10). Communities with higher low-income rates were primarily occupied by immigrants and racialized people (City of Toronto, 2010 p. 10). There is also a concentration of single-parent families in social housing (City of Toronto 2010, p. 10). The homeless population in Toronto, according to April 2013 data, was 5,253 people (City of Toronto 2017c). Unfortunately, a more recent survey of the number of homeless people and an updated poverty profile has not been undertaken so more accurate figures are unavailable. Despite this, poverty and homelessness are constantly discussed within Toronto's context.

In 2011, the non-governmental organization United Way Toronto looked into unaffordability in the city and its correlation with building disinvestment (United Way Toronto 2011, p. 2). Poor living conditions are an indication of one's inability to cope with severe weather conditions in relation to their housing as well as personal health. Sub-par housing worsens both physical and mental health issues, which is concerning as poverty rises in Toronto (Toronto Public Health 2011, p. 8). *The Opportunity Equation* Report, which was also done by United Way Toronto, stated that income inequality is increasing in Toronto more rapidly than in the province and country as a whole (United Way Toronto 2015, p. 10). Research has also shown that the city is quite segregated according to income levels (Hulchanski 2010, p. 3). It is worth noting that the city acknowledges that Toronto is both the "provincial capital of income inequality" and the "national capital of working poverty" (City of Toronto June 2015, p. 1). Poverty is clearly a problem in the city, and this further diminishes people's ability to cope with the negative effects of climate change.

WoodGreen, a community services organization, released a report in 2014 about the devastating ice storm that hit Toronto in 2013. Data was gathered from 360 older adults living in the East York area and they were surveyed about their experiences during the ice storm (WoodGreen Community Services 2014, p. 2). According to the report, "42% of 360 seniors in East York said they did fair to very poorly during the ice storm" (WoodGreen Community Services 2014, p. 2). Income was also surveyed, showing that those who did poorly were mostly low-income individuals (WoodGreen Community Services 2014, p. 4). WoodGreen's case study further established the effect of income and its relationship to how people deal with conditions caused by climate change. As climate change worsens, the already-significant number of people severely affected will increase without proper action by the municipal government to enhance people's access to resources during difficult situations.

Services for homeless individuals in Toronto are sorely stretched, as shelters are overcrowded and basic needs are often not met (Ontario Coalition Against Poverty 2016, p. 1). As a result, homeless individuals have died during extreme cold. In January 2015, 2 homeless men, aged 55 and 60, were found dead on the streets of Toronto (Martin 2015). To mitigate this issue, the city has an *Out of the Cold* program, discussed below. Homelessness has become a more pressing issue in Toronto as the city becomes more unaffordable. Those who are living in precarious situations and/or have unstable income sources are at risk of becoming homeless in Toronto or being pushed out of the city. Housing affordability and homelessness are connected issues. Having a home constitutes very basic protection from the environment, especially important in light of climate change.

The City's climate response so far

Toronto's *Climate Driver Study* outlined weather projections up until 2050, emphasizing changing "extremes." These projections were to help the city with departmental and service planning (SENES Consultants Ltd. 2012, p. 8). The number of heat waves is predicted to increase; rainfall will increase by 50% to 80% during the summer months with more intense rainstorms; and there will be a fourfold increase in the number of days where the humidex is over 40°C (SENES Consultants Ltd. 2011, p. 12). The study laid out the scientific groundwork for concern for vulnerable communities. It affirmed the claims from Toronto Public Health (2011) as they acknowledged that low-income groups are disproportionately vulnerable to extreme heat due to where they live in the city (Toronto Public Health 2011, p. 8).

The November 2016 city staff report, *Resilient City—Preparing for a Changing Climate—Status Update and Next Steps*, discusses climate change adaptation through a brief *equity lens*, since consideration for vulnerable populations is a concern for the city. It states that

> Climate change adaptation and resilience actions should provide multiple community benefits and recognize: 1. the need for a fair distribution of benefits and burdens across all segments of society, to help reduce poverty and create a diversity of job opportunities and; 2. the need to consider generational impacts and propose actions that do not result in unfair burdens for future generations.
>
> (City of Toronto Chief Corporate Officer 2016a, p. 4)

Despite mentioning those who are vulnerable, no targets or specific strategies are outlined. Instead, one can look to the *TO Prosperity: Toronto Poverty Reduction Strategy* of 2015 to find indirect responses to climate change vulnerability via housing strategies and social service programs (City of Toronto, 2015). However, climate change is not mentioned in this report at all—ignoring how climate change only perpetuates poverty and the negative health and social externalities that come with extreme weather. *TransformTO*, Toronto's climate change strategy adopted in 2017, advocates inclusivity and accessibility in civic engagement during the implementation of the city's climate change plan (City of Toronto Chief Corporate Officer, 2016b, p. 1). It also sets ambitious targets for greenhouse gas (GHG) reduction and retrofitting (City of Toronto, 2017e. See also City of Toronto Chief Corporate Officer, 2016b).

Other City of Toronto initiatives for vulnerable populations include:

1. *Air-conditioned public spaces and cooling centres.* During times of extreme heat warnings, the city recommends visiting these sites if one is not in possession of an air conditioner. These spaces include libraries, shopping malls, and community centres—a combination of privately owned public spaces and city-owned space (City of Toronto 2019a).
2. *Warming Centres & Shelter Services.* There are emergency shelter services available, 24-hour drop-ins, housing help, and street outreach during extreme cold (City of Toronto 2017a).
3. *Out of the Cold Program.* This program, funded by the city, provides an additional average of 100 spaces a night during times of extreme cold from

mid-November to mid-April. This program is operated by faith groups (City of Toronto 2017a).

4. *Extreme Weather Portal.* This platform helps Torontonians get emergency-ready by providing guides for residents and education on extreme weather. It outlines where to check for updates, the components of an emergency kit, and information on power outages (City of Toronto 2017b).

It's not enough

According to the Toronto Public Health report *Health Benefits of a Low-Carbon Future*, "health" is "a complete state of physical, mental and social well-being, and not merely the absence of disease or infirmity" (Toronto Public Health 2016, p. 24). This broad and comprehensive definition increases the threshold for required action to protect the health of all.

The current steps taken during severe weather events are crucial to assisting vulnerable populations. However, these do not address systemic equity and poverty issues which are inherently needed for climate justice. A 2016 Ontario Coalition Against Poverty report reveals that shelters are short of space, health supplies, and adequate sleeping conditions (Ontario Coalition Against Poverty 2016, p. 4 and 7). The report also found that shelter standards were not fully complied with in any of the city's "warming centres" or *Out of the Cold* facilities (Ontario Coalition Against Poverty 2016, p. 4). The shelter system has been criticized by the media as many locations did not have enough or sufficient restroom facilities, mats were the primary option for sleeping at all locations, pillows were only offered at two locations, and individuals were placed only a few inches to a couple of feet apart from each other, increasing the risk of illness and disease transmission (Ormsby 2016). Shelters do not fulfil the broad definition of health adopted by the city. Cooling centres are public buildings, shopping malls, public swimming pools, and other air-conditioned spaces used to service those who are marginalized in extreme heat conditions. The city threatened to cut funding for cooling centres in 2017, even though they are a lifeline for many vulnerable people whose homes are not air-conditioned.

In 2017, City Council approved $330,000 from the Operating Budget to support three short-term strategies related to climate change (City of Toronto 2017e, p. 11). The operating budget covers spending pertaining to services that affect day-to-day living, such as public health services, garbage collection, and city roads (City of Toronto 2018a, p. 1. See also City of Toronto, 2018b). One of TransformTO's briefing notes expresses that in fact $1.6 million would be needed for a half year of funding in 2017 (City of Toronto Chief Corporate Officer 2017, p. 1), so the amount set out in the 2017 budget was not even a quarter of what was needed. For 2018, $6.7 million would have been required for TransformTO's full year budget (City of Toronto 2017d, p. 5). However, the city only committed $2.5 million in the final vote at City Council: $2.3 million for key priorities and $0.2 million for increases to fleet services (City of Toronto

Budget Committee 2018). Partial funding means that only certain parts of the plan are prioritized to be implemented. This is problematic because all proposed strategies work in harmony together to maximize TransformTO's effect on GHG emissions and advancing social equity.

TransformTO, if properly funded, would not only decrease the city's output of GHG emissions, but Toronto's people would also experience community benefits. TransformTO's mandate is connected to strategies pertaining to equity and neighbourhood improvement. However, without dedication and political will, there is no guarantee that services for marginalized populations will be improved. Issues pertaining to shelters and poverty may not be seen as climate change adaptation issues to some members of City Council, regardless of the clear interconnectedness between social and environmental issues. Adaptation in Toronto has primarily been understood as related to the longevity of physical infrastructure, pushing vulnerable people to the periphery of the conversation.

The initiatives that the City of Toronto provides are helpful to vulnerable people in the city in times of crisis, but these programs do not address the long-term systemic problems that result from lack of investment to meet people's underlying needs for housing, healthcare, childcare, eldercare, and employment opportunities. The "justice" aspects of climate policies are thorny but important.

A brief discussion of ideological underpinnings

The lack of consideration for social issues, such as climate change adaptation in vulnerable communities, increasing poverty in the city, and increasing unaffordability, highlights Toronto's neoliberal problem. To put it simply, neoliberalism is characterized by reducing taxation for social services to facilitate the accumulation of profit, promotion of individual liberty, self-interest, free-trade, private property rights, and entrepreneurial interests (Harvey 2007, p. 22). Toronto is in cost-saving mode to make up for gaps in funding as a result of the history of federal and provincial downloading of social responsibilities and services (Fanelli 2014, p. 5). This means that there is not much left for investment in the public sector. Further, Toronto's City Council is not investing enough in its own constituents. Cost-cutting reflects neoliberal ideals as it promotes the divide between upper and lower classes (Harvey 2007, p. 29). The consequence of growing income inequality and the lack of investment in climate change adaptation—which includes various social services that enhance people's ability to survive—means higher income people are more advantaged when it comes to dealing with climate change.

There has to be strong political will to take on these challenges because of resistance from vested, moneyed interests. The ideological underpinnings of political will are connected to a grassroots theory of change that builds power from the bottom-up. Establishing power at the grassroots level comes from capacity and knowledge building—equipping people with the tools to challenge power in the city. While power relations in the city are multi-faceted and complex, it is important to look critically at the relationship between Toronto's constituents and City Council. When it comes to climate change adaptation in the city, City Council's policies show that people who are historically disadvantaged are becoming further marginalized. Neoliberal motivated policies do not favour constituents who need assistance the most—especially those facing poverty and homelessness.

Linked justice challenges

Transit-related inequities worsen some people's access to other services, such as emergency weather facilities. For example, if Toronto Public Health decides that the cooling centre program is not effective, these services might be cut in the future. Further, cutting the number of cooling centres and other emergency centres in the city means that some people will have to travel farther for relief from extreme weather conditions. Ridership on public transit and Wheel-Trans (transit services for people with mobility challenges) has increased, services have been cut, and overcrowding continues to be an overwhelming problem (Wilson 2016). For those people who are already at the periphery of the city, it might be even more difficult to access transit in severe weather conditions.

The City of Toronto needs updated information about poverty and its interaction with other services that are crucially important for climate change adaptation. Climate change must be included in other public-sector service discussions. Neglecting to recognize its importance does a disservice to those who are most affected since climate justice is directly interconnected with vulnerability. With increasing unaffordability in the city, justice issues will become more pressing as extreme weather events become more common.

Toronto residents may well ask, "Who is Toronto for if services cannot cater to those who are most at risk? Is Toronto for everyone?" The short answer is that Toronto's climate policies have effectively left many of its residents unaccounted for. The previous discussion on the city budget shows where the priorities are, which illustrate the values of those in power. While Toronto's City Council has some champions advocating for the issues mentioned here, many people affected are not part of the conversation and many councillors do not see these issues as pressing matters. Unfortunately, Toronto's allocations in the budget do not put social issues, such as housing affordability, health services, public transit, and climate change—all interrelated budget items both tangibly and ideologically—at the forefront of politics and policy. The people who are affected the most are those who are most vulnerable in Toronto.

What we can do

Municipal government is the level of government that is closest to the people, which means that we have the power to put pressure on local governments to support policies that will make our towns and cities more sustainable and equitable. The decisions made at the local level regarding climate change affect our day-to-day lives. Decisions about water, infrastructure, electricity, transit, public health, waste management, parks, and recreation services are all made at the local level. The following are some ways to take action for climate justice at the local level:

Learn more about municipal government processes. Civic literacy is significant; local governments all operate differently. Knowledge of local government operations and processes is important so that one can participate meaningfully in the opportunities where public input is needed, whether it is by joining in a public consultation by expressing views, presenting at a committee meeting, or making a phone call to a city councillor. By taking steps to understand how decisions are made and your role within the process, you can begin to understand the reasons behind the city's current state, and how one's involvement can contribute to a liveable and more equitable city. While resources are available online, local non-profit organizations such as Progress Toronto work to ensure that Toronto's politicians are held accountable through educating the public on how they vote at City Hall on the issues that affect people directly (Progress Toronto n.d.). Understanding municipal processes and government operation is democratic power that can be harnessed by constituents.

Connect with issue-based campaigns. This can be done with organizations or politicians through letters, phone calls, petitions, or meetings. While it may seem mundane, it has an overwhelming impact when enough people come together. NoJetsTO, a non-profit organization that was dedicated to protecting Toronto's waterfront, educated and mobilized thousands of constituents to reject the Toronto island airport expansion, which would worsen gridlock and pollution (Battersby 2015). In 2015, the plans were taken off the table completely because of the citizen-led movement (Battersby 2015).

Do the "small" things. If you're reading this book, you are probably aware of the small lifestyle changes you can make in your life to reduce carbon emissions, such as recycling and using LED light bulbs. These small changes are often tied to municipal services like waste management or electricity. A prime example is handling household waste more mindfully. Up to 86% of waste can be diverted (Toronto Environmental Alliance 2016a, p. 26). Proper waste management at home can reduce greenhouse gases significantly on a larger scale, and help ensure that the city's waste system runs more efficiently. Methane gas, a very potent greenhouse gas, is released in landfills when organic waste is not exposed to oxygen in the process of breaking down (Toronto Environmental Alliance 2016b, p. 7). Twenty per cent of Canada's methane emissions come from landfills (Government of Canada, 2017).

Engage with local organizations who are already engaged in the issues. Volunteering with local organizations that work to ensure your city is green and healthy is one of the best ways to learn about your community and help reduce the city's climate impact. Citizen science—projects that involve community members in gathering scientific data—is a meaningful and often fun way of contributing to the organizations working hard to make the city better. It is also an opportunity to be more active in your own community. The INHALE project, which was created by the Toronto Environmental Alliance (TEA) and Environment Hamilton, is an initiative to crowd-source air quality data using mobile air monitors (Inhale project n.d. a). This project was inspired by Pittsburgh's Group Against Smog and Pollution (GASP) and their Bicycle Air Monitoring Project (Group Against Smog and Pollution n.d. b). Other cities such as London, San Francisco, and Brooklyn also have citizen science air quality projects (Inhale Project n.d.b). Collecting this data creates conversations about air quality and brings people together. Mobile air monitoring was made easy to incorporate into daily life; individuals could attach the air monitors to their bikes, strollers, and backpacks during their daily errands and free time.

Contribute to creating an adaptive community. During the Toronto ice storm in December 2013, there were many stories about neighbours helping neighbours (WoodGreen Community Services 2014, p. 3). As a result of power outages, many residents could not request assistance by phone; some had no heat; food spoiled in refrigerators; elevators and appliances did not function. A local network of expertise and collaboration can ensure support for a myriad of emergency situations, especially for differently-abled people, elderly people, and those with small children. Actions can include getting together with community members to translate available resources for newcomers where English is not their first language. Toronto's Extreme Weather Portal was translated from English into ten other languages (City of Toronto 2017f). Information and assistance must be available to everyone.

Conclusion

Making a conscious effort to reduce personal emissions, encourage local politicians to support sustainable and equitable policies, and engage with our neighbours and local organizations can have a global impact. This chapter provides examples of changes that are grounded in grassroots action in light of city budget allocations that do not do enough to improve the lives of Toronto's more marginalized people. Toronto is an example of a former leader in municipal environmental policy. To get back on the right track, people have to work together to challenge Toronto's policies and the decisions being made at City Council.

Of course, cities vary in terms of historical, economic, cultural, social, and political contexts. However, the tools and actions that need to be taken seriously are related. In the North American context, municipalities have the potential to influence about 52% of nationally emitted GHGs (Robinson and Gore 2005, p. 105). Further, the actions we take in cities impact national GHG output. When looking at the global picture, the tangible impact of local action is very clear at the international scale: actions at the local level snowball into a greater impact because they challenge the power relationships in place.

This chapter calls people everywhere to critically think about your municipality in the context of how decisions are made and how you can participate in a meaningful and effective way. If you are able to think critically about the municipal processes and be civically literate, you can empower yourself and others to effect positive change that would make your city a more livable, sustainable, equitable, and climate-resilient place.

References

Alamenciak, T. (2014). "East-end TCHC building remains without heat as pipes continue to burst." *Toronto Star*. www.thestar.com/news/gta/2014/01/10/eastend_tchc_building_remains_without_heat_as_pipes_continue_to_burst.html (Accessed March 2019).

Battersby, S.J. (2015). "No expansion for Toronto island airport." *Toronto Star*. www.thestar.com/news/gta/2015/11/13/no-expansion-for-toronto-island-airport-transport-minister-says.html (Accessed March 2019).

Casey, L. (2014). "Toronto community housing complex goes two weeks without heat after ice storm." *Toronto Star*. www.thestar.com/news/gta/2014/01/07/toronto_community_housing_complex_goes_two_weeks_without_heat_after_ice_storm.html (Accessed June 2017).

CBC News (2010). "Rob Ford: 'Transit city is over.'" *CBC News*. www.cbc.ca/news/canada/toronto/rob-ford-transit-city-is-over-1.926388 (Accessed March 2019).

City of Toronto (2000). *Clean, Green and Healthy: A Plan for an Environmentally Sustainable Toronto*. Toronto: Environmental Task Force.

City of Toronto (2008). "Ahead of the storm: Preparing Toronto for climate change. Toronto. www.climateneeds.umd.edu/pdf/ahead_of_the_storm.pdf (Accessed March 2019).

City of Toronto (2010). "Poverty profile." Toronto. https://communitydata.ca/sites/default/files/poverty_profile_2010%20v4.pdf (Accessed June 2017).

City of Toronto (2013). "Toronto now the fourth largest city in North America." Toronto. http://wx.toronto.ca/inter/it/newsrel.nsf/bydate/88678E26C2B5C0DE85257B25007655DD (Accessed March 2019).

City of Toronto (2015). "TO Prosperity: Interim poverty reduction strategy." Toronto. www.toronto.ca/legdocs/mmis/2015/ex/bgrd/backgroundfile-81607.pdf (Accessed March 2019).

City of Toronto (2017a). "Extreme cold weather alerts." Toronto. Available at: www1.toronto.ca/wps/portal/contentonly?vgnextoid=d81ed4b4920c0410VgnVCM10000071d60f89RCRD&vgnextchannel=cfa2d62869211410VgnVCM10000071d60f89RCRD (Accessed February 2017).

City of Toronto (2017b). "Extreme weather portal." Toronto. www1.toronto.ca/wps/portal/contentonly?vgnextoid=f33dd8f3e3214510VgnVCM10000071d60f89RCRD (Accessed February 2017).

City of Toronto (2017c). "Quick facts about homelessness and social housing in Toronto." Toronto. www1.toronto.ca/wps/portal/contentonly?vgnextoid=f59ed4b4920c0410VgnVCM10000071d60f89RCRD&vgnextchannel=c0aeab2cedfb0410VgnVCM10000071d60f89RCRD (Accessed February 2017).

City of Toronto (2017d). "2018 Operating Budget Briefing Note TransformTO 2018 Operating Budget Request." Toronto. www.toronto.ca/legdocs/mmis/2017/bu/bgrd/backgroundfile-109987.pdf (Accessed March 2019).

City of Toronto (2017e). "TransformTO: Climate Action for a Healthy, Equitable and Prosperous Toronto – Report #2 – The Pathway to a Low Carbon Future." City of Toronto Website, April. www.toronto.ca/wp-content/uploads/2017/10/99b9-TransformTO-Climate-Action-for-a-Healthy-Equitable-and-Prosperous-Toronto-Report-2-The-Pathway-to-a-Low-Carbon-Future-Staff-Report-April-2017.pdf (Accessed March 2019).

City of Toronto (2017f). "Preparedness guides now translated into multiple languages." Toronto. Available at: www1.toronto.ca/wps/portal/contentonly?vgnextoid=228ff2b43a364510VgnVCM10000071d60f89RCRD&vgnextchannel=4b3307ceb6f8e310VgnVCM10000071d60f89RCRD (Accessed February 2017).

City of Toronto (2018a). "City council approves balanced budget that invests in key service priorities for Toronto residences and businesses." Toronto. http://wx.toronto.ca/inter/it/newsrel.nsf/7017df2f20edbe2885256619004e428e/f32220d1caeb33198525823300132a69?OpenDocument (Accessed 26 March 2019).

City of Toronto (2018b). "Understanding the Toronto City budget." Toronto. www.toronto.ca/wp-content/uploads/2017/11/97f7-A170XXXX_Budget_Basics_Understanding-final-web.pdf (Accessed March 2019).

City of Toronto (2019a). "Places to cool down." Toronto. www.toronto.ca/community-people/health-wellness-care/health-programs-advice/extreme-heat-and-heat-related-illness/places-to-cool-down-2/#otherPlaces\ (Accessed March 2019).

City of Toronto (2019b). "Biography—David Miller." Toronto. www.toronto.ca/explore-enjoy/history-art-culture/mayors-councillors-reeves-chairmen/mayor-reeve-chairman/david-miller/ (Accessed March 2019).

City of Toronto Budget Committee (2018). "2018 capital and operating budgets." Toronto. http://app.toronto.ca/tmmis/viewAgendaItemHistory.do?item=2018.BU41.1 (Accessed March 2019).

City of Toronto Chief Corporate Officer (2016a). "Resilient city—Preparing for a changing climate status update and next steps." Toronto. www.toronto.ca/legdocs/mmis/2016/pe/bgrd/backgroundfile-98049.pdf (Accessed March 2019).

City of Toronto Chief Corporate Officer (2016b). "TransformTO: Climate Action for a Healthy Equitable, and Prosperous Toronto – Report #1." Toronto. www.toronto.ca/legdocs/mmis/2016/pe/bgrd/backgroundfile-98039.pdf (Accessed 26 March 2019).

City of Toronto Chief Corporate Officer (2017). "TransformTO Short-term Strategies Financial Estimates." Toronto. www.toronto.ca/legdocs/mmis/2017/bu/bgrd/backgroundfile-99799.pdf (Accessed March 2019).

Fanelli, C. (2014). *Under Pressure: How Public Policy Is Constraining Ontario Municipalities.* Toronto: Canadian Centre for Policy Alternatives. www.policyalternatives.ca/sites/default/files/uploads/publications/Ontario%20Office/2014/10/Under%20PressureFINAL.pdf (Accessed March 2019).

Government of Canada (2017). "Municipal solid waste and greenhouse gases." Government of Canada. Available at: www.ec.gc.ca/gdd-mw/default.asp?lang=En&n=6f92e701-1 (Accessed March 2019).

Grosvenor Group (2014). "Resilient Cities Research Report." Grosvenor Group Limited. www.grosvenor.com/news-views-research/research/2014/resilient%20cities%20research%20report/ (Accessed March 2019).

Group Against Smog and Pollution (n.d.). "Bicycle air monitoring project." Group Against Smog and Pollution website. http://gasp-pgh.org/projects/bam/ (Accessed February 2017).

Harvey, D. (2007). "Neoliberalism as Creative Destruction." *The Annals of the American Academy of Political and Social Science*, 610(1), pp. 22–44.

Hulchanski, D. (2010). *The Three Cities within Toronto: Income Polarization Among Toronto's Neighbourhoods, 1970–2005*. Toronto: Cities Centre: University of Toronto.

Inhale Project (n.d. a). "About." Inhale Project. www.inhaleproject.ca/about (Accessed March 2019).

Inhale Project (n.d. b). "Citizen science resources." Inhale Project. www.inhaleproject.ca/citizen_science_resources (Accessed March 2019).

Martin, S. (2015). "Deaths of 2 men during cold snap prompt call for action." *CBC News Website*, 5 January. www.cbc.ca/news/canada/toronto/deaths-of-2-men-during-cold-snap-prompt-call-for-action-1.2891428 (Accessed March 2019).

Ontario Coalition Against Poverty (2016). "Out in the cold: The crisis in Toronto's shelter system." Ontario Coalition Against Poverty.

Ormsby, M. (2016). "Packed Toronto shelter system poised for breakdown: Report." *Toronto Star Website*, 15 February. www.thestar.com/news/gta/2016/02/15/packed-toronto-shelter-system-poised-for-breakdown-report.html (Accessed March 2019).

Progress Toronto (n.d.). "Our story." Toronto. www.progresstoronto.ca/our-story/ (Accessed 26 March 2019).

Robinson, P.J. and Gore, C. (2005). "Barriers to Canadian municipal response to climate change." *Canadian Journal of Urban Research*, 14(1), pp. 102–120.

SENES Consultants Ltd. (2011). "'7.0 What does it all mean?,' in Toronto's future weather & climate driver study." *City of Toronto Website*, December. www.toronto.ca/legdocs/mmis/2013/pe/bgrd/backgroundfile-55152.pdf (Accessed March 2019).

SENES Consultants Ltd. (2012). "Toronto's Future Weather & Climate Driver Study: Outcomes report." *City of Toronto Website*, 30 October. www.toronto.ca/wp-content/uploads/2018/04/982c-Torontos-Future-Weather-and-Climate-Drivers-Study-2012.pdf (Accessed March 2019).

Toronto Community Housing (2015). "2015 resident survey." *Toronto Community Housing Website*, 8 May. www.torontohousing.ca/news/whatsnew/Documents/Nielsen%20TCHC%20-%202015%20Resident%20Survey%20-%20Final%20Report.pdf (Accessed March 2019).

Toronto Environmental Alliance (2009). "Toronto pesticide by-law." *Toronto Environmental Alliance Website*, 8 April. www.torontoenvironment.org/toronto_pesticide_by_law (Accessed March 2019).

Toronto Environmental Alliance (2016a). "Zero waste Toronto." *Toronto Environmental Alliance Website*, February. https://d3n8a8pro7vhmx.cloudfront.net/toenviro/pages/1636/attachments/original/1456190964/TEA_-_Zero_Waste_Toronto_Report_-_2016.pdf?1456190964 (Accessed March 2019).

Toronto Environmental Alliance (2016b). "Organics First: Setting Toronto on the Path to Zero Waste." *Toronto Environmental Alliance Website*, June. https://d3n8a8pro7vhmx.cloudfront.net/toenviro/pages/1775/attachments/original/1465851365/Organics_First_-_TEA_Report_-_June_2016.pdf?14658513659 (Accessed March 2019).

Toronto Public Health (2011). "Protecting vulnerable people from health impacts of extreme heat." *City of Toronto Website*. www.climateontario.ca/doc/ORAC_Products/TPH/Protecting%20Vulnerable%20People%20from%20Health%20Impacts%20of%20Extreme%20Heat.pdf (Accessed March 2019).

Toronto Public Health (2016). "Health benefits of a low-carbon future." *City of Toronto Website*. www.indeco.com/wp-content/uploads/2018/10/Low-C-Report-TPH-2016-06-23-6-4-AODA.pdf (Accessed March 2019).

United Way Toronto (2011). "Poverty by postal code 2: Vertical poverty, declining income, housing quality and community life in Toronto's inner suburban high-rise apartments." *United Way Toronto & York Region Website*. www.unitedwaytyr.com/document.doc?id=89 (Accessed March 2019).

United Way Toronto (2015). "The opportunity equation." *United Way Toronto & York Region Website*, February. www.unitedwaytyr.com/document.doc?id=286 (Accessed March 2019).

Wilson, B. (2016). "City budget: Day 2 budget committee review." *Social Planning Toronto Website*, 19 December. Available at: www.socialplanningtoronto.org/2017_city_budget_day_2_budget_committee_review (Accessed February 2017).

WoodGreen Community Services. (2014) "Ice storm 2013." *WoodGreen Community Services Website*, December. www.woodgreen.org/Portals/0/PDFs/Ice%20Storm.pdf (Accessed March 2019).

6

THE RIGHT TO REMAIN

Community-led responses to land dispossession in the context of global and local climate injustice

Meagan Dellavilla

Rising sea levels are swallowing up shores. Droughts are drying up once fertile agricultural land. The well-documented consequences of climate change are becoming increasingly hard to ignore, and for many communities, increasingly difficult to withstand. As climatic conditions become increasingly erratic, the threat of displacement continues to plague a rising number of communities. Since the 1980s, the highly contested concept of climate-induced migration has complicated conversations on border security, diplomatic relations, and climate debt reparations (White 2011). Yet despite gaining traction in political spheres, the proposed solutions to climate-induced, or climate-compelled, migration routinely ignore the first desire of impacted communities: their wish to remain, or, at the very least, have the free will to decide when they are to depart from the places where their ancestors rest, where their relations to the winged, finned, two-legged, and four-legged ones, and the lands and waters, are rooted, and their traditions are established.

This chapter aims to contribute to this dialogue by first examining the far-reaching and entangled impacts of land dispossession, specifically on Indigenous communities. Through this lens, I seek to reaffirm scholar Kyle Whyte's (2017) assertion that the disproportionate exposure to *environmental bads* "represents only one dimension of the structure of environmental [or climate] injustice against Indigenous peoples." Following this logic—that achieving climate justice, therefore, demands more than mere re-distribution of the associated "bads"—the paper concludes by recounting transferable, grassroots strategies employed by communities near and far in their relentless efforts to reassert their sovereignty in their fight for climate justice.

> It is more than being connected or attached to the land, we are part of the land, it is part of us and we are part of it … the water, the air, all of it runs

> through our veins and souls. To be here is to live, to be elsewhere is to die to who we are.
>
> (*Maldanado et al. 2013*, n.p.)

As illustrated by the quote above, the loss of land is often devastating to the social, economic, and cultural constructs of a community. This devastation, as demonstrated below, manifests in a number of interconnected ways. Though the experiences of Indigenous communities are emphasized here, I acknowledge that land dispossession has and will continue to also impact those not necessarily indigenous to their lands. While social roles and conceptions of land connection may vary between settler and Indigenous populations (as they also do amongst Indigenous peoples), there is a relatively universal understanding of "home," and many shared challenges that one is forced to endure when involuntarily removed from that place. This chapter, in part, seeks to speak to that sentiment.

However, the experiences of Indigenous communities remain central to this analysis for three interrelated reasons. First, colonization sought to eliminate Indigenous communities. Survival, therefore, demanded rapid, widespread adaptation. As articulated by Whyte (2016) and McGregor (2018), the very existence of Indigenous peoples on Turtle Island today implies that they bring both knowledge and experience in adapting to forces, such as climate change, that seek to dismantle and decimate their (and others') presence. Second, it is becoming increasingly obvious that proposed solutions to climate change that are rooted in colonial, capitalist, patriarchal ways of thinking often perpetuate the underlying causes of the climate crisis. In the case that they do bring about change, it is rare that it happens in a way that flattens hierarchies or challenges our interlaced systems of oppression. Therefore, it seems critical that Indigenous knowledge systems become central to the climate justice conversation. Finally, given that the larger discourse of climate (in)justice is founded upon the disproportionate ways in which climate change is and will continue to be experienced, it is important to reiterate that without intervention, those least responsible for the climate crisis will continue to bear the brunt of its effects. A growing body of literature that cites the reoccurring and accumulating impact of the climate crisis on Turtle Island's original inhabitants confirms this claim (Scott 2008; Sandlos and Keeling 2016; Wiebe and Konsmo 2014).

Prior to exploring community-based interventions to resist land dispossession, I survey the myriad ways that one's removal from their land (physical or not) impacts individual and communal well-being. While a few coastal and Northern nations in the United States and Canada are currently weathering the rage of rising sea levels, the majority of people indigenous to Turtle Island, particularly those residing in the Great Lakes watershed, are battling erratic weather conditions alongside another component of the climate crisis: routine exploitation perpetrated by the culprits of the said crisis—namely, the extractives industry, and the regulatory and economic systems that embolden its presence. As articulated by various land protectors, changing climatic conditions coupled with this chronic contamination complicate the continuation of land-oriented ceremonies and practices, including but not limited to hunting, fishing, and medicine gathering (B. Gray and V. Gray 2017, personal communication; Konsmo and Pacheco 2016). Disruption to these significant and sacred practices mirrors the cultural and communal diaspora that ensues when one is forced or compelled to leave their home because of changing climatic conditions. The following examples are not offered in an effort to homogenize communities' experiences but rather to illustrate parallels in such experiences—an arguably important element in building nation-to-nation solidarity.

The entangled impacts of land dispossession

Socio-cultural—Land dispossession, specifically in Indigenous communities, can increase women's social vulnerability. In many Indigenous communities, women are perceived as the keepers and teachers of traditional knowledge. However, when women are unable to continue engaging in traditional practices, as is common in cases of land dispossession, they are displaced "from their roles and positions in their societies" (Kuokkanen 2008, p. 223). This, in turn, may cause their social status and value to diminish, "making them more vulnerable to marginalization and exclusion" (Kuokkanen 2008, p. 223). This disruption to individual identity often further manifests as a disruption to intergenerational cultural continuation. When a community loses access to their ceremonial sites, medicines, and foods, the ability to share cultural and healing practices with the next generation is consequently compromised. This can trigger a communal and cultural fracturing whereby language and tradition are compromised and the subsequent void is filled by "unsafe" coping practices (e.g. alcohol and drug use). As a member of Akwesasne First Nation, a reserve located on the St. Lawrence River that has been negatively impacted by industry, explains,

> It's not our way of life ... [because] we have to make the decision between that way of life—that culture, that tradition—and our health ... We're trying to fill those voids, [replace] those things that we're missing. And one of the ways that so many people are reaching for that is ... with drinking and drugs.
>
> *(Konsmo and Pacheco 2016, p. 29)*

Economic—When forces beyond a community's control impact its ability to proceed with long-practised land-oriented livelihood strategies, self-sufficiency often decreases and the community is forced into the global economic system. In many cases, internationally influenced strategies of "economic stimulation" (e.g. mining; entrance into the market economy) have proven to exacerbate both global climate challenges and the vulnerability of the local community. For example, in Bangladesh, where rising sea levels have significantly interfered with stability, security, and quality of life, World Bank loans have pushed "farmers away from rice and towards export-driven shrimp farming on a massive scale" (Klein 2010, p. 60). In turn, such practices have devastated mangrove networks, which traditionally act as natural barriers to cyclones and avert substantial land erosion, making the community more susceptible to the fury of increasingly violent cyclones (Klein 2010). Similar circumstances have routinely played out in the Great Lakes watershed. When a community's subsistence practices are severed, by way of, for example, contamination or decline in wildlife populations, survival often requires entry into wage-labour. Though employment with the corporations perpetuating such atrocities may seem counter-intuitive, decades of destruction make it so that few, if any, alternatives exist.

Health—It has been suggested that men's mental health, in particular, may be vulnerable to environmental changes such as those described above (Kukarenko 2011). When these changes affect the places and resources critical to men's livelihoods, the masculine identity is inevitably challenged. Consequently, this has been documented to decrease perceptions of self-worth and increase mental stress amongst Indigenous men (Vinyeta, Whyte and Lynn 2015). This psychological burden has also been linked to the increased likelihood of domestic violence, thereby further compromising the health and safety of women and children (Kukarenko 2011). Further, researchers are continuing to note the parallels between the ways in which Indigenous land is treated, and the ways in which human bodies are impacted by this treatment (Konsmo and Pacheco 2016). For example, in Aamjiwnaang First Nation, a reserve located in Canada's so-called "Chemical Valley," 40% of the female population has reported experiencing at least one miscarriage or stillbirth (Macdonald and Rang 2007). Nationwide, in contrast, The National Institute of Child Health and Human Development (2010) suggests that across Canada approximately 17% of women experience a miscarriage, and stillbirths occur in 0.5% of pregnancies. Furthermore, the Aamjiwnaang community has been experiencing a documented decline in the ratio of male to female newborns for many years (Mackenzie et al. 2005). It is hypothesized that the declining rate of male births is due to chronic exposure to a group of endocrine-disrupting chemicals (Scott 2008). Moreover, such industrial chemicals often resist being broken down by the body and are stored in the fat cells (Cook 2003). As such, it has been suggested that,

> The only known way to excrete large amounts of these contaminants is 1) through pregnancy, when they cross the placenta, or 2) during lactation, when they move out of storage in fat cells and show up in breast milk.
>
> *(Konsmo and Pacheco 2016, p. 26)*

This implies that succeeding generations inherit a toxic body burden from their mothers, resulting in maternal feelings of guilt and vulnerability. As midwife Katsi Cook summarizes, "In this way, we, as women, are the landfill" (Cook as quoted in Konsmo and Pacheco 2016, p. 20). Moreover, even if or when relocation is independently considered, residents, for example, of Aamjiwnaang, namely women, are very aware that they will carry this toxic load with them (Gray 2016). Perhaps best described as a form of "slow violence" (Nixon 2013), communities are now beginning to question the compounding intergenerational impacts of this cumulative exposure.

Finally, it is important to ground the above impacts in historical context. As Whyte (2016) explains, colonialism "can be understood as a system of domination that concerns how one society inflicts burdensome anthropogenic environmental change on another society" (p. 5). Since their arrival, settler communities have been dismantling Indigenous peoples' way of life by employing a strategy of "containment" that runs counter to the traditionally employed adaptable and dynamic systems of social and political organization (Whyte 2016). In North America, this was most evidently demonstrated through the forcible relocation of Indigenous peoples from their original territories to government allocated plots of lands. As mentioned above, while adaption to environmental change is not a new feat for Indigenous communities, the means of recourse are continually stunted by the colonial configuration of forced containment. Thus, it is fair to conclude that the forms of displacement described above should not be perceived as novel, but rather a continuation of old patterns, or a case of what Whyte (2016) describes as "colonial déjà vu."

Community-led responses to land dispossession in the context of global climate injustice

According to the United Nations High Commissioner for Refugees (UNHCR), 22.5 million individuals worldwide were displaced due to a climate or weather-related event between 2008 and 2015 (Yonetani 2015, p. 20). The International Organization for Migration (IOM) predicts this number could reach 200 million by 2050 (International Organization on Migration, 2019). But in fact, as this chapter argues, the act of leaving one's home is often at odds with what many communities have rightfully declared is their desired response to contemporary forms of land dispossession.

As articulated in the UN Declaration on the Rights of Indigenous People (2007), climate-induced displacement severs the physical and spiritual ties, as well as the rights Indigenous peoples have to their traditionally occupied land. As illustrated above, the loss of connection to the land often acts as the catalyst for a number of challenges plaguing today's society. In response, organizations and individuals are employing strategies to empower communities and enable their right and ability to remain on their traditionally occupied land.

For example, in late 2009, the EarthLore Foundation,[1] located in South Africa, hosted a gathering for Indigenous peoples from around the world. During

1 Additional information on how and where the EarthLore Foundation works can be found at http://earthlorefoundation.org.

this ceremony, Colombian Elders from an Amazonian community shared a long-practised and refined participatory community-mapping tool, referred to as eco-cultural mapping, with the Tshidzivhe community of Venda, South Africa, as well as visitors from Indigenous communities of present-day Russia ("Eco-cultural mapping" 2015). This process—which encourages community members to first identify what the community once was; next, how it presently is; and finally, their vision for the future—has proven to be "a powerful path for restoring the confidence of communities in their traditions and in defending their rights to their ancestral territory and traditional culture" ("Eco-cultural mapping" 2015). It not only prompts and encourages Elders to apply their traditional ecological knowledge to the present-day challenges brought on by climate change and colonialism, but also directs the community toward a collective vision.

Once that collective vision is established, the EarthLore Foundation offers support to communities in reaching their goals. In the past, communities have opted to place a large focus on protecting biodiversity and cultural diversity through the revitalization of seeds and food sovereignty. In addition to increasing resiliency in the face of climate change, such efforts also enable greater community cohesion, allow for the recuperation of complementary gender roles, and strengthen community ecological governance. The EarthLore Foundation has been able to support such efforts through facilitating agro-ecology trainings, seed fairs, and community exchanges of knowledge and experience. Further, the EarthLore Foundation also supports Elders in restoring sacred sites and the ceremonies associated with these spaces.

This often involves paralegal training, whereby the EarthLore Foundation assists community members in understanding their constitutional rights and their means of exercising those rights. In the case of Venda, South Africa, the EarthLore Foundation worked collaboratively with the local "protectors" of the sacred natural sites to formally register three sacred community forests with the South African Heritage Resources Agency (SAHRA) and then the World Conservation Monitoring Centre. Registering these sites not only offers a higher level of protection from encroaching development initiatives, but also "firmly [establishes] the customary rights of the local communities" over each site ("Sacred Lands" 2016). This work exemplifies the potential of centring local knowledge and practices in the face of shifting climatic conditions and ever-growing threats from mining and industrial exploitation. Further, the strategies and tools employed by the EarthLore Foundation and their community partners remain transferable and adaptable to community needs.[2]

Prolonged displacement, particularly as a result of climate change, is not often regarded as an immediate concern in the Great Lakes watershed. In fact, it has

2 Participatory mapping projects are similarly employed throughout the Great Lakes watershed. One such example, facilitated by Carly Dokis and Paige Restoule, is currently underway with Dokis First Nation (Dokis & Restoule 2017).

even been suggested that southern Canada may stand to "benefit" from climate change—by way of extended growing seasons and the opening up of shipping routes due to melting ice in the Arctic (Klein 2010). Yet, such "benefits" come at a high cost—not only to the exploited communities of South Africa, but to Turtle Island's own peoples, and disproportionately to the First Peoples.

As a result of colonialism, and the consequent inequality and exploitation, First Nations, Inuit, and Metís communities have horrifically and unjustly borne the brunt of Canada's industrial expansion. This is perhaps most apparent through the widespread contamination of air, water, and land. Though such contamination has not always resulted in the physical relocation of entire communities, it has, as illustrated above, significantly challenged the relationship Indigenous communities traditionally share with their surroundings, resulting in impacts that run parallel to those experienced in more conventionally conceived of cases of displacement.

Recognizing the interrelated challenges associated with contemporary land treatment, the Women's Earth Alliance and the Native Youth Sexual Health Network teamed up to create the Violence on the Land, Violence on our Bodies initiative. In 2016, after listening extensively to the ways in which impacted communities are healing themselves and their lands, the team released a report and toolkit. The resulting resource aims to "center the experiences and resistance efforts of Indigenous women and young people in order to expose and curtail the impacts of extractive industries" on their peoples and lands (Konsmo and Pacheco 2016). As co-authors Konsmo and Pacheco (2016) explain,

> This framework is an alternative to mainstream responses that often see the bodily impacts of environmental destruction as being solved by increased policing or criminalization, rather than community-based solutions developed by those most often impacted.
>
> *(p. 6)*

The toolkit, which "offers both guidance and support for developing and strengthening culturally-rooted, nation-specific responses" to environmental violence,[3] strives to assist Indigenous communities in identifying the connections between the way their bodies and the land is being exploited by the extractives industry (Konsmo and Pacheco 2016). For example, the Environmental Violence Assessment tool (included in the toolkit) invites communities to document the specific ways violence is manifesting across a number of areas of impact (e.g. Indigenous Governance; Drug and Alcohol Use) in their community. It

3 Defined as: "the disproportionate and often devastating impacts that the conscious and deliberate proliferation of environmental toxins and industrial development (including extraction, production, export and release) have on Indigenous women, children and future generations, without regard from States or corporations for their severe and ongoing harm" (Konsmo & Pacheko, 2016 p. 62).

then allows the community to name the true perpetrator of the violence and to identify appropriate action plans. Similar to the eco-cultural mapping process, this helps to orient and organize impacted communities around a shared resistance strategy (Eco-Cultural Mapping, 2015). Other activities, for instance, encourage contemplation on the connection between Free, Prior and Informed Consent (FPIC) over one's body and over the land. Recognizing the feelings of pain and grief that may surface as a community employs such tools, it is suggested that nation-specific medicines be made available to participants. Examples of this type of support are also included in the toolkit.

Looking ahead

It is understood that as environmental conditions worsen due to climate change, parts of the world may disappear into the sea or become entirely uninhabitable. In such cases, participatory relocation and resettlement strategies need to be considered. Yet, (arguably a larger) part of the global and local response to the current and ensuing injustice of land dispossession needs to continue to address the desire and the right of communities to maintain (or regain) their traditionally occupied spaces and places. In small but growing pockets of the world, such strategies are gaining traction. It is now up to us, as scholars, policymakers, advocates, and impacted community members, to piece our supportive efforts around such community-led responses, for when a community is enabled to reclaim their land, their ability to adapt to pending climate change (amongst other) challenges will be significantly strengthened—perhaps pushing us one small step closer to achieving climate justice.

References

Cook, K. (2003) "Women Are the First Environment." *Indian Country Today Media Network*. Available at: http://indiancountrytodaymedianetwork.com/2003/12/23/cook-women-are-first-environment-89746

Dokis, C. & Restoule, P. (2017) "A World Covered in Stories." Challenging Canada 150: Nipissing University. Available at: https://challengingca150.nipissingu.ca/paper-abstracts/day-4-friday/

"Eco-cultural mapping," (2015) The EarthLore Foundation [Online]. Available at: http://earthlorefoundation.org

Gray, V. (2016, October 17) Guest Speaker—Environmental Law and Justice Course. Toronto, ON: York University.

Gray, V. & Gray, B. (2017) Personal communication.

International Organization on Migration (2019) *Migration, Climate Change and the Environment*. Available at: www.iom.int/complex-nexus#estimates. Accessed July 2019.

Klein, N. (2010) "Paying Our Climate Debt" in Sandberg, L.A. & Sandberg, T., eds. *Climate Change –Who's Carrying the Burden*, Ottawa, ON: Our Selves, Our Books, p. 60.

Konsmo, E. & Pacheco, A.M.K. (2016) "Violence on the Land, Violence on our Bodies," [online]. Available at: http://landbodydefense.org/uploads/files/VLVBReportToolkit_2017.pdf

Kukarenko, N. (2011) "Climate change effects on human health in a gender perspective: Some trends in Arctic research," *Global Health Action*, 4, p. 7913.

Kuokkanen, R. (2008) "Globalization as racialized, sexualized violence," *International Feminist Journal of Politics*, 10(2), pp. 216–233. Available at: https://doi.org/10.1080/14616740801957554

MacDonald, E. & Rang, S. (2007, October) "Exposing Canada's Chemical Valley: An Investigation of Cumulative Air Pollution Emission in the Sarnia, Ontario Area." Ecojustice [online]. Available at: www.environmentalhealthnews.org/ehs/news/2012/2007-study.pdf

McGregor, D. (2018) "Taking Responsibilities for the Water." Ottawa, ON: Indigenous Women's Leadership Forum.

Mackenzie, C., Lockridge, A. & Keith, M. (2005) "Declining sex ratio in a first nation community," *Environmental Health Perspectives*, 113, p. 10. Available at: https://doi.org/10.1289/ehp.8479.

Maldanado, J.K., Shearer, K., Bronen, R., Peterson, K. & Lazrus, H. (2013) "The impact of climate change on tribal communities in the US: Displacement, relocation, and human rights," *Climatic Change*, 120(3), pp. 601–614. Available at: https://doi.org/10.1007/s10584-013-0746-z.

National Institute of Child Health and Human Development. (2010) "Research on Miscarriage and Stillbirth." Available at: www.nichd.nih.gov/womenshealth/research/pregbirth/miscarriage_stillbirth.cfm

Nixon, R. (2013) *Slow Violence and the Environmentalism of the Poor*. Cambridge, MA: Harvard University Press.

"Sacred Lands." (2015) The EarthLore Foundation. Available at: http://earthlorefoundation.org

Sandlos, J. & Keeling, A. (2016) "Toxic legacies, slow violence, and environmental injustice at Giant Mine, Northwest Territories," *Northern Review*, 42, pp. 7–21. Available at: http://journals.sfu.ca/nr/index.php/nr/article/view/566

Scott, D. (2008) "Confronting chronic pollution: A socio-legal analysis of risk and precaution." *Osgoode Hall Law Journal*, 46(2), pp. 293–343. Available at: https://digitalcommons.osgoode.yorku.ca/cgi/viewcontent.cgi?referer=https://www.google.com/&httpsredir=1&article=1196&context=ohlj

"United Nations Declaration on the Rights of Indigenous Peoples." (2017) "United Nations. Department of Economic and Social Affairs." Available at: www.un.org/development/desa/indigenouspeoples/wp-content/uploads/sites/19/2018/11/UNDRIP_E_web.pdf

Vinyeta, K., Whyte, K. & Lynn, K. (2015) "Climate change through an intersectional lens: Gendered vulnerability and resilience in indigenous communities in the United States." United States Department of Agriculture. Available at: www.fs.fed.us/pnw/pubs/pnw:gtr923.pdf

White, G. (2011) "Climate-Induced Migration: An Essentially Contested Concept" Chapter 1 in White, G. ed. *Climate Change and Migration: Security and Borders in a Warming World*. New York, NY: Oxford University Press, pp. 16–28.

Whyte, K. (2016) "Is it Colonial Déjà Vu? Indigenous Peoples and Climate Injustice," *Humanities for the Environment: Integrating Knowledges, Forging New Constellations of Practice*. Available at: https://static1.squarespace.com/static/55c251dfe4b0ad74ccf25537/t/5830ca4ef7e0ab3a3c8af727/1479592529139/Colonial+Deja+Vu%2C+IP+Climate+Justice+11-19-16.pdf,pg. 5

Whyte, K. (2017) "Indigenous Food Systems, Environmental Justice, and Settler-Industrial States" in Rawlinson, M. & Ward, C., eds. *Global Food, Global Justice: Essays on Eating Under Globalization*, Cambridge Scholars Publishing, p. 153.

Wiebe, S. & Konsmo, E. (2014) "Indigenous Body as Contaminated Site? Examining the Struggles for Reproductive Justice in Aamjiwnaang" in Paterson, S., Scala, F. & Sokolon, M.K., eds. *Fertile Ground: Exploring Reproduction in Canada*, Montreal, QC: McGill-Queens University Press, pp. 325–358.

Yonetani, M. (2015) "Global Estimates 2015: People Displaced by Disasters." Available at: www.internal-displacement.org/sites/default/files/inline-files/20150713-global-estimates-2015-en-v1.pdf

7

INTERNATIONAL ADVOCACY FOR CLIMATE VICTIMS IN BANGLADESH

Nowrin Tabassum

The excessive carbon emissions of the Western-industrialized countries are responsible for the creation of global warming, but zero/low carbon-emitting countries have been suffering from the severe impacts of climate change which they did not cause. The complex cross-border impacts of global warming raise the questions: How can climate justice be established globally in a situation where low carbon-emitting countries have a minimal role in producing global warming, but they have been suffering the most? Do we need to focus on the local impacts of climate change and then transmit that local information into global forums for establishing climate justice? What kind of activism can bridge knowledge of local sufferings together with knowledge about global climate change, its causes, and ways to address those causes? What is the path by which we envision steps toward climate justice taking place?

Activism can take many forms. It can be violent protests against an unlawful activity, or a peaceful campaign for establishing social and political justice on certain issues. In this chapter, I discuss how activism can also consist of sharing knowledge—the knowledge of local sufferings—and connecting that knowledge to global forums of action. The knowledge sharing can educate global forums about who has been suffering in the local areas, what is the reason for their vulnerabilities, how to respond to their sufferings, and who can take what actions in response. In this knowledge-sharing process, diasporas—people living abroad, far from their countries of origin and heritage—can be useful in collecting knowledge of local sufferings in their original homeland and transmitting the knowledge to global audiences or activist groups in their adopted countries, and in this way help to connect local issues with global activism. The strength of diaspora is that the members of a diaspora can speak their local languages, understand local conditions, and gain easy access to information in their home country. They can collect information from the local climate change-affected people and bring details about the victims' sufferings to the attention of activists in their host countries.

Simply by being a member of the Bangladeshi diaspora—now living on the shores of Lake Ontario—I have become a member of the group of activists working for global climate justice. I collect local information about people's sufferings and transmit the information to activist groups in the Great Lakes Watershed. The venues I choose for sharing local information about climate change's impacts on people in Bangladesh are mainly academic forums.

Why academic forums? I have a story about choosing academic forums. I came to Canada in 2012 as a permanent resident and I am now a doctoral candidate in the Department of Political Science, McMaster University, Hamilton, Ontario. I wanted to focus my doctoral research and my professional career on climate change impacts in Bangladesh because I am personally aware of how serious these impacts are, and how under-reported the situation is. My academic research gives me access to academic forums—organized in the Great Lakes watershed—which pinpoint socio-political injustices and prescribe remedies for these problems. The academic forums include workshops, seminars, international conferences, academic publications, and knowledge-sharing talks, organized mainly by universities in collaboration with government agencies, international institutions, and non-governmental organizations. So, the forums are excellent places to share information with governmental, non-governmental, and international actors. For this reason, I have chosen these academic forums to share information. This book is such a forum in which I am sharing information on climate-affected people in Bangladesh.

This chapter is the true story of Bangladeshi-born climate refugees who have already been displaced from their homes due to climate change-induced disasters, and the response to their sufferings at the national and international levels.

It is also my story. Because I speak Bengali, I was able to interview people in Bangladesh during my research fieldwork, who were, in particular, slum-dwellers, government officials, and climate scientists. As a "foreigner," based in Canada, I was able to ask questions and pursue topics that local academics might not have for various political and career-related reasons. As a Hijab-wearing woman, I had access to many Muslim families and I could speak to the women of the families and include more diverse views that are represented in most research on Bangladesh, which is largely conducted by men.

In this chapter, I outline my review of the existing literature on climate refugees, as well as my research findings, and I conclude by presenting two opposing views: one from the viewpoint of Bangladeshi climate scientists and policymakers, and the other from the perspective of climate victims who have been uprooted from their living places due to climate change-induced disasters. The former viewpoint suggests implementing donor-funded resilience projects for making climate refugees resilient in dealing with the effects of climate change, and the latter view demonstrates that the victims cross national borders by ignoring the resilience projects and official immigration processes. I also reflect on questions of access to knowledge, power, and equity, and how the viewpoint of the storyteller matters. It seems to me that my story may indicate how the global diaspora makes possible new stories and solidarities to confront climate change.

Review of existing literature on climate refugees and the case of Bangladesh: Who are climate refugees?

Many studies on the socio-economic impacts of climate change conclude that there is a direct connection between the damaging effects of climate change and subsequent population movements. Examples of this literature are Docherty and Giannini's (2009) analysis on sea level rise and climate refugees, Biermann and Boas' (2010) discussion on climate refugees and their protection regime, Guzman's (2013) book on the human cost of climate change, and the 1990 First Assessment Report of the Intergovernmental Panel on Climate Change (IPCC). The way the literature frames the connection between climate change and population movement is this: climate change has uncontrollable and detrimental effects on the environment and ecology, which in turn have the potential to make lands uninhabitable, leading to deterioration in the living conditions of their inhabitants; as a result, the inhabitants migrate from their homes in search of new livelihoods (Guzman 2013, pp. 11–18, 63–71). Some of the most vulnerable countries to climate change are small island nations and countries at lower altitudes, these being more prone to inundation as a result of climate change-induced sea level rise. Among those at risk are the Pacific Island countries (such as Fiji, Kiribati, Samoa, Solomon Islands, Tonga, Tuvalu, and Vanuatu), Bangladesh, and the Maldives (Guzman 2013, pp. 54–96). Climate change is also causing droughts and water shortages in Yemen, Syria, the entire Arabian Peninsula and Persian Gulf Coast, and Northern Africa (also many parts of Sub-Saharan Africa), and people in these areas are prone to leave their homes for lack of water (Chellaney 2013, pp. xxi, 161–165).

The authors who accept that climate change can cause population movement are divided regarding the issue of how to label the people thereby uprooted. Methmann and Oels (2015: 52–58) observe that the early science and policy documents from the 1980s and 1990s did indeed label the climate change-induced uprooted people as climate refugees; academic and non-academic literature did likewise. For example, Frank Biermann and Ingrid Boas—Professor of Environmental Policy at the University of Amsterdam and Professor of Climate Governance at Wageningen University, respectively—stated that climate refugees can be defined as:

> people who have to leave their habitats, immediately or in the near future, because of sudden or gradual alteration in their natural environment related to at least one of three impacts of climate change: sea-level rise, extreme weather events, and drought and water scarcity.
>
> *(Biermann and Boas 2009, p. 67)*

For Biermann and Boas (2009), the people can be internally displaced people (IDPs), or they can also cross their national borders to take shelter in a foreign country. However, Bonnie Docherty and Tyler Giannini (2009), two distinguished lecturers at Harvard Law School, disagree with the definition above. For them, "refugees" must cross national borders and IDPs are not refugees; but climate change and its effects should not be restricted into only three categories:

sea-level rise, extreme weather events, and drought and water scarcity (Docherty and Giannini 2009, pp. 367–372). For Docherty and Giannini (2009), the displaced person should fulfil the following requirements to be called a climate refugee:

a. The migration must be forced migration
b. Their relocation can be temporary or permanent
c. They must move across national borders
d. The cause of migration must be consistent with the disruption of climate change
e. It can be a sudden or gradual environmental disruption, and
f. A "more likely than not" standard for human contribution to the disruption

(Docherty and Giannini 2009, p. 372)

However, no international/regional organization or country treats climate refugees as one of the recognized categories of refugees. Remarkably, in 2005, the United Nations Environmental Program (UNEP) published a colour-coded map on their website entitled "50 Million Climate Refugees by 2010". The publication of the map can be interpreted as the UNEP, a unit of the United Nations, accepting that climate refugees are one of the recognized categories of refugees; however, in 2010, the UNEP deleted the map from its server (*Wall Street Journal* 2011, p. 1). The deletion of the map indicates that the UNEP was not confident about treating climate change-induced uprooted people as one of the recognized categories of refugees.

Although the term climate refugee has no legal recognition worldwide, in this chapter I consider all sorts of climate change-induced uprooted people as climate refugees; they may be both IDPs and cross-border migrants.

Who are Bangladesh's climate refugees?

Among all the countries endangered by climate change, Bangladesh may potentially generate the maximum number of climate refugees because of its huge and dense population (almost 1,300 people per square kilometre), and because several effects of climate change are evident in the country simultaneously (Biermann and Boas 2009, pp. 8–17). Examples include the following:

River bank erosion

Global warming is quickly melting the Himalayan glaciers, which flood the major river basins of Bangladesh flowing into the Ganges-Brahmaputra-Meghna river delta. This overflow is causing massive riverbank erosion, which uproots people. I contacted the Water Development Board and the Ministry of Environment and Forests—the two official bodies concerned in Bangladesh—to collect statistics on climate change-induced riverbank erosion and the subsequent population movement in Bangladesh. The Board and Ministry did not provide me with the information. However, academic documents suggest that global warming-induced riverbank erosion has already uprooted 100,000 people (Dastagir 2015, p. 49; Naser 2012, pp. 63–64; Leckie et al. 2011).

Sea level rise

Melting glaciers of the Himalayas and thermal expansion of the Indian Ocean have caused sea levels to rise, leading to the submersion of coastal land, which results in population movement. The 1990 First Assessment Report of the IPCC, for example, states:

> In coastal lowlands such as in Bangladesh, China and Egypt, as well as in small island nations, inundation due to sea-level rise and storm surges could lead to significant movement of people.
>
> *(IPCC 1990, p. 3)*

The 2000 World Bank Report 21104-BD, entitled *Bangladesh Climate Change and Sustainable Development*, describes sea level rise as reaching 10 cm by 2020, which will inundate 2% of the country's land area (World Bank 2000, p. 40). The document also explains that the situation for Bangladesh will continue to worsen, with a sea level rise of as much as 25 cm by 2050, causing land inundation of up to 4% of the country (including 40% of the Sundarbans, the location of the largest mangrove forests of Bangladesh), and an increase in storm surges (World Bank 2000, p. 40). The same document also estimates that sea level rise will be 100 cm in 2100, which will inundate 17.5% of the country's land area, including the entire Sundarbans, with the associated increase in storm surges causing the displacement of 20 million people (World Bank 2000, p. 40).

Bangladesh's Bhola Island, almost 1,500 square kilometres in size, has been reported to have half of its landmass submerged already, displacing half a million people (Docherty and Giannini 2009, p. 356). Another example is Kutubdia Island, almost 100 square kilometres in size, which is likely to be submerged due to rising sea levels, threatening its 125,000 inhabitants.

However, I could not collect any exact data from government sources on how many people are already displaced or about to be displaced due to the submersion. Many of the residents of this island have emigrated to the Chittagong Hill Tracts where the land is above sea level. The Chittagong Hill Tracts is an area where many aboriginal people live.[1] According to one of my interviewees, the emigration of the islanders to the Hill Tracts has been reported to cause social unrest in the area. The interviewee added that the aboriginal people view the emigration as land grabbing by the islanders whereas the islanders state that they have bought the lands and so they have rights to live in the area.

1 Aboriginal people in India and Bangladesh, the continent's earliest inhabitants, often face discrimination on ethnic grounds. See Cultural Survival (2000).

Severe cyclones and tropical depressions

The Bay of Bengal is producing more severe storms and tropical depressions due to the rising temperature. From 1986 to 2009, "the Bay of Bengal produced on an average 5.84 storms per year and global warming, and climate change can increase the numbers of storms up to 7.35 per year" (Chowdhury et al. 2012, p. 20). The warmer temperature also produces new breeding zones for severe cyclones (Chowdhury et al. 2012, p. 20). Examples of such areas are the latitudes between 15° and 19° North: the zone absorbs the warmer weather faster and gathers more strength and moisture, producing severe cyclones (Chowdhury et al. 2012, pp. 12–20). Bangladesh experienced the severe cyclone Aila in 2009: this was the first severe cyclone to have originated from the latitude 15° to 19° North (Chowdhury et al. 2012, pp. 12–20). Bangladesh had not experienced such a storm since records began; its cause was the unprecedented warm weather conditions (Chowdhury et al. 2012, pp. 12–20). Cyclone Aila in 2009 displaced more than 70,000 families, permanently destroying their homelands (Dastagir 2015, p. 49). Thus, the increased number of severe cyclones create climate refugees.

In addition to the severe cyclones, the number of less severe cyclones, such as tropical depressions, has also increased in recent years. The Bangladesh Meteorological Department maintains a chart of the early warning signals for cyclones and tropical depressions: it announces the signal numbers in the media to alert people to take precautionary measures as needed. Among the signals, the *Local Cautionary Signal No. III* (LC III) is used to inform people that a tropical depression with a wind speed of 40–50 km/hr will soon take place. Since 1998, the Bangladesh Meteorological Department has announced LC III more frequently.[2] In the three consecutive years 2007, 2008, and 2009, there were more than 20 tropical depressions each year, whereas in the previous years they had not exceeded 15. This indicates that the Bay of Bengal is warming up; the seawater and air are expanding, which results in more tropical depressions. A climate scientist and IPCC member I interviewed in Bangladesh remarked:

> When the Bangladesh Meteorological Department announces Local Cautionary Signal No. III, the fishermen in coastal Bangladesh (particularly, Chittagong), are prohibited to go fishing because the tropical depression makes the sea rough for fishing in the open sea, which can also be life-threatening. In this circumstance, if any fisherman goes fishing when the signal is on, the government of Bangladesh will not take any responsi bility for losses and damages that the fisherman encounters. The increased numbers of tropical depressions already cut the numbers of days for fishing,

2 See the increased number of signals for tropical depressions from 1998 to 2009 on the Bangladesh Meteorological Department website http://bmd.gov.bd/p/Signals/ (Data for tropical depressions between 2009 and 2017 are not available.)

> and as a result, they cannot earn [a living] by fishing when the sea is rough. Consequently, they have a lesser amount of catch and a lesser amount of income. The situation forces them to live in starvation due to lack of earnings. To avoid the starvation, the fishermen either take risks of their life, ignore the Cautionary Signal and go fishing, or alternatively, they migrate from the coastal areas to the highlands.

In my view, these fishermen must also be counted among the climate refugees in Bangladesh because the increase in climate-change-induced depressions forces them to risk their lives to seek an income, or else to depart from their homelands.

International advocacy and Bangladesh's climate refugees

As mentioned earlier, no international/regional organization or country has classified climate refugees as one of the recognized categories of refugees. The Western-industrialized countries remain against recognition of the concept of climate refugees because it would be an implicit source of blame on them for their excessive carbon emissions. Australia, for example, is the highest per capita carbon dioxide emitter in the world, and never recognizes climate refugees (Karasapan 2015, p. 3): the Australian Labour Party proposed in 2006 to accept climate refugees from the Pacific Island countries, but this was rejected by the Australian government (Biermann and Boas 2010, p. 66). In 2007, the Australian Green Party tabled a bill at the Australian parliament named the Migration Amendment Bill 2007 to recognize climate refugees (Parliament of Australia 2007), but this effort also was not successful. By 2015, New Zealand and Australia had rejected 17 applications from the Pacific Island countries seeking climate refugee status (O'Brien 2015, pp. 7–10).

The non-recognition of climate refugees indicates that no specific protection law exists for the people displaced by climate change. For this reason, no country is bound to provide a haven to these individuals. How do the climate refugees survive in this world without having a protection regime?

Some authors note that, in contrast to the idea that refugees are helpless, climate refugees may seem to be active agents, not passive victims, because they can raise their voices against the fossil fuel-based Western neoliberal economic system—which is the main cause of their uprootedness—and they can take their own responsibility (McNevin 2006, p. 136; Rajaram 2002, pp. 251–254; Hartmann 2010, pp. 238–242).

On the other hand, some climate scientists and policymakers from climate-affected countries, who were adamant in demanding refugee status for climate victims at many international conferences in the early 2000s, have surprisingly made a complete U-turn from their previous position (Methmann and Oels 2015, p. 52). They now argue that climate-affected people should be "resilient" in facing climate change and should learn how "to prepare for the unavoidable effects of climate change" (Methmann and Oels 2015, p. 52). At the national government level, the Pacific islands and low-lying countries have stopped using the term climate refugees and replaced it with that of "climate change-induced migration" in their official government documents and policy papers. It is worth noting that these countries also have accepted many Western donor-funded climate change adaptation and resilience projects to attempt to curb the disastrous effects of climate change.

The position of the Bangladeshi government, its major climate-oriented non-governmental organizations, and some Bangladeshi climate scientists seems to follow the trend—not using the term climate refugees and replacing it with that of climate change-induced migration. During the 1990s and early 2000s, they stood for the radical position of holding high carbon emitters responsible for compensation and providing shelter (McAdam 2011, p. 6), while more recently they have moved away from this position. I wondered what drove these climate scientists, policymakers, and the government of Bangladesh to shift from their previous more radical position regarding climate refugees.

In order to find out the answer, I visited Bangladesh from October 2016 to March 2017. I conducted interviews with some climate scientists and policymakers in order to understand their view on climate change victims. I also visited some of the victims who had migrated, in order to grasp their opinions regarding the shifting of responsibility and lack of advocacy by policymakers. My visit to Bangladesh provided me with opposing views: one came from the climate scientists and the policymakers whom I interviewed, and the other one from the victims who had migrated due to climate-induced disasters.

The "official" view

According to the climate scientists and policymakers who spoke to me, Bangladesh will not recognize the climate victims as "refugees" unless the international community classifies them as such. The following discussion clarifies why.

Bangladesh's economy is heavily dependent on foreign funds which are donated by the Western countries, who are mainly high carbon emitters. These funds are channelled into Bangladesh through, in particular, the World Bank and Asian Development Bank (ADB). The Western countries, as mentioned above, do not support recognition of climate refugees, although these countries, as per the 1992 United Nations Climate Change Conference, have agreed to provide funds to climate change adaptation projects in developing countries in order to curb the effects of climate change (UNFCCC 1992, article 4). These projects are intended to help the developing countries transfer technology and technological know-how, and build massive infrastructure projects intended to address climate change (Martin 2010, pp. 400–403; Naser 2012, pp. 114; Rai et al. 2014, pp. 527–543; UNFCCC 1992, article 4). The funded projects include (i) building sea-walls and embankments to protect against sea-level rise and riverbank erosion, (ii) coastal forestation projects which can deter the severity of cyclones,

(iii) planting saline-tolerant crops to reduce food scarcity caused by saline intrusion of arable lands, and (iv) technological support for introducing the "green economy" and reducing their own carbon emissions (Huq 2011, pp. 56–69; Karim and Mimura 2008, pp. 498; Martin 2010, pp. 403; Rai et al. 2014, pp. 527–543; World Bank 2006, pp. 14–24). The idea in implementing these projects is to prevent climate change effects so that there will be no climate change refugees (Bettini and Gioli 2015, p. 2; Huq 2011, pp. 56–69; Karim and Mimura 2008, p. 498; Martin 2010, p. 403; World Bank 2006, pp. 14–24).

Bangladesh has received four large funds from foreign donors for the implementation of the above-mentioned projects: (i) Bangladesh Climate Change Resilience Fund (BCCRF) ($170 million), (ii) Global Environmental Facility's (GEF) Least Developed Countries Fund (LDCF) (information on the amount of money is not publicly available), (iii) Pilot Programme for Climate Resilience (PPCR) of the Climate Investment Fund (CIF) (approximately $400 million), and (iv) the Green Climate Fund (GCF) (approximately $100 million) (Rai and Smith 2013, p. 12). However, after analysing the actual budgets of the projects, I have found that there exist unexplained anomalies regarding the amount of money spent on the projects. One project—Climate Resilient Participatory Afforestation and Reforestation Projects (CRPARP)—reportedly exceeds $75 million whereas the World Bank's website shows that the amount is $35 million.

According to my interviewees, while receiving such lucrative funding, Bangladesh must support the prescriptions (i.e. implement climate change resilience projects) suggested by the donors. According to one of my interviewees, who is a university professor and also works as a policy consultant for the government of Bangladesh,

> The government of Bangladesh does not adopt any policy, including decisions on climate change adaptation projects, which is not supported by the donors. As a result, Bangladesh cannot take any decision at the national level by bypassing the donors. If Bangladesh tries to do so, the donors withdraw their climate change-related funds from Bangladesh, and consequently, Bangladesh faces severe economic losses. The issue of climate refugee is not supported by the donors: the World Bank and developed countries. Therefore, the actors who were very vocal in international forums for recognizing the climate refugees and influencing the donors remained silent at the national level and did not press for adopting a policy related to the issue of climate refugees.

However, the climate-induced disasters and the subsequent migration are already taking place. As a result, turning a blind eye to the problem means that those affected by such changes do not have access to any resources to help with migration and resettlement. The irony is that, as fast as they are built, the donor-funded adaptation projects are being destroyed by some effects of climate change: (i) polders and embankments (under construction) are severely damaged by increased attacks of cyclones, coastal flooding, and sea level rise; (ii) forestation efforts are washed away and destroyed by frequent cyclones and floods; (iii) the saline concentration in the land is levels beyond the tolerance threshold of saline-tolerant crops; and (iv) excessive salt water has penetrated drinking water channels causing health problems such as spontaneous abortions in pregnant women (Rawlani and Sovacool 2011, p. 860; Yamamoto and Esteban 2014, p. 56). The failure to recognize the severity of climate change in Bangladesh in forced efforts to appease donors has done little to curb the number of climate refugees in Bangladesh.

The migrants' view

While I was in Bangladesh, I searched for the locations where climate change victims have made new homes. I spoke with people, surveyed newspaper articles, and read online publications of the United Nations High Commission for Refugees, the United Nations Framework Convention on Climate Change (UNFCCC), and the World Bank. I concluded that many migrants were widely dispersed throughout Bangladesh, except in two areas (some neighbourhoods in Dhaka city and a village in the Chittagong Hill Tracts) where there was a high concentration of climate refugees. Reports of the United Nations Development Programme, the *National Geographic*, and the *Guardian* mostly mention the Korail slum area in Dhaka city as being a place where climate refugees live (McPherson 2015; Shachi 2015). Other slums cited in local newspapers in Bangladesh, where climate victims live, are Badda and Bhola (in Mirpur) in Dhaka city. Climate scientists and policymakers in Bangladesh whom I interviewed recommended that I visit the Korail and Bhola neighbourhoods to observe the lifestyle of these refugees.

I visited these slums to gain first-hand knowledge about the people affected in November and December 2016. I found that the large Korail slum houses more than 100,000 inhabitants. The Badda slum area has approximately 3,000 residents, while the total population of the Bhola slum is about 5,000. All the slums contain both climate victims and other migrants.

I visited the Bhola slum three times, and then, in early January 2016, I received a death threat over the phone that stated that if I visited the slum again, I would be killed. In trying to make sense of this, I realized that I had already noticed something strange about this slum. The Bhola slum was established in 1970. After a devastating cyclone in 1970 in Bhola Island, people from the cyclone-affected area moved to the slum in Dhaka city and established their community there. The 1970 cyclone has not been identified as a climate change-caused cyclone. So, the people who moved there cannot be called climate refugees or climate change-induced displaced people. When I visited the slum, I found something odd. First of all, it is a very organized and developed slum. The men who live in the slum have secured jobs in the garment factories of the neighbourhood. The women in the slum wear gold jewellery (such as earrings and nose-pins, which seemed to be expensive). Children's education in the slum is provided free—funded by World Vision. The slum-dwellers have a contract with the government of Bangladesh which states that the dwellers own the houses and they do not have to pay any rent or property taxes to the government. I tried but failed to obtain a copy of this contract. To me, the slum-dwellers did not seem to be displaced but rather owners of houses in the slums. So, projecting the slum-dwellers as displaced persons was misleading and wrong.

I also came to know from local Bangladeshi newspapers and a few of my interviewees that some research organizations (such as ICCCAD, the International Centre for Climate Change and Development) have provided funds to develop the slum and to make the slum "accessible and secured" for foreign researchers. ICCCAD, funded by the United Kingdom Department for International Development (DFID), conducts research on building communities' resilience to climate hazards and collects traditional wisdom for climate change adaptation and monitoring (Independent Commission for Aid Impact 2011, p. 22).

ICCCAD hosts many international conferences and workshops on climate change and supports many international scholars who conduct field studies in the Bhola slum regarding climate change-induced migration (see www.icccad.net/tag/bhola/). The whole slum gave me a feeling that it was established by the people who migrated from Bhola Island but later on that it was artificially developed for displaying (showcasing) fake climate change-induced uprooted people for some specific audiences. In making the artificial slum, I found both national (such as ICCCAD) and international (such as IOM, International Organization for Migration) actors to be involved. Maybe my suspicions about the slum disturbed some people who were involved in funding/showcasing the slum, and perhaps they wanted to scare me and stop my visiting the slum by means of the phone threat. They were a bit successful. The issue of the phone call hampered my field research after I informed my advisors at McMaster University. The University gave me three options and advised me to choose one:

(i) Leave Bangladesh and return to McMaster as soon as possible. (The School of Graduate Studies would adjust the damage caused by my early return from the field research).
(ii) Stay in the field but follow some conditions prescribed by McMaster University (i.e. use an International SOS for emergency contact, not visit the slums anymore, etc.).
(iii) Take a full-time off-campus leave (if I did not want to follow the above-mentioned conditions but wanted to stay in the field).

I chose the second option, stayed in the field, and did not visit the slums anymore.

However, my previous visits to other slums gave me some ideas. I discovered that climate victims do not subscribe to the limitations imposed by statehood and national boundaries. According to them, the "state" is an elite club which is produced to serve the interests of the elites by ignoring the well-being of the non-elites and poor. Their definition of elite includes those who are privileged, well-educated, have a secure job, and do not suffer from climate-induced calamities. In addition, the climate victims do not care about state immigration systems and border security forces (which are definitely shaped and managed by the elites, as they stated). Therefore, many relatives of the climate victims, as they told me, have crossed national borders without a passport or visa, and gone to India. They chose their destination in India because it was closer to their home than another part of Bangladesh.

Since 2004, India has been constructing a 2,500 mile long wire fence along the Bangladeshi border to prevent a mass influx of migrants from Bangladesh (Buckley 2014, p. 200; Guzman 2013, p. 62). For this reason, it was a surprise for me to understand how they could manage to cross the long wire fence without the proper documentation—visa and passport. One victim explained to me that they did not cross the border through the fence, but instead dug an underground tunnel connecting their village to one in India and then crossed the border through the underground tunnel (underneath the fence) by avoiding the border security of both countries—Bangladesh and India. They managed to escape from climate-induced disasters by migrating to another place without the help or permission of any authority. Therefore, it seems to me that the climate victims are not passive victims of climate change awaiting rescue from any authority; instead, they are active agents who do what they must do to escape the climate-affected areas, seeking to secure their livelihood.

After I returned to Canada, I started to write my dissertation. I am also presenting papers at conferences, workshops, and seminars, and I am writing articles on my research. Every time I present my findings in public forums, I learn more about how we, the researchers who live in the Western world, can easily be misguided in non-Western situations by fake field sites or artificially created field sites, such as the Bhola slum. Manchester University (UK) researchers worked in the Bhola slum, identifying it as a slum where climate change-induced displaced people live (see http://blog.gdi.manchester.ac.uk/gdipotgan-rich-mix/). As far as I know, they did not realize or question how the field site was produced for manufacturing fake knowledge about the slum and its residents.

In addition, we researchers who live in the Western world have a misconception that climate-affected people in poorer countries view the Western developed countries as saviours for their climate change-induced problems. In practice, climate change affected people are brave enough to cross national borders by ignoring any border security measures and find their new homes in foreign lands. We believe that the donor-funded climate change adaptation projects can be solutions for climate refugees; however, in some cases, perhaps many, this does not work in practice.

I have spoken to other Bangladeshi-Canadian researchers (based at the University of Toronto, York University, and Lethbridge University) who also have similar views. We Bangladeshi-born researchers in Canada find that there exist huge communication gaps between the Western world (such as Canada) and the non-Western world (such as Bangladesh) which generate these misconceptions. It is challenging for the researchers to eliminate the misconceptions because there is no institutional support from either side—Bangladesh or Canada—or public support to reduce the misconceptions. So, my future plan is to do research on climate change-related *fake news, fake information and fake field sites*, and bring these issues to conferences, seminars, workshops, and in some published writings, so people can be aware of this.

As a diaspora researcher, I feel responsible for continuing to try to generate information that will help both my country of birth and my new country to address climate change effectively. The stakes are very high.

References

Bettini, G., and Gioli, G. (2016). Waltz with development: insights on the developmentalization of climate-induced migration. *Migration and Development, 5*(2), 171-189.

Biermann, F. and Boas, I. (2009) Protecting climate refugees: the case for a global protocol. E*nvironment: Science and Policy for Sustainable Development, 50*(6), 8–17.

Biermann, F. and Boas, I. (2010) Preparing for a warmer world: towards a global governance system to protect climate refugees. *Global Environmental Politics, 10*(1), pp. 60–88.

Buckley, M. (2014). *Meltdown in Tibet: China's Reckless Destruction of Ecosystems from the Highlands of Tibet to the Deltas of Asia.* St. Martin's Press: Macmillan.

Chellaney, B. (2013) *Water, Peace, and War: Confronting the Global Water Crisis.* Lanham, MD: Rowman & Littlefield.

Chowdhury, S.R., Hossain, M.S., Shamsuddoha, M. and Khan, S.M.M.H. (2012) *Coastal Fishers' Livelihood in Peril: Sea Surface Temperature and Tropical Cyclones in Bangladesh.* Dhaka, Bangladesh: Center for Participatory Research and Development.

Cultural Survival. (2000) *Are there Indigenous people in Asia?* Available at: www.culturalsurvival.org/publications/cultural-survival-quarterly/are-there-indigenous-peoples-asia (Accessed 25 March 2019).

Dastagir, M.R. (2015) Modeling recent climate change induced extreme events in Bangladesh: a review. *Weather and Climate Extremes*, 7, pp. 49–60.

Docherty, B. and Giannini, T. (2009) Confronting a rising tide: a proposal for a convention on climate change refugees. *Harvard Environmental Law Review, 33*, pp. 349–403.

Guzman, A.T. (2013) *Overheated: The Human Cost of Climate Change.* Oxford, UK: Oxford University Press.

Hartmann, B. (2010) Rethinking climate refugees and climate conflict: rhetoric, reality and the politics of policy discourse. *Journal of International Development, 22*(2), pp. 233–246.

Huq, S. (2011) Lessons of climate change, stories of solutions. *Bulletin of the Atomic Scientists*, *67*(1), pp. 56–59.

Independent Commission for Aid Impact. (2011) *The Department for International Development's Climate Change Programme in Bangladesh* [Online]. Available at: www.oecd.org/countries/bangladesh/49092047.pdf (Accessed: 27 March 2019).

Karasapan, O. (2015) *Refugees: Displaced from the Paris climate change agreement?* [Online]. Available at: http://reliefweb.int/report/world/refugees-displaced-paris-climate-change-agreement (Accessed: 27 March 2019).

Karim, M.F. and Mimura, N. (2008) Impacts of climate change and sea-level rise on cyclonic storm surge floods in Bangladesh. *Global Environmental Change, 18*(3), pp. 490–500.

Leckie, S., Simperingham, E. and Bakker, J. (2011) 'Climate change and displacement reader', *The Ecologist*, 18 April [Online]. Available at: www.theecologist.org/blogs_and_comments/commentators/other_comments/854868/bangladeshs_climate_displacement_nightmare.html (Accessed: 27 March 2019).

LIMUN (London International Model United Nations). (2017) *Study Guide–UNEP* [Online]. Available at: https://limun.org.uk/FCKfiles/File/LIMUN_HS_2018_Study_Guides/LIMUN_HS_UNEP.pdf. (Accessed: 27 March 2019)

Martin, S. (2010) Climate change, migration, and governance. *Global Governance, 16*, pp. 397–414.

McAdam, J. (2011). Swimming against the tide: Why a climate change displacement treaty is not the answer. *International Journal of Refugee Law*, 23 (1). 2–27.

McNevin, A. (2006) Political belonging in a neoliberal era: The struggle of the sans-papiers. *Citizenship Studies, 10*(2), pp. 135–151.

McPherson, P. (2015) 'Dhaka: The City Where Climate Refugees are Already a Reality', *The Guardian*, 1 December [Online]. Available at: www.theguardian.com/cities/2015/dec/01/dhaka-city-climate-refugees-reality (Accessed: 27 March 2019).

Methmann, C. and Oels, A. (2015) From 'fearing' to 'empowering' climate refugees: governing climate-induced migration in the name of resilience. *Security Dialogue, 46*(1), pp. 51–68.

Naser, M.M. (2012) *Protecting Climate Change Induced Displacement in Bangladesh: Legal and Policy Responses* (Doctoral dissertation, Macquarie University).

O'Brien, E. (2015) *An Islander's Bid to be the World's First Climate-refugee* 30 March [Online]. Available at: www.bloomberg.com/news/articles/2015-03-30/an-islander-s-bid-to-be-the-world-s-first-climate-refugee (Accessed: 27 March 2019).

Parliament of Australia. (2007). *Migration (Climate Refugees) Amendment Bill 2007.*

Rai, N., Huq, S. and Huq, M.J. (2014) Climate resilient planning in Bangladesh: a review of progress and early experiences of moving from planning to implementation. *Development in Practice, 24*(4), pp. 527–543.

Rai, N. and Smith, B. (2013) Climate Investment Funds: Pilot Programme for Climate Resilience (PPCR) in Bangladesh–a status review. IIED Country Report.

Rajaram, P.K. (2002) Humanitarianism and representations of the refugee. *Journal of Refugee Studies, 15*(3), pp. 247–264.

Rawlani, A.K. and Sovacool, B.K. (2011) Building responsiveness to climate change through community based adaptation in Bangladesh. *Mitigation and Adaptation Strategies for Global Change, 16*(8), pp. 845–863.

Shachi, S. (2015) 'Climate Refugees and a Collapsing City', *Inter Press Service*, 25 November [Online]. Available at: www.ipsnews.net/2015/11/climate-refugees-and-a-collapsing-city/ (Accessed: 27 March 2019).

The World Bank. (2006) *Managing Climate Risk Integrating Adaptation into World Bank Group Operations*. Washington, D.C.: World Bank.

UNFCCC. (1992) United Nations Framework Convention on Climate Change [Online]. Available at: https://unfccc.int/files/essential_background/background_publications_htmlpdf/application/pdf/conveng.pdf (Accessed: 27 March 2019).

Wall Street Journal. (2011) "Climate Refugees, Not Found: Discredited by reality, U.N.'s prophecies go missing." April 21, p. 1.

Yamamoto, L., and Esteban, M. (2014). *Atoll island states and international law: climate change displacement and sovereignty*. Springer Science & Business Media.

8

REFUGEE SPONSORSHIP AND CANADA'S IMMIGRATION POLICY IN TIMES OF CLIMATE CHANGE[1]

Michaela Hynie, Prateep Kumar Nayak, Teresa Auntora Gomes, and Ifrah Abdillahi

Introduction

Global events, including climate change and related conflicts, have consequences for population flow and resettlement patterns, both within and between countries. A salient example is Canada's response to the recent political crisis in Syria. Between November 2015 and February 2017, Canada resettled over 40,000 Syrian refugees. This was a significant change in policy; as a result, refugees, protected persons, and their dependents represented 20% of the 296,346 new permanent residents in 2016, double the usual numbers (IRCC, 2017). While Canada was applauded for its effort to provide durable solutions to the protracted conflict in Syria, the number resettled represents a tiny fraction of the 5.6 million Syrians who sought asylum across international borders in the first 7 years of the conflict, and the 6.6 million who have been displaced within the country to date (UNHCR 2018), emphasizing how great the need is for more, and more accessible solutions

The Syrian refugees resettled in over 360 communities across Canada (CIC 2018). However, a map of settlement communities shows that they were most densely clustered around the Southern Great Lakes and along the St. Lawrence Seaway, with about 10% settling in Montreal and Toronto (CIC 2018). This pattern of settlement reflects Canada's different resettlement programs. Canada accepts refugees as permanent residents through two pathways: resettled refugees

1 This chapter draws from the authors' longer report, *Environmental Displacement and Environmental Migration: Blurred Boundaries Require Integrated Policies*, which includes more detailed references to the work of other authors on this topic and was made possible by a Social Sciences and Humanities Research Council of Canada grant to the first author. That report is available at: https://refugeeresearch.net/rrn_node/environmental-displacement-and-environmental-migration/

and in-land claims. Normally, approximately half of the refugees and their dependents who settle in Canada enter through each of these paths. Resettled refugees are selected and approved by Canadian officials overseas from among those who have been referred by the United Nations High Commissioner for Refugees (UNHCR). These are individuals who have been assessed as satisfying the United Nations 1951 Refugee Convention criteria for refugees: people with a well-founded fear of persecution on account of their social or political status, who cannot or will not be protected by their home state. In some cases, such as Syria, where there is wide-ranging violence and mass migration, country of origin alone is grounds for asylum, although all resettled refugees must pass additional assessment for approval by Canadian officials.

Resettled refugees are eligible for 1 year of financial and settlement support, although in exceptional cases this may be 6 or 12 months longer. Resettled refugees who are sponsored by the government (Government Assisted Refugees, or GARs) receive financial and settlement support through government-funded agencies. Those refugees facing exceptional vulnerability, such as difficult medical conditions, large families, or single parenthood, are accelerated for resettlement under the GAR program (UNHCR 2018b). These refugees settle in the urban centres where the government settlement programs are run; during the time of the Syrian refugee resettlement, this included 37 communities across Canada, with Toronto and other centres around Southern Ontario settling the largest numbers. Those who are settled under the private sponsorship pathway (Privately Sponsored Refugees, or PSRs) receive financial and settlement support from family, faith groups, NGOs, and groups of five or more Canadian citizens. PSRs can be named and are then approved, and thus this program is often used to support family reunification among those forcibly displaced (Labman 2016). Most PSRs were sponsored through faith communities, but also by community organizations and groups of private citizens (Groups of Five), and while many organizations supported Canadian residents by sponsoring their family members, there were others for whom the sponsored newcomers were strangers. PSRs settle in the communities of their sponsors, and many PSRs were sponsored by the pre-conflict Syrian-Canadian community, who resided primarily in Montreal and Toronto (IRCC 2011), hence the concentration of resettled Syrian refugees in and around Toronto and Montreal. A small number also come through the blended-private/government sponsorship program (Blended Visa Office-Referred, or BVORs), where financial support is shared between government and private sponsors but settlement support is entirely private. Canada is increasing the emphasis on this blended pathway for future resettlements (IRCC 2017).

In the Syrian resettlement initiative, 21,876 resettled as GARs, 14,274 as PSRs, and 3,931 as BVORs (CIC 2018). As these numbers reveal, private sponsorship provides opportunities to provide settlement to an increased number of forced migrants; the private sponsorship model is based on a principle of additionality such that the number accepted under private sponsorship is supposed to add to, rather than replace, Canada's commitments to resettle through government sponsorship (Labman 2016). Private sponsorship also offers a means by which citizens can engage in the process of creating new Canadian citizens (Macklin et al. 2018). The engagement and empowerment of Canadian citizens in the resettlement process may contribute to Canadians' relatively pro-migration stance. Relative to other Organisation for Economic Co-operation and Development (OECD) countries, Canadians have quite positive attitudes towards immigrants and refugees (Environics Institute for Survey Research 2018; IPSOS 2017). This may be partially due to Canada's history as a settler nation, but also its perception of itself as a champion of refugee resettlement, and the opportunities it creates for individuals and communities to choose to participate in the resettlement of refugees (Hynie 2018).

However, Canadian attitudes are more divided than many acknowledge (Environics Institute for Survey Research 2018), with some believing Canada has gone too far in welcoming refugees. This is particularly true with respect to refugees making in-land claims. Changes in US immigration rules by the Trump administration, and their ripple effects, have a direct impact on Canada. One response to the changes in US immigration policy is that an increased number of individuals and families have been claiming refugee status outside of regular border crossings or airports, which is the usual location of claims in Canada. The discourse around these claimants often fails to highlight that their entry, while irregular because it is not through a regular border crossing or entry point, is not illegal. Under international law and the UN Convention, to which Canada is a signatory, people have the right to cross borders to claim asylum and have their claims heard. But in Canada, asylum seekers are typically viewed far less positively than resettled refugees, and are often portrayed as economic migrants who are manipulating the immigration system (e.g. Diop 2014; Molnar 2016). The increased, and inaccurate, discussion of "illegal asylum seekers" and misplaced fears around the burden imposed by the entry of fewer than 30,000 asylum seekers (28,100 in 2018) in a country that hosts 400,000 landed immigrants and temporary foreign workers every year may also be having negative spillover effects on attitudes towards resettled refugees.

The anxiety currently being expressed about the inability to manage Canadian migration is paralleled by a negative global discourse about migration more generally (Hynie 2018). It reveals that the current global refugee system is failing to acknowledge and respond to the reality of current migration patterns. The current refugee regime was adopted in response to the forced migrations and statelessness that arose in Europe following World War II and has not kept pace with changing political realities (Barnett 2002). Those risking their lives to cross the Mediterranean to reach Europe in this past decade, for example, are part of a rise in mixed migration flows of people seeking asylum due to a variety of humanitarian conditions: starvation, exploitation, poverty — conditions often linked to environmental change (Crawley et al. 2016). Indeed, the international community has been struggling with its lack of readiness to respond to the needs of these forcibly displaced migrants. In contrast, the International Displacement

Monitoring Centre (Bilak et al. 2016) notes that 25.4 million people have been displaced by natural disasters every year since 2006, double the number displaced by conflict alone. This number does not take into account migration due to slower-onset environmental changes, which is more difficult to estimate than disaster-related migration, or the role of environmental change in influencing migration indirectly. Nonetheless, there are few national or global agreements on international environmental migration (Popp 2014).

The broader issues of the current patterns of forced migration are not only the complexity of "push" factors of extreme weather plus economic pressures plus violence/war and cultural breakdown. There are questions of how global policies and practices can balance people's "right to remain", how to respond to the tendency for the "best and brightest" to migrate and to be welcomed in destination countries, the need for fairer global wealth and income distribution, and the neoliberal contrast between capital flowing freely across borders while labour faces barriers. Recent efforts to craft and adopt the Global Compact on Migration and the Global Compact on Refugees mark the increasing awareness that migration is a response to rapid global changes that require coordinated international responses, with climate-driven migration poised to become perhaps the greatest challenge to existing systems.

Environmental refugees?

There has been an ongoing debate about applying the term "environmental refugee" or "climate refugee" to those forcibly displaced by environmental change, with environmental scientists generally more in favour of the terms, and migration theorists generally opposed (e.g. Gill 2010; Piguet 2013; Stavropoulou 2008). Support for the language of "environmental refugees" has been reinforced through the leading role taken by the office of the United Nations High Commissioner on Refugees (UNHCR) in moving the agenda forward on recognizing migration as a response to climate change, and in encouraging states to recognize and offer protection to those displaced by environmental changes (McAdam 2014). However, those displaced by environmental factors are not eligible to receive protection under the international refugee regime. Those forced across international borders by environmental change do not meet the legal requirements to be considered "Convention Refugees" (i.e. those meeting the 1951 Convention Relating to the Status of Refugees definition of being forced to migrate as the result of conflict or persecution and a lack of state protection).

Moreover, the distinction between forced displacement and voluntary environmental migration is rarely clear. The term "environmental refugee" is therefore starting to lose ground in favour of "environmental migrant." The International Organization for Migration (IOM) defines environmental migrants as

> persons or groups of persons who, for reasons of sudden or progressive changes in the environment that adversely affect their lives or living

> conditions, are obliged to have to leave their habitual homes, or choose to do so, either temporarily or permanently, and who move either within their territory or abroad.
>
> *(IOM 2014, p. 3)*

Regardless of the terminology, current international instruments are not inclusive enough to recognize environmental migrants or people otherwise displaced by environmental drivers to provide legal protection and respite. There is also no international consensus on who deserves protection under which conditions of international environmental migration (Laczko and Aghazarm 2009; Zetter 2009). The lack of legal and policy frameworks to deal with international environmental migration is particularly alarming given the large numbers of people affected, a number that is expected to multiply in the near future.

Environmental changes as drivers of migration

Environmental changes, conflict, livelihood opportunities, and other drivers of migration are interconnected, making it difficult to clearly state the ultimate factors that determine the decision to migrate (Black 2001; Black et al. 2011; Lilleor and Van den Broeck 2011; Martin 2013). This, combined with a lack of clear international legal definitions pertaining to forced environmental displacement, contributes to a lack of data, making it difficult to estimate the number forcibly displaced by environmental change per se (Black 2001). A common estimate, however, is that 200 million people will be displaced by environmental changes by 2050 (Brown 2007; Stern 2006, p. 3). The rate and extremity of environmental change are rapidly increasing, and an equally increasing number of people are being affected (EM-DAT 2016). Moreover, the rates of environmental migration from all causes are expected to escalate and exacerbate other drivers of migration such as conflict and dwindling livelihood opportunities. Environmental change is therefore unquestionably the biggest influence in patterns of migration worldwide.

A major shift in the framing of environmental migration has been away from trying to isolate the independent impact of the environment on migration pathways to emphasizing the importance of recognizing environmental change as one of many drivers of migration, both as an independent driver of migration and as a factor that amplifies or reduces the impact of other drivers. In this chapter, we use the term "environment" rather than the narrower term "climate" to acknowledge that climate change is a subset of environmental changes that may include natural disasters, such as volcanic eruptions, or other anthropogenic changes, such as dams or deforestation. Climate change can also interact synergistically with these other environmental changes, increasing each other's impact. For example:

1. Changes in temperature and rainfall can result in decreased agricultural yields and lack of access to clean water but also increase the use of agricultural strategies that further deplete the land.

2. Natural disasters such as volcanic eruptions or tidal waves can create immediate threats to life as well as long-term threats to livelihoods, and their effects can be exacerbated by anthropogenic changes like deforestation.
3. Environmental degradation can create or exacerbate conflicts over limited resources, and conflict is one of many forces leading people to inhabit ecosystems that are highly vulnerable to environmental change.

Thus, for example, there has been considerable debate about whether and how much environmental change contributed to the conflict in Syria (e.g. Gleick 2014; Kelley et al. 2017; Selby et al. 2017). Going forward, linking patterns of environmental migration and displacement with the increasing trends in the uncertainty, variability, and unpredictability associated with environmental and climate change processes require scholarly and policy attention. As suggested above, this continues to remain uncharted territory.

Even subjective assessments of one's own motivations may overlook the ultimate role of environment. When asked about their reasons for migrating, those leaving conditions of environmental hardship can perceive more proximal drivers, such as livelihoods, as the primary deciding factor, making it difficult to attribute specific environmental conditions to the cause of migration at the individual level (Black et al. 2011). There is also the challenge of recognizing that environmental changes tend to affect individuals and communities differently. Vulnerability theories emphasize that the impact of any specific environmental changes depend on the affected individuals' and communities' ability to adapt (e.g. McLeman and Smit 2005). Thus, the likelihood that migration is a necessary adaptation strategy in situations of environmental change depends on the resilience and vulnerability of specific individuals, populations, and ecosystems.

Environmental migration is often internal, within a country's borders. Internal displacement is particularly dominant for the poor. The poor have fewer resources to invest in other adaptation strategies and so may need to rely more on migration to adapt to environmental change (Bierman and Boas 2010). However, although migration might become the only possible adaptation in the face of increasingly severe environmental degradation, severity of environmental changes can also decrease international migration. The resources required for longer distance migration are no longer available, and this again is felt most severely by the poor, since they have the fewest resources available for long-distance migration (Kniveton et al. 2009; Laczko and Aghazarm 2009).

Where environmental migration is international, it is typically regional, over the nearest border, and follows along pre-existing migration corridors. Although Canada is far from the countries most seriously impacted by climate change, Canada's diversity and immigration history mean that there are already diaspora communities from many climate-affected countries in Canada. The potential and impact of these diasporic communities can be seen in the Syrian resettlement initiative, where the pre-conflict Syrian community coordinated to privately sponsor large numbers of Syrian refugees. Many others will come through other migration pathways, including, if necessary, irregular migration paths, since the risks inherent in irregular migration are less than the certainty of the negative consequences such as starvation that are inherent in situations of severe climate-induced changes. For these and other reasons, Canada should take leadership on policies that recognize the place of environmental migration at the intersection of environment, immigration, development, security, and human rights by developing models of intersectoral policy and governance, while demonstrating that commitments to humanitarian principles, human rights, and sustainable development can be consistent with domestic goals for economic growth and security.

Environmental migration, adaptation, and justice

Policies on environmental migration have tended to focus on trying to reduce the influence of environmental change on migration, by trying to reduce the rate of change, and building the adaptive capacity of communities (Foresight 2011). These responses tend to be linked to a class of arguments that assume that migration of this scale is undesirable, that it will lead to increased conflict, that maintaining secure borders is essential, but that doing so may be impossible (Castles 2010; Popp 2014; Zetter 2009). These arguments have been raised in response to refugee migration also, and have been particularly salient in recent years (Hynie 2018). Mitigating environmental change and promoting adaptation that can keep people safely in their homes and support their "right to remain" are important national and international priorities, but migration may at times be the best or only possible adaptation response (Laczko and Piguet 2014; Martin 2013). Indeed, in recent years, environmental migration is being reframed from a failure of adaptation that should be prevented to a resilient response that has traditionally been used by many communities around the world throughout human history.

Canadian national and international environmental adaptation policies should continue to encourage mitigation and adaptation but also explicitly include support for migration as an adaptation strategy. A focus on the vulnerability of those exposed to disasters and environmental degradation, and international humanitarian obligations in the face of this vulnerability, is frequently invoked in the context of environmental disasters where migration is more clearly forced. The poorest people in both high- and low-income countries are the most affected by environmental change, and the most vulnerable to these changes; the poor have limited resources to mitigate change or adapt when it happens, and a greater likelihood of living in settings susceptible to environmental risk (Assan and Kumar 2009; Leichenko and Silva 2014; IPCC 2014). This perspective typically looks to international legal tools and agreements for the protection of people displaced by conflict as a model. However, as noted above, it is not clear that these tools are appropriate in their current form.

From the perspective of restorative justice, high-income countries have made the greatest contribution to global environmental effects; poorer countries are not only less responsible for the environmental changes that are disproportionately experienced by their populations, but also have fewer resources to invest in other adaptation strategies that could reduce the impact of these changes (Bierman and Boas 2010). Moreover, despite the urgency that high-income countries express about limiting migration, most forced migration, whether due to conflict or environmental change, is between low-income countries rather than from low- to high-income countries; low-income countries host 80% of refugees worldwide (UNHCR 2017). This draws attention to the international community's responsibility for shouldering the costs of mitigation and adaptation in poorer regions and countries and is a frequent issue in international discussions around both climate change and forced migration.

Recognizing environmental migration links the related policy objectives of sustainable development, climate change adaptation, humanitarian responses, security, human rights, and disaster risk reduction (IOM 2014; McAdam 2014). In this view, environmental migration is an inevitable part of these multiple policy domains and therefore requires comprehensive and coordinated international planning. Many regions have already developed agreements on international migration, with the African Union leading the way in incorporating environmental issues in its regional migration agreement (Popp 2014). This agreement recommends that environmental concerns be incorporated into migration policies for both internal and international migration, whether voluntary or forced, improved data collection, and increased mitigation strategies for environmental change. Other regions, like the European Union, have shown little appetite for developing policies at the nexus of environment and migration.

The emerging discourses on environmental migration and climate refugees may have a direct bearing on the success of the UN Sustainable Development Goals (SDGs). Goal 13 pertaining to climate action highlights "urgent action to combat climate change and its impacts" through (1) strengthening resilience and adaptive capacity to climate-related hazards and natural disasters, (2) integrating climate change measures into national policies, strategies, and planning, (3) improving education, awareness, and human and institutional capacity on climate change mitigation, adaptation, impact reduction, and early warning, and (4) promoting mechanisms for raising the capacity for effective climate change-related planning and management. Each of these action areas, including plans and strategies, can be carefully crafted to include issues of climate displacement and environmental migration as a core focal area within the climate action agenda. The onus is now on the respective nation states to move this agenda further. Barring environmental migration and climate displacement issues from the climate action agenda will have ripple adverse effects on the achievement of a host of other SDGs that range from deeply social issues (e.g. gender, equality, equity, health, and well-being) to important economic issues (e.g. hunger, poverty, income, consumption, growth and production) to grossly political subjects (e.g. peace, justice, partnerships, and institutions). The onus is now on the respective nation states to move this agenda further and Canada's leadership role, given its historical position on issues of immigrants and refugees, will be important.

Canada has regained a leadership role in the realm of forced migration with its response to the Syrian conflict. Moreover, Canada's unique model of private sponsorship is currently being considered by several other nations, and is being promoted as a promising alternative pathway for dealing with conflict-induced migration. Canada could leverage and extend its current influence on environmental forced migration. Canada's most important role at this juncture may be to work with other states to encourage movement on regional agreements, support international organizations like the Internal Displacement Monitoring Centre (IDMC) in their work on collecting data and sharing knowledge on internal displacement due to migration change, and encouraging adherence to current policies, such as the UN Guiding Principles of Internal Displacement (1998). Canada's support for regional and international policies governing internal environmental displacement and migration should be strengthened as part of our commitment to global environmental and humanitarian goals.

One possible solution to environmental change that is sensitive to the cyclical and temporary nature of most environmental migration is through ensuring protected temporary migration pathways. Temporary migration is not necessarily the best adaptation. It can increase inequities in sending countries. Temporary migration can be a more successful strategy for those who already have more resources, while being less successful, and even creating new vulnerabilities, for those already marginalized (e.g. Robson and Nayak 2010). It may also leave communities less resilient, in the absence of key community members, and make an entire community politically voiceless and disempowered (Adger et al. 2002; Nayak 2014). Undertaken with the rights of migrants and sending communities as core values, however, bilateral agreements offer a means of developing temporary migration policies in partnership with other countries and their citizens to simultaneously protect the rights of migrants and address the vulnerability of those communities left behind.

Canada's approach to temporary foreign workers has been celebrated as a model in international circles, providing greater protection for migrant worker rights than in many other countries. However, much more could be done to ensure that these programs protect the rights of the workers, including removing restrictions on freedom of movement, ensuring opportunities for integration into communities, and protecting workers from exploitation by employers (Hennebry and Preibisch 2012). By virtue of its "model" programs, Canada is also a recognized leader in this area of policy and can leverage this reputation to encourage adherence to migrant workers' rights in other countries, and to transform our own temporary migrant worker programs into ones that meet the humanitarian needs of the environmentally displaced, respect the rights of workers, and still fill the domestic market needs. Canada can build on its status as a "best practices" model for temporary migration by endorsing the UN Convention on Migrant Workers and Families (United Nations, 1990), and leading in revising national policies to better protect the rights of international migrants, including temporary and cyclical migrants within Canada. Bilateral labour agreements for temporary and cyclical migration could then be expanded to address humanitarian, environmental, and livelihood needs, and revised to further ensure protection for migrants' rights.

Conclusion

The international community has collaborated on its response to conflict-induced migration through the 1951 Convention on the Rights of Refugees, and a global refugee regime operates to protect those forcibly displaced by conflict. While there are flaws in this regime, it still stands as an example of an international humanitarian system that protects the rights of the vulnerable. In contrast, although the international community has become more aware of environmental migration as a form of adaptation, there has been little success in developing international policies to protect and plan for those who are forced or compelled to migrate. To some extent, the lack of planning or protection for environmental migrants has been driven by the emphasis on security in the responses to environmental migration, consistent with a general increase in the reframing of immigration in security terms (Aikin et al. 2014). It may also be an unintended consequence of the evocation of mass migration as a threat in order to motivate action supporting climate mitigation strategies by high-income countries. The current erosion of refugee rights in high-income countries is evidence of how vulnerable international human rights agreements are in the face of these changes.

Migration and environment are policy realms that elicit high levels of concern, both among political actors and among the public. Developing strategies to integrate these issues will be challenging, domestically and on the international stage. Nonetheless, although environmental migration can be unpredictable, it is inevitable (Foresight 2011). Proactive responses are necessary to ensure that this migration maximizes the resilience and adaptive capacity of individuals, communities, social systems, and ecosystems. The absence of policy frameworks to support and protect individuals who are compelled to migrate by environmental change will not prevent migration. Rather, it will force vulnerable people into precarious and irregular migration (Long and Rosengaertner,2016). Eventually, environmental change will affect us all, and being prepared with policies that protect those affected now will also ensure we can protect a wider range of individuals, communities, and countries in the future.

References

Adger, W. N., Kelly, P. M., Winkels, A., Huy, L. Q. & Locke, C. (2002) Migration, remittances, livelihood trajectories and social resilience. *Ambio*, 31(5), 358–366.

Aiken, S. J., Lyon, D. & Thorburn, M. (2014) Introduction: 'Crimmigration, surveillance and security threats': A multidisciplinary dialogue. *Queen's Law Journal*, 40(1), 1–8.

Assan, J. K. & Kumar, P. (2009) Policy arena: Livelihood options for the poor in the changing environment. *Journal of International Development*, 21(3), 393–402.

Barnett, L. (2002) Global governance and the evolution of the international refugee regime. *International Journal of Refugee Law*, 14(2/3), 238–262.

Biermann, F. and Boas, I. (2010) Protecting climate refugees: The case for a global protocol. *Environment*, 50(6), 8–16.

Bilak, A., Cardona-Fox, G., Ginnetti, J., Rushing, E. J., Scherer, I., Swain, M., Walicki, N. & Yonetani, M. (2016). GRID 2016: Global report on internal displacement. Geneva, Switzerland: The Internal Displacement Monitoring Centre. Available from: www.internal-displacement.org/assets/publications/2016/2016-global-report-internal-displacement-IDMC.pdf.

Black, R. (2001) Environmental refugees: Myth or reality? (Vol. 34). Geneva, Switzerland: UNHCR.

Black, R., Adger, W. N., Arnell, N. W., Dercon, S., Geddes, A. & Thomas, D. (2011) The effect of environmental change on human migration. *Global Environmental Change*, 21, S3–S11.

Brown, O. (2007) Climate change and forced migration: Observations, projections and implications. A background paper for the Human Development Report 2007/2008—*Fighting Climate Change: Human Solidarity in a Divided World*. Geneva, Switzerland: Human Development Report Office, UNDP.

Castles, S. (2010) Understanding global migration: A social transformation perspective. *Journal of Ethnic and Migration Studies*, 36(10), 1565–1586.

Crawley, H., Duvell, F., Signoa, N., McMahon, S. & Jones, K. (2016) Unpacking a rapidly changing scenario: Migration flows, routes and trajectories across the Mediterranean. Unravelling the Mediterranean Migration Crisis (MEDMIG), *Research Brief No. 1*. Available from: www.compas.ox.ac.uk/media/PB-2016-MEDMIG-Unpacking_Changing_Scenario.pdf

Diop, P. M. (2014) The "bogus" refugee: Roma asylum claimants and the discourses of fraud in Canada's Bill C-31. *Refuge*, 30(1), 67–80.

EM-DAT. (2016) Disaster trends. Available from: www.emdat.be/disaster_trends/index.html

Environics Institute for Survey Research. (2018) Canadian public opinion about immigration and minority groups. *Focus Canada—Winter 2018*. March 22.

Foresight. (2011) *Foresight: Migration and Global Environmental Change; Final Project Report*. London, UK: The Government Office for Science.

Gill, N. (2010) "Environmental refugees": Key debates and the contributions of geographers. *Geography Compass*, 4(7), 861–871.

Gleick, P. H. (2014) Water, drought, climate change, and conflict in Syria. *Weather, Climate and Society*, 6, 331–340.

Hennebry, J. L. and Preibisch, K. (2012) A model for managed migration? Re-examining best practices in Canada's seasonal agricultural worker program. *International Migration*, 50, e19–e40.

Hynie, M. (2018) Canada's Syrian refugee program, intergroup relationships and identities. *Canadian Ethnic Studies*, 50(2), 1–13.

Immigration, Refugees and Citizenship Canada (IRCC). (2017) *Annual Report to Parliament on Immigration*. Available from: www.canada.ca/en/immigration-refugees-citizenship/corporate/publications-manuals/annual-report-parliament-immigration-2017.html#sec1_1

Intergovernmental Panel on Climate Change (IPCC). (2014) *Climate Change 2014: Impacts, Adaptation, and Vulnerability*. New York, NY: Cambridge University Press.

International Organization for Migration (IOM). (2014) *IOM Perspectives on Migration, Environment and Climate Change*. Geneva, Switzerland: International Organization for Migration. Available from: http://publications.iom.int/books/iom-perspectives-migration-environment-andclimate-change

IPSOS. (2017) *Global views on immigration and the refugee crisis*. Available from: www.ipsos.com/sites/default/files/ct/news/documents/2017-09/Global_Advisor_Immigration.pdf

Kelley, C., Mohtadi, S., Cane, M., Seager, R. and Kushnir, Y. (2017) Commentary on the Syria case: Climate as a contributing factor. *Political Geography*, 60(1), 245–247.

Kniveton, D., Smith, C., Black, R. and Schmidt-Verkerk, K. (2009) Challenges and approaches to measuring the migration-environment nexus. In F. Laczko & C. Aghazarm (Eds.), *Migration, Environment and Climate Change: Assessing the Evidence* (pp. 41–111). Geneva, Switzerland: International Organization for Migration.

Labman, S. (2016) Private sponsorship: Complementary or conflicting interests? *Refuge*, 32(2), 67–80.

Laczko, F. and Aghazarm, C. (2009) Introduction and overview: Enhancing the knowledge base. In F. Laczko and C. Aghazarm (Eds.), *Migration, Environment and Climate Change: Assessing the Evidence* (pp. 7–40). Geneva, Switzerland: International Organization for Migration.

Laczko, F. and Piguet, E. (2014) Regional perspectives on migration, the environment and climate change. In E. Piguet & F. Lacko (Eds.), *People on the Move in a Changing Climate: The Regional Impact of Environmental Change on Migration* (pp. 1–20). New York: Springer.

Leichenko, R. and Silva, J. A. (2014) Climate change and poverty: Vulnerability, impacts and alleviation strategies. *WIREs Climate Change*, 5, 539–556.

Lilleør, H. B. and Van den Broeck, K. (2011) Economic drivers of migration and climate change in LDCs. *Global Environmental Change*, 21, S70–S81.

Long, K. and Rosengaertner, S. (2016) *Protection Through Mobility: Opening Labor and Study Migration Channels to Refugees*. Washington, D.C.: Migration Policy Institute.

Macklin, A., Barber, K. Goldring, L., Hyndman, J., Korteweg, A., Labman, S. and Zyfi, J. (2018) A preliminary investigation into private refugee sponsors. *Canadian Ethnic Studies*, 50(2), 35–58.

Martin, S. F. (2013) Environmental change and migration: What we know. *Migration Policy Institute*, Policy Brief No. 2.

McAdam, J. (2014) Creating new norms on climate change, natural disasters and displacement: International developments 2010–2013. *Refuge*, 29(2), 11–26.

McLeman, R. and Smit, B. (2005, June) Assessing the security implications of climate change related migration. Paper presented at Human Security and Climate Change: An International Workshop, Oslo.

Molnar, P. (2016) The boy on the beach: The fragility of Canada's discourses on the Syrian refugee 'crisis'. *Contention: The Multidisciplinary Journal of Social Protest*, 4(1–2), 67–75.

Nayak, P. K. (2014) The Chilika Lagoon social-ecological system: An historical analysis. *Ecology and Society*, 19(1), 1. http://dx.doi.org/10.5751/ES-05978-190101.

Piguet, E. (2013) From "primitive migration" to "climate refugees": The curious fate of the natural environment in migration studies. *Annals of the Association of American Geographers*, 103(1), 143–162.

Popp, K. (2014) Regional policy perspectives. In E. Piguet & F. Lacko (Eds.), *People on the Move in a Changing Climate* (pp. 229–253). New York, NY: Springer.

Robson, J. P. & Nayak, P. K. (2010) Rural out-migration and resource-dependent communities in Mexico and India. *Population and Environment*, 32(2), 263–284.

Selby, J., Dahi, O. S., Fröhlich, C. & Hulme, M. (2017) Climate change and the Syrian civil war revisited. *Political Geography*, 60, 232–244.

Stavropoulou, M. (2008) Drowned in definitions? *Forced Migration Review*, 31, 11–12.

Stern, N. (2006) *Stern Review: The Economics of Climate Change*. London, UK: HM Treasury.

UNHCR. (2017) *Global Trends: Forced Displacement in 2016*. Geneva, Switzerland: United Nations High Commissioner for Refugees. Available from: www.unhcr.org/statistics/unhcrstats/5943e8a34/global-trends-forced-displacement-2016.html

United Nations. (1990) *International convention on the protection of the rights of all migrant workers and members of their families*. Available from: www.ohchr.org/EN/ProfessionalInterest/Pages/CMW.aspx

Zetter, R. (2009) The role of legal and normative frameworks for the protection of environmentally displaced people. In F. Laczko & C. Aghazarm (Eds.), *Migration, Environment and Climate Change: Assessing the Evidence* (pp. 385–440). Geneva, Switzerland: International Organization for Migration.

9

OUT OF CREDIT

Climate finance in the face of climate debt

Alicia Richins

On April 25, 2009, the Plurinational State of Bolivia, led by its first Indigenous president, Evo Morales, and supported by 49 other developing nations, submitted to the United Nations Framework Convention on Climate Change (UNFCCC) a statement on the theoretical concept of climate debt, which charges developed countries with the responsibility to make up for their historically unequal use of atmospheric emissions space. The statement included the following words:

> We call on developed countries to commit to deep emission reductions, … to reflect their historical responsibility for the causes of climate change, and to respect the principles of equity and common but differentiated responsibilities in accordance with the UNFCCC. … The excessive past, current and proposed future emissions of developed countries are depriving and will further deprive developing countries of an equitable share of the much diminished environmental space they require for their development and to which they have a right. By over-consuming the Earth's limited capacity to absorb greenhouse gases, developed countries have run up an "emissions debt" which must be repaid to developing countries by compensating them for lost environmental space, stabilizing temperature and by freeing up space for the growth required by developing countries in the future.
>
> *(Plurinational State of Bolivia 2009, p. 3)*

This chapter aims to conceptualize and explain climate debt—a central concept in climate justice—as effectively a form of climate justice accounting, and to provide an overview of the climate financing solutions which are arguably being employed to pay those debts. While climate debt is usually focused at the global scale, the final section of this chapter considers climate debt at the national/local level, and introduces climate financing measures which could be employed at that scale.

What is climate debt?

Greenhouse gas (GHG) emissions, which have been accumulating in the atmosphere since the Industrial Revolution, play a major role in "anthropogenic interference with the climate system" (Plurinational State of Bolivia 2009, p. 3). Climate debt thus is based on the idea of a fair share of emissions across the global population; in the words of researcher Damon Matthews, it is "the amount by which national carbon contributions have exceeded a hypothetical equal per-capita share over time" (Matthews 2016, p. 60). As this definition suggests, the per-capita shares of the human contribution to global climate change in reality have been far from equal, reflecting underlying historical disparities in economic development among nations. The strides in industry and development that have facilitated present levels of global wealth and trade are also responsible for the current crisis of impending climate instability, and these levels of wealth (and emissions) are, for the most part, concentrated in a small number of developed countries.

Meanwhile, the larger proportion of the global population in developing countries, due to the mere chance of geography, overwhelmingly faces the brunt of adverse effects associated with the changing climate. Today we already face more frequent and intense droughts, floods, fires, and extreme weather events; in addition are water stress, adverse impacts on agriculture, and threats to ecosystems, coastlines, and infrastructure (Plurinational State of Bolivia 2009, p. 3). Most of these are concentrated in developing states of the Global South, which also continue to contend with poverty, disease, hunger, and increasing levels of national debt. Thus, we have a situation where 20% of the global population is generally responsible for 75% of greenhouse gas emissions (Plurinational State of Bolivia 2009, p. 3), while the remaining 80% of the population faces the largest burdens of climate change, with markedly fewer resources and weaker systems and infrastructure to face the looming challenge, having only contributed around one quarter of the emissions that are creating the problem.

Since the adoption of the Kyoto Protocol in 2005, nation-states have committed to binding emissions-reduction targets. These targets are subject to regular review and evaluation through mechanisms overseen by the UNFCCC. When the Paris Climate Agreement entered into force in 2016, it brought together for the first time "all nations into a common cause to undertake ambitious efforts to combat climate change and adapt to its effects, with enhanced support to assist developing countries to do so" (UNFCCC 2019d), charting a new course in the global climate effort. Central to this effort was the aim of containing the global temperature rise this century to 2°C above pre-industrial levels. In late 2018, the Intergovernmental Panel on Climate Change (IPCC), the UN body charged with assessing the science related to climate change, released a new report to much fanfare in support of restricting climate change to 1.5°C above pre-industrial levels (IPCC 2018). The IPCC researchers concluded that warming that surpasses 1.5° will lead to unforeseeable and catastrophic impacts and disasters. Their findings cite impacts on developing states—small-island developing states (SIDS) in particular—who are most vulnerable to changes in climate.

It is in acknowledgement of this restricted emissions space that Bolivia and others have sought to charge a climate debt to account for the share of global emissions required by developing countries to meet their "first overriding priorities" (Plurinational State of Bolivia 2009, p. 4) such as poverty eradication and economic and social development. In their submission to the UNFCCC, climate debt is comprised of two forms of debt:

1. Emissions debt—the unequal share of emissions contributions over time
2. Adaptation debt—the costs, damages, and lost opportunities faced by developing countries due to climate change

A NOTE ON ECOLOGICAL AND HISTORICAL DEBT

Climate debt forms part of a broader ecological and historical debt that reflects the "heavy environmental footprint, excessive consumption of resources, materials and energy and contribution to declining biodiversity and ecosystem services" (Plurinational State of Bolivia 2009, p. 4) of developed countries. Broader than the purely ecological debt, historical debt also accounts for historical loss of population due to slavery and war, unpaid labour in the instances of slavery and indentureship, biopiracy, resource extraction and plunder in the Global South, and its associated pollution and degradation of ecosystems (Ross 2013).

These broader ecological and historical debts form the theoretical basis for the concept of climate debt (Pickering and Barry 2012, p. 669). These debts are both intergenerational and international, reflecting both the temporal and spatial nature of the concept: it is accrued by past/current populations and owed to current/future generations, and is accrued by the populations of some nation-states and owed to those of others. In international climate negotiations—"the only existing means of resolving political conflict over the collective action problems posed by human induced climate change" (Pickering and Barry 2012, p. 667)—climate debt claims tend to focus on the international dimension of these debts.

While worthy of acknowledgement, ecological and historical debts do not readily lend themselves to quantification. Thus, the primary benefit of a focus on climate debt extracted from the more extensive historical and ecological debts is the relative ease of calculation. Atmospheric emissions estimates are easier to measure than any of these other historical debts. It is also relatively easier to trace lines of responsibility for climate debt in order to justify its payment. Hence we proceed with climate debt, while recognizing its place within a much larger frame of ecological and historical debt.

Quantifying climate debt

According to the work undertaken by Damon Matthews on climate debt calculations, and as widely published online (Merchant 2015), the United States owes the world an estimated USD $4 trillion in climate debt. Counting from 1990—since this was when a verifiable link between carbon emissions and climate change was first cited (in the first report of the Inter-Governmental Panel on Climate Change), so this is thus the least contestable measure of the debt—the United States has accounted for 32% of the cumulative global climate debt from 1990–2010. Canada is also a significant debtor country at 3.9% of the world climate debt, meaning that Canada owes the world USD $500 billion (Matthews 2016, p. 60).

As explained in the introductory section of this chapter, these calculations are based on emissions accounting of an equal per-person share of emissions into our common-pool climate resource. In this way, the climate debts represent the accumulated difference between the actual temperature change caused by each country and their per capita "fair share" of global temperature change. Matthews further specifies his accounting through a focus only on national carbon dioxide emissions due to fossil fuel combustion: Carbon accounts for approximately 60% of global greenhouse gas emissions, with the rest taken up by other gases such as methane, nitrous oxide, and aerosols. Unlike carbon and the other gases, aerosols have a cooling effect on our atmosphere, and the rate of that cooling is more or less equal to the heating effect of the methane and nitrous oxide. Thus, carbon emissions form a reasonably accurate indicator of broader greenhouse gas emissions, as the others cancel each other out (Matthews 2016, p. 61).

From 1990 to 2013, the carbon debt totals 250 billion tonnes of carbon dioxide, equivalent to 40% of global emissions over this period. This is an example of *production*-based emissions accounting, excluding any account of the international transfer of emissions associated with the export and import of goods and services—which is one of several ways to calculate the debt. Other possibilities would be to calculate it based on *consumption* of fossil fuel products and related emissions, or even on emissions generated by a country based on all of the global emissions of the companies who hold their headquarters there. Whatever the methodology, carbon debts are a useful tool for examining broader climate debt as they are easy to calculate, carry lower uncertainty, and can be monetized using estimates of the cost of climate damages from carbon dioxide emissions (Matthews 2016, pp. 62–63). Matthews assures us of coming improvements in national emissions estimates that will make climate debt calculations even more robust and policy relevant.

Debt justification

In its formal submission to the Ad-Hoc Working Group on Long-term Cooperative Action of the UNFCCC (quoted at the beginning of the chapter),

Bolivia cited the mandate of the Bali Action Plan (UNFCCC 2019b), which calls for:

> Measurable, reportable and verifiable nationally appropriate mitigation commitments or actions, including quantified emission limitation and reduction objectives, by all developed country Parties, while ensuring the comparability of efforts among them, taking into account differences in their national circumstances.
>
> *(UNFCCC 2007, cited in Plurinational State of Bolivia 2009, p. 1)*

Focusing on the comparability of efforts, the submission calls for debt calculations that take into account both the per-capita emissions of developed countries (historic and current), and the share of global emissions that developing countries still require in order to address their most urgent priorities of economic and social development (UNFCCC 2007, cited in Plurinational State of Bolivia 2009, p. 1). This position inherently acknowledges a "right to development" of developing countries (Bauer et al. 2008), to secure a basic level of economic prosperity and social well-being despite the possible attendant emissions; requiring historical emitters to leave room for those who need it most.

To ethically and politically analyze this concept of climate debt, Pickering and Barry provide a useful framework. They identify five core elements of the climate debt idea: (1) moral responsibilities; (2) characterization of these responsibilities as debts; (3) the content of the responsibility; (4) identity of debtors and creditors; (5) form of repayment required (Pickering and Barry 2012, p. 670). While all of these elements are contested in some way by climate debt detractors, the second and third elements are those most often specifically challenged in the international climate debt arena.

On the ethical front, the concept of moral responsibilities stands up to criticism, as the atmosphere is widely accepted as a global commons which should be subject to a shared responsibility for fair use. While it is generally accepted in international climate discourse that the debtors and creditors are developed and developing countries respectively, these are, in the words of Pickering and Barry, "broad-brush approximations" that fail to take into account the changes over time in wealth and emissions across countries, as well as wealthy individuals in so-called developing countries who have exceeded their fair emissions allotment. Thus, the climate debt concept would benefit from a more nuanced identification of debtors and creditors.

The question of the form of repayment required also raises the distinction between emissions and adaptation debts. For instance, emissions debt proposals call for the reallocation of future emissions rights in order to offset past over-emittances, while adaptation debt proposals call for financing to support developing countries with their own emissions reductions in order to meet their own per capita entitlements (Pickering and Barry 2012, pp. 670–2, 76–77).

However, as the authors point out, the most fundamental objections to the concept of climate debt arise from the elements of the characterization of the responsibilities as debts, and the question of the content of the debt responsibility. First, "guiltless responsibility" arguments assert that debt only makes sense in the case of culpability, whereas the greenhouse effect was unknown to emitters until relatively recently. However, responsibility is not synonymous with culpability. Rather, we have a case of unjust enrichment for which climate debt assigns not blame, but a responsibility for cost-bearing. Similarly, the second category of objection—"excusable ignorance," contends that since countries were, until relatively recently, ignorant of the effects of their emissions, then they should not be held responsible for those emissions nor their effects. Nevertheless, excusable ignorance would no longer apply once knowledge of the risks associated with greenhouse gas emissions became available. Many propose then that climate debt be calculated from the year 1990, when the IPCC published its first report (as also acknowledged in Matthew Damon's calculations of carbon debt in the previous section). This would provide for a discounting of the "total" amount owed to uphold this principle (Pickering and Barry 2012, pp. 672–5).

Finally, the intergenerational objection claims it implausible to hold countries responsible in the present for actions undertaken by past generations. As Pickering and Barry point out, the prevalence of sovereign debt is a prime example to the contrary, as the obligation to repay loans can be carried by successive generations. "Odious debts," acquired illegitimately, or appropriated unjustly, or subject to changing distributions of resources, can be cancelled; however, in this scenario where developed countries have amassed great wealth through the use of global common resources in excess of their fair share, an "odious debt" argument would be implausible. Moreover, current generations continue to benefit from the historical emissions of earlier generations in the form of the higher standards of living provided for by industrialization (Pickering and Barry 2012, pp. 675–6).

Thus, we can confirm the moral coherence of the argument for climate debt. Political plausibility, however, is a separate matter. While supported by many developing countries and international advocacy organizations, and having been made highly visible in the media, calls for climate debt repayments have largely been excluded from the resulting agreements (Pickering and Barry 2012, p. 670). Problems of measurement form a major political obstacle in informing climate negotiations (p. 677). Fortunately, as pointed out earlier, a focus on accounting from 1990—when sound country-level emissions data first became available—simplifies the political challenge of quantification. While there are a variety of methodological choices available (of which Damon Matthews provides just one) to account for emissions since 1990, an agreement to take this as the base year clears up the politics if not the math.

The other political obstacle identified by Pickering and Barry is that of rhetorical emphasis. They argue that the frame of climate debt is a form of "bonding rhetoric," which aims to motivate people who are already of similar mind, helping to mobilize dispossessed groups and bring wider public attention to their political concerns. This bonding rhetoric has an adversarial emphasis, dividing the world into climate debtors (developed countries) and creditors (developing countries), and contrasting with the usual debt rhetoric of developing countries owing debts to developed countries. The authors posit that in order to overcome the mistrust that pervades climate negotiations, bridging rhetoric is required to facilitate meaningful and fair agreements through the adoption of "mutually acceptable frames for collective action" (Pickering and Barry 2012, p. 679). Since the concept of climate debt seems incapable of embodying a bridging rhetoric, Pickering and Barry explore other modes of promoting collective responsibility for climate action. Their prime example is that of the carbon budget: while compatible with climate debt, it does not require its focus on the two groups of debtors and creditors, but rather a broader collective undertaking of "balancing the budget" (p. 680).

While a carbon budget does not talk explicitly of debtors and creditors, the accounting for climate debt is implicit in its deficits and surpluses, and it involves a responsibility for emissions—past, present, and future. Thus, despite the apparent political expediency of using language other than "climate debt," the existence of climate financing mechanisms like those reviewed in the next section proves the political plausibility of emission and adaptation debts, through commitments to deep emission reductions and financial funding for foreign adaptation measures.

Global climate finance

Climate finance refers to local, national, and transnational financing that seeks to support mitigation and adaptation efforts to address climate change (UNFCCC 2019c). In recognition of the responsibility of developed countries, many of them have committed, through the UNFCCC, to provide financial resources for mitigation and adaptation efforts in developing countries. This section provides a brief overview of the global climate finance initiatives currently underway as a form of climate adaptation debt repayment—though they are defined in terms of collective responsibility rather than debt accounting.

Since the inception of the UNFCCC in 1994, the Global Environmental Facility (GEF) was formed as an international partnership of 183 countries, international institutions, civil society organizations, and private businesses to address environmental issues. To date, the GEF has provided more than USD $17.9 billion in grants and mobilized a further $93.2 billion in co-financing for over 4,500 projects in 170 countries (GEF 2018a). The financial contributions, which form the GEF Trust Fund, are replenished every 4 years by the GEF's 39 donor countries, and administered by its Trustee, the World Bank. The GEF project areas include protected areas, sustainable landscape and seascape, sustainable forest management, sustainable land management, GHG emission reduction, integrated water resources management, safe disposal of hazardous chemicals, and adaptation to climate change.

The GEF also manages the Special Climate Change Fund (SCCF) and the Least Developed Countries Fund (LDCF), both established in 2001. Open to all developing countries, the SCCF funds climate change adaptation and technology transfer activities with 77 projects in 79 countries and almost USD $350 million in voluntary contributions (GEF 2018c). The LDCF, on the other hand, is exclusive to the least developed countries (LDCs), and is designed to address their special needs and to help them prepare and implement their National Adaptation Programs of Action (NAPAs). Holding the largest portfolio of adaptation projects in LDCs, the Fund has provided over USD $1.16 billion in grants for 250 projects, including the formulation of NAPAs for 51 countries to identify urgent and immediate adaptation needs (GEF 2018b).

The Adaptation Fund was also established in 2001 under the Kyoto Protocol. Its purpose is to finance tangible adaptation projects in developing countries that are Parties to the Kyoto Protocol that are particularly vulnerable to the adverse effects of climate change (United Nations Framework Convention on Climate Change 2019a). It is financed with a share of proceeds from the Clean Development Mechanism (CDM), which allows emission reduction initiatives in developing countries to earn Certified Emission Reduction (CER) credits, among other sources of funding. This Fund is managed by the Adaptation Fund Board (AFB), composed of 16 members and 16 alternates who meet at least twice per year.

Finally, the Green Climate Fund (GCF) was established at the 2010 Conference of the Parties (COP16) to limit or reduce GHG emissions in developing countries, and to help vulnerable communities adapt to the unavoidable impacts of climate change. Administered in the form of grants, loans, equity or loan guarantees, the GCF to date has funded 93 projects totalling USD $4.6 billion, through which an estimated 1.4 billion tonnes of carbon dioxide emissions have been avoided and 272 million people supported with increased resilience (GCF 2019). GCF's approach has three distinct features: (1) a balanced portfolio between mitigation and adaptation efforts; (2) unlocking private finance through its Private Sector Facility (PSF); and (3) country ownership, by which developing country partners exercise ownership of funding, allowing them to integrate it within their own national action plans.

The long-term goal of the global climate finance process is to jointly mobilize US $100 billion per year by 2020 to address the needs of developing countries (UNFCCC 2019c). Confirmed with the Paris Agreement, there is still more work to be done as this goal is set to be re-evaluated before the 2025 Conference of Parties. With the overwhelming focus on assisting developing countries with the mitigation and adaptation challenges related to climate change, one can in fact chart a clear path from climate debt to the climate finance mechanisms currently at play across the globe. Given the earlier-cited estimate of USD $4 trillion in US climate debt, however, these mechanisms are orders of magnitude too small to address the overall scale of global climate debt.

Acting locally

Local/national climate debt also accrues within developed countries such as Canada, the United States, and Australia, where Native and other marginalized communities suffer at the hands of fossil fuel emission-intensive industries (Luginaah et al. 2010, Vice Staff 2013). And as pointed out earlier, the true climate debtors are a combination of corporations and individuals who are most responsible for the high levels of emissions that created the crisis. The average citizen in Toronto or Texas has not derived significant financial gains from the profits of the petrochemical industry. Rather, the beneficiaries are the shareholders, owners, and officers of capitalist extractive industries who (especially since 1990) sought the means of extraction, refinement, manufacturing, and sale.

These corporations themselves are responsible for various social, economic, and environmental abuses at home and abroad (McSheffrey 2017; Beaumont 2017; Dean 2013). Thus, in addition to global climate financing, some efforts at recovering some of those profits to re-invest into economic, social, and climate mitigation and adaptation efforts within developed countries are also required. Many propose tax systems that allow countries to charge the true emitters and redistribute their wealth at home and abroad. Perhaps the most daring proposal yet is that of recently elected US Congresswoman Alexandria Ocasio-Cortez and Congressman Ed Markey: the Green New Deal (GND) (Ocasio-Cortez et al. 2019). While still in its developing stages, the resolution adopting the GND was introduced to Congress in early 2019. In very broad strokes, it outlines high-level goals such as deep emissions reductions, large investments in clean energy jobs and infrastructure, and guarantees of health, housing, and employment for all citizens. It is at its core a new vision for the economy, one that is based 100% on renewable energy, but that is also based on major social and economic reforms. While the details and actual policies are yet to be mapped out, supporters of the GND look forward to new tax policies and budgeting priorities to fund the undertaking.

These and other established initiatives to bring about large emissions reductions within developed countries are an essential factor in the broader framework of climate debt, addressing the emissions debt at home, while the UN system attempts to address the global adaptation debts head-on.

References

Bauer, P. et al. (2008) 'The greenhouse development rights framework: The right to development in a climate-constrained world,' *Cambridge Review of International Affairs* 21(4): 649–669. DOI: 10.1080/09557570802453050

Beaumont, H. (2017) 'Women from Papua New Guinea bring rape complaints to Canadian mining company's door,' *MiningWatch Canada*, 26 April [online]. Available at: https://miningwatch.ca/news/2017/4/25/women-papua-new-guinea-bring-rape-complaints-canadian-mining-company-s-door

Dean, D. (2013) '75% of the World's Mining Companies Are Based in Canada,' *Vice*, 9 July [online]. Available at: www.vice.com/en_ca/article/wdb4j5/75-of-the-worlds-mining-companies-are-based-in-canada

Green Climate Fund (2019) *About the Fund* [online]. Available at: www.greenclimate.fund/who-we-are/about-the-fund

GEF (2018a) *About Us* [online]. Available at: www.thegef.org/about-us

GEF (2018b) *Least Developed Countries Fund* [online]. Available at: www.thegef.org/topics/least-developed-countries-fund-ldcf

GEF (2018c) *Special Climate Change Fund* [online]. Available at: www.thegef.org/topics/special-climate-change-fund-sccf

Intergovernmental Panel on Climate Change (2018) *Special Report: Warming of 1.5°C* [online]. Available at: www.ipcc.ch/sr15/

Luginaah, I. et al. (2010) 'Surrounded by Chemical Valley and 'Living in a Bubble': The Case of the Aamjiwnaang First Nation, Ontario,' *Journal of Environmental Planning and Management* 53(3): 353–370.

Matthews, D. (2016) 'Quantifying Historical Carbon and Climate Debts Among Nations,' *Nature Climate Change*, 6:60–64. DOI: 10.1038/nclimate2774

McSheffrey, E. (2017) 'Mining Violence Survivors Demand Justice in Toronto,' *MiningWatch Canada*, 25 April [online]. Available at: https://miningwatch.ca/news/2017/4/25/mining-violence-survivors-demand-justice-toronto

Merchant, B. (2015) 'The U.S. Owes the World $4 Trillion for Trashing the Climate,' *Vice News*, 15 September [online]. Available at: https://motherboard.vice.com/en_us/article/bmj97q/the-us-owes-the-world-4-trillion-for-trashing-the-climate

Ocasio-Cortez, A. et al. (2019) 'Recognizing the Duty of the Federal Government to Create a Green New Deal.' US Congress Resolution, 7 February [online]. Available at: https://ocasio-cortez.house.gov/sites/ocasio-cortez.house.gov/files/Resolution%20on%20a%20Green%20New%20Deal.pdf

Pickering, J. and Barry, C. (2012) 'On the Concept of Climate Debt: Its Moral and Political Value,' *Critical Review of International Social and Political Philosophy* 15(5): 667–685. DOI: 10.1080/13698230.2012.727311

Plurinational State of Bolivia (2009) 'Commitments for Annex I Parties under paragraph 1(b)(i) of the Bali Action Plan: Evaluating developed countries' historical climate debt to developing countries' submitted to AWG-LCA of the UNFCCC [online]. Available at: http://climate-debt.org/wp-content/uploads/2009/11/Bolivia-Climate-Debt-Proposal.pdf

Ross, A. (2013) 'Climate Debt Denial,' *Dissent Magazine*, Summer [online]. Available at: www.dissentmagazine.org/article/climate-debt-denial

United Nations Framework Convention on Climate Change (2019a) Adaptation Fund [online]. Available at: https://unfccc.int/process/bodies/funds-and-financial-entities/adaptation-fund

UNFCCC (2019b) *Bali Road Map Intro* [online]. Available at: https://unfccc.int/process/conferences/the-big-picture/milestones/bali-road-map

UNFCCC (2019c) *Introduction to Climate Finance* [online]. Available at: https://unfccc.int/topics/climate-finance/the-big-picture/introduction-to-climate-finance

UNFCCC (2019d) *The Paris Agreement* [online]. Available at: https://unfccc.int/process-and-meetings/the-paris-agreement/the-paris-agreement

Vice Staff (2013) 'The Chemical Valley,' *Vice News*, 8 August [online]. Available at: www.vice.com/en_us/article/4w7gwn/the-chemical-valley-part-1

PART II

Personal action and local activism

For many people, growing personal awareness and concern about climate change lead to the question, "What am I doing to address this problem; how can my own actions help to bring about a small step in the right direction?" While the scale of the challenge can seem daunting, individual choices do matter, both in their collective impact and as part of the educational and cultural shifts which lead to snowballing policy change. For example, using car seatbelts is now accepted and enforced by law, and the linked reduction in injuries and deaths in automobile accidents has increased the welfare for all and saved billions in insurance and health costs. There was a time, however, when people rebelled against using seatbelts, claiming it was a personal choice not to use them; a combination of accelerating evidence, public interest research, activism led by people whose families had been affected, and widespread educational campaigns eventually led to policies which then seemed progressive but now are "normal."

Other examples of relatively rapid policy change sparked by individual actions and activism include changes in cigarette advertising and public smoking regulations, divestment as a contributor to the end of apartheid in South Africa, the removal of lead from gasoline in many places, and bans on DDT and other pesticides and food additives. There are many other examples.

Social action to build interpersonal connections at the local level hastens such processes because it helps people understand others' struggles and experiences as related to their own. This section's chapters focusing on individual choice and social action on food, money management, housing and community development show how what happens at the local level can give people insights into similar livelihood issues globally—and their relation to climate justice.

10

THE FOSSIL FUEL DIVESTMENT MOVEMENT

A view from Toronto

Aaron Saad

The movement to remove investments (or to *divest*) from fossil fuel companies has grown enormously since kicking off around 2012 following a landmark essay by climate campaigner Bill McKibben (2012). It has now become global in scope, and not a moment too soon. The world has entered an unnerving stall with regards to climate action. The 2015 Paris Climate Agreement was intended to coordinate global fossil fuel emissions reductions to keep the increase in global average temperatures to 1.5°C, or at least well below 2°C. As part of the agreement, each country independently selected its own emissions reductions targets. Predictably, many of the targets were unambitious (Climate Action Tracker 2017). If not improved upon, they place the world on a pathway to warm by a calamitous 3°C or more relative to pre-industrial times by 2100 (United Nations Environment Program 2018, p. 21). More worrying still is that countries seem off track to meet even these already inadequate targets (Victor et al. 2017). Under the Donald Trump administration, the United States has pulled out of the agreement altogether, creating potentially far-reaching harm (Saad 2018).

Divestment might be one tool to help pull the world out of this morass. By tightly combining a powerful economic argument about the financial risks of continued investment in fossil fuel companies with an urgent climate justice argument about the moral consequences of continued fossil fuel use, the divestment movement seeks to stigmatize and erode the political power of the industry most responsible for driving the climate crisis.

This chapter serves as an introduction to fossil fuel divestment. It discusses how the movement emerged from a relatively new way of looking at the climate crisis, explains the climate justice-based rationale behind the divestment strategy, overviews the growth of the movement and the effect divestment is likely to have on investment returns, and concludes with some personal reflections based on my time as an activist in Toronto's university divestment movement.

A new way of looking at climate change: The carbon budget

Divestment grew out of a new way of looking at the problem of climate change. Climate change goals have tended to be communicated, overly abstractly, in terms of percentage reductions relative to a baseline year by a certain deadline (e.g. "Our country must lower emissions by X% relative to 1990 emissions by the year 2030"). But the "carbon budget" approach that began picking up steam in 2009 with the publication of two papers in the journal *Nature* (Meinshausen et al. 2009; Allen et al. 2009) offers a fresh way of looking at the problem, one that emphasizes the need to bring emissions down to zero as rapidly as possible.

A *carbon budget*, put simply, indicates the total amount of carbon that can ever be emitted to offer a decent chance of preventing global average temperatures from rising by a certain amount. The lower we wish to keep the rise in temperature and the more certain we want to be of not exceeding that temperature, the lower the total budget must be.

Due to differences in assumptions about the role of non-carbon dioxide greenhouse gases and other matters, climate scientists have produced a few different carbon budgets (Peters 2017). For the purposes of illustration, let us use the figures from a 2018 special report by the Intergovernmental Panel on Climate Change (IPCC). The authors calculate that for a likely (66%) chance of keeping temperature rise to 2°C relative to pre-industrial times, the carbon budget at the beginning of 2018 was 1,170 gigatons of carbon dioxide (GtCO2) (IPCC 2018, p. 108). Past that point, no more carbon can be added without risking far more dangerous climate change than the world is already seeing. (There are uncertainties that potentially change the size of the carbon budgets. For example, carbon released from melting permafrost could reduce the budgets by 100 GtCO2 over the 21st century (IPCC 2018, pp. 107–108)).

Once a budget is established, two insights follow. First, it becomes possible to estimate the time until the budget is expended. In 2018, annual global emissions stood at around 42 GtCO2. At that rate, the carbon budget for a likely chance of 2°C warming will be spent by 2045. But given the severe climate change effects the world is currently experiencing at only 1°C of warming, letting the rise in global average temperature approach 2°C is dangerous. As an important 2018 paper argued, a temperature rise of around 2°C could be enough to trigger self-reinforcing cascading biogeophysical feedbacks in the earth system that would push the world into a "Hothouse" state (Steffen et al. 2018). It would therefore be much safer to attempt to hold warming to 1.5°C. For an even chance of doing so, the carbon budget stood at just 580 GtCO2 (and 420 GtCO2 for a two-thirds chance) at the beginning of 2018 (IPCC 2018, p. 108). At current emission rates, *that budget will be exhausted by 2031*. What this means is that we live in a world that is *highly* "carbon constrained."

The second insight emerges through comparing the remaining carbon budget against how much carbon fossil fuel companies hold. Oil Change International (2016) issued a report that was unique in taking into consideration the carbon dioxide that would be emitted from burning just the fossil fuels in "developed reserves"--that is, only from the gas, oil, and coal in the fields and mines that are *already operating*. The amount it found was striking—942 GtCO2, far in excess of the 580 GtCO2 budget for 1.5°C.

But the industry holds much more than those developed reserves. If burned, the fossil fuels in developed and still-to-be developed reserves would together release just over 2,500 GtCO2 (Oil Change International 2016, p. 15, fig. 2). Another study puts the amount even higher at 2,900 GtCO2 (McGlade and Ekins 2015). And the industry has every intention of finding more reserves. By one estimate, fossil fuel companies invested $674 billion in 2013 in developing and exploring for more reserves (Carrington 2015).

What carbon budgets make clear is that there is no reason society should accept the search for new fossil fuel resources: we have already found more than can ever be used. In fact, some of the fields *currently operating* will need to be shut down if we are to constrain warming to 1.5°C. It is precisely these truths that the fossil fuel industry is resisting, however. The reason is straightforward: the stock value of fossil fuel companies is determined based on the assumption that they will bring to market the fossil fuels in their reserves; the greater the reserves they hold, the greater their stock value. It is a logic that yields a business model requiring them to find and produce new fossil fuel sources in perpetuity and to ensure that our society remains addicted to them, even at the cost of a habitable climate.

Changing the story: The rationale behind divestment

It is from this context that a new current in the climate movement emerged. What carbon budgets were revealing was that the fossil fuel industry was driving the climate crisis for business reasons, and that some way of disrupting its influence was necessary. A movement began to grow around the idea that selling off fossil fuel company stocks and refusing to ever invest in the industry again could do just that.

The rationale behind that idea is not as simple as is sometimes assumed. There is a misconception that divestment is aimed at directly harming fossil fuel companies economically—though there are early indications that this is happening (McKibben 2018)—and that it will not work because some other buyer will simply acquire the stocks that divesting institutions sell off. But this is the wrong way to look at it—at least in the short term. The main thrust of the strategy behind divestment is to *change the way society perceives fossil fuel companies*. Our society, particularly in Canada, still tends to see them primarily as employers, providers of vital energy resources, and sensible investments. Governments continue to give them enormous subsidies, by one estimate reaching US$5.3T (6.5% of global gross domestic product) in 2015 when taking into account governments' failure to adopt policies to make fossil fuel prices reflect their environmental costs (Coady et al. 2017). Major banks remain invested in even the most carbon-intensive sectors of the fossil fuel industry. Between 2014–2016, 37 of the largest private banks in the world invested $290 billion in extreme fossil fuels like tar sands, deepwater oil, coal, and liquid natural gas exports. Four of the top seven banks investing in tar sands were Canadian (RBC, CIBC, Bank of Montreal, and TD) (Rainforest Action Network 2017, pp. 3, 17).

The divestment movement harnesses two powerful arguments to shake up this state of affairs. The first is moral. As mentioned above, fossil fuel companies' business model itself absolutely requires wrecking a liveable climate. More egregiously, to protect that business model they have recklessly delayed action on climate change by funding organizations that create and disseminate climate change denier propaganda; as a series of 2015 exposés showed, Exxon-Mobil has been doing so even though its own scientists have understood for decades that climate change is a real threat (Inside Climate News 2015; Jennings et al. 2015; Jerving et al. 2015). Fossil fuel companies fund the election campaigns of politicians who promise to resist or rescind climate and environmental legislation (Center for Responsive Politics no date; Gaworecki 2016), or, should that legislation come to a vote, they pressure governments to kill it. From 2000–2016, the major fossil fuel suppliers and users outspent environmental and renewable energy groups in lobbying government by a ratio of 10 to 1 in the United States (Brulle 2018). They have even co-opted and undermined high-level negotiations under the United Nations Framework Convention on Climate Change intended to coordinate a rapid global shift away from fossil fuels (Corporate Accountability 2017). In their need for ever more oil, coal, and gas, they have exposed frontline communities, many of them Indigenous or predominantly people of colour, to the pollution involved in extracting, transporting, and refining fossil fuels.

To continue to invest in the industry is to tacitly accept these practices and to continue to provide fossil fuel companies what is referred to as "social license"– the approval society grants them to freely carry on with their usual business. It is this social license that the movement aims to progressively strip away by convincing more and more people and institutions to divest. Each divestment commitment acts as a condemnation of the moral transgressions the industry commits, a declaration that fossil fuel companies are engaging in acts too opprobrious to associate with any longer. This is the reason divestments by major respected institutions (like universities) are especially important; due to their stature, their divestment commitments sound this message at the loudest of volumes.

Each victory also works in concert with the others, and with enough of them it might be possible to delegitimize an entire sector of the global economy. Delegitimized, it will be easier to impose stricter regulation on the industry, like imposing a steeper price on carbon pollution through carbon taxes or even taking some of their profits to invest in a renewable future.

But not everyone is persuaded by moral reasoning alone. So the divestment movement deploys a second argument, an economic one. Carbon budgets revealed that the way fossil fuel companies are valued—based on the size of their reserves—is deeply flawed. As the world gets more serious about climate change, those companies will be holding what are called "stranded assets," assets they cannot sell because no one can use them without making our climate increasingly volatile, and those holdings will become worthless. Some analysts, like Bank of England governor Mark Carney (Clark 2015), have begun to suggest that continued investment in fossil fuels in a carbon-constrained world opens up massive financial risk. There is increased talk of an economic "carbon bubble" due to burst as the world wakes up to the reality the carbon budgets reveal.

The movement has good reason to believe this strategy can work. It takes direct inspiration from the successful divestment campaign against apartheid South Africa. In the mid-20th century, the white South African political leadership institutionalized a system of racial segregation that stripped black South Africans of a host of rights. By the 1980s, people around the world, horrified by the injustices being committed by the South African state, looked for a way to apply pressure on the regime from outside. One of the tactics they turned to was divestment, withdrawing investments from businesses and banks that had dealings with South Africa, and turning the regime into a global pariah. The movement started small before spreading through campuses and to municipalities. In the United States, 155 campuses, 26 state governments, and 90 cities divested from apartheid South Africa.

As Archbishop Desmond Tutu (2010), a leading figure in anti-Apartheid struggles, writes,

> In South Africa, we could not have achieved our freedom and just peace without the help of people around the world, who through the use of non-violent means, such as boycotts and divestment, encouraged their governments and other corporate actors to reverse decades-long support for the Apartheid regime. Students played a leading role in that struggle, and I write these words of encouragement for student divestment efforts cognizant that it was students who played a pioneering role in advocating equality in South Africa and promoting corporate ethical and social responsibility to end complicity in Apartheid.

Movement growth and impact on returns

The fossil fuel divestment movement is the fastest growing divestment movement in history. The end of 2018 saw an important landmark as the number of institutions that had divested either fully or partially from fossil fuels reached 1,000. (Partial divestment occurs when institutions stop investing in the most environmentally destructive kinds of fossil fuels, like coal or the Alberta tar sands.) Together, these institutions command US$8 trillion in assets. This represents a truly massive growth from the 181 institutions, representing $50 billion in assets, that had divested by September 2014. Notable institutions that have divested include the *Guardian* newspaper and media company, the British Medical Association, the Rockefeller Brothers Fund, the City of Oslo, the City of New York, the Republic of Ireland, and the Norwegian Sovereign Wealth

Fund. Canada saw its first university fully divest from fossil fuels in 2017 with the successful campaign launched by the organization "ULaval sans fossiles" at Université Laval in Quebec (Simard 2017).

It is becoming easier than ever for Canadians to divest. In 2013, Genus Capital Management became the first Canadian company to offer a carbon-free investment fund. Vancouver-based credit union Vancity announced in 2015 that one of its affiliated mutual funds (the IA Clarington Inhance Global Equity SRI Class fund) was no longer investing in fossil fuels.

Data is still emerging on the financial performance of carbon-free investment funds, but available studies show that no major financial penalty is incurred by switching to them (Genus 2016; Patsky 2017). It is likely that fossil-free funds will outperform fossil fuel-invested funds by even greater margins in the future. The world is changing, after all. Solar energy costs have plummeted dramatically and have become competitive with fossil fuels. Governments are announcing increasingly ambitious climate targets. Scotland is eyeing a target of 100% of electricity to be generated by renewable sources by 2020. Vancouver and Victoria plan on deriving 100% of the energy used in those cities from renewables by 2050. Hawaii is aiming for 100% clean energy by 2045. In 2018, California committed to sourcing 100% of its electricity from zero-carbon sources by 2045 (for more commitments, see Go 100% Renewable Energy). Scotland announced plans to end the sale of new diesel- and gasoline-powered vehicles by 2032; the United Kingdom and France plan to do the same by 2040. Governments all over the world are adopting policies to raise the price of carbon in order to better reflect its social and environmental costs (World Bank 2015). If they can realize their climate goals, and if movements can build on this progress to demand even more ambitious climate policy, the fossil fuel industry will not be long for this world.

Reflections on the university divestment movement in Toronto

I joined the fossil fuel divestment movement in 2014 when I helped launch a divestment campaign at York University. What I observed happening in the larger Toronto university divestment movement was eye-opening, and I would like to share that here, as I do not believe those observations are unique to Toronto's divestment scene; I suspect they are happening throughout the movement.

Divestment was an important first entry point into environmental politics and social justice for many students. Climate change presents a problem on such an unprecedentedly large scale that people can find themselves overwhelmed by it and unable to see how they can make a difference. But divestment takes that complex, unwieldy global problem and relocalizes it so that people can connect and recognize opportunities to work together to make institutions more democratically accountable and more socially and environmentally responsible. It distils the often complex science of climate change into these easy-to-understand terms: there is a limited amount of carbon we should ever emit and the fossil fuel industry is set on seeing all of it and more spewed into the air. Divestment establishes clear and achievable goals and is embedded in a logic showing how eroding the social license of fossil fuels can have a broad impact on the climate crisis.

Divestment created spaces for the spread and exchange of important ideas. Students could not discuss the unchecked power and influence of the fossil fuel industry without also considering the role of neoliberal capitalism in driving the climate crisis. Nor could they truly talk about the development of fossil fuel sites and pipeline infrastructure in North America near or on Indigenous lands without drawing links to the continuing legacies of colonialism. Divestment opened opportunities for many students to engage for the first time in organizing and activism, and in ways that expanded their worldviews. Students who once believed they were too nervous to speak in public suddenly found themselves making presentations in massive lecture halls. At other times, they were holding important discussions about how to build a movement that was inclusive, accessible, anti-racist, and decolonized. They examined privilege and power *within* the movement by considering matters like who speaks, who is spoken over, and who stays silent in meetings. Students who were women, LGBTQ, Indigenous, and of colour took on leadership roles. Much more work along these lines needed to be done, as I'm sure the participants would acknowledge, but it was an important start.

Divestment also teaches valuable lessons about institutional power and to whom it is, or may be, accountable. To see campus campaigns rejected by university administrations is to understand the dangers of institutions standing outside of democratic control. Sadly, it was this last experience that those of us in the divestment groups at the University of Toronto and York University (Saad 2017) went through in 2016 and 2017, respectively, when our university administrations rejected our divestment proposals.

The University of Toronto and York decisions were hardly unique in the Canadian university landscape. Dalhousie rejected divestment in 2014, McGill in 2016, and the University of Guelph in 2019. The University of Calgary *preemptively* rejected divestment in 2015 (Hussein 2015). Others have found ways to sidestep it. In 2016, the University of British Columbia rejected a proposal to divest its $1.4 billion endowment fund in favour of establishing a $10 million low-carbon investment fund. Concordia made similar moves in 2014. (Efforts by schoolteachers to divest the $175 billion Ontario Teachers' Pension Plan have also been similarly frustrated (Sparks 2017).)

The consequence is that Canadian universities lag behind the rest of the world. Fortunately, others have stepped up. As of August 2017, the UK is leading the world with 54 (or one-third) of its universities divesting either fully or partially. In Australia, 14% of universities have divested, in Sweden 15%, in New Zealand 25%, and in Ireland 29%. Canada's figure of 1% (Morgan 2017) is shameful in comparison. As many in the movement have pointed out, this is especially egregious given that universities are supposed to be preparing youth for the future, not profiting from companies destroying that future. More scholarship will need to emerge on what factors explain this recalcitrance. To what degree might Canadian universities, squeezed for funds in a neoliberal era, have become overreliant on sponsorships from the fossil fuel industry? The governing boards of Canadian universities are composed to a large degree of people with private-sector backgrounds (Canadian Association of University Teachers 2016). What effect might that have on perceptions of fiduciary duties or the use of divestment as a political tool?

Whatever the reason, the struggle for divestment is not dead and will need to escalate in the coming years.

References

Allen, M. et al. (2009) 'Warming caused by cumulative carbon emissions towards the trillionth tonne', *Nature* 458, pp. 1163–1166.

Brulle, R.J. (2018) 'The climate lobby: a sectoral analysis of lobbying spending on climate change in the USA, 2000 to 2016', *Climatic Change* 149, pp. 289–303.

Canadian Association of University Teachers. (2016) 'Do you know who sits on your board?' *Canadian Association of University Teachers*, December [online]. Available at: www.caut.ca/bulletin/2016/09/do-you-know-who-sits-your-board (Accessed 18 March 2019)

Carrington, D. (2015) 'Leave fossil fuels buried to prevent climate change, study urges', *Guardian*, 7 January [online]. Available at: www.theguardian.com/environment/2015/jan/07/much-worlds-fossil-fuel-reserve-must-stay-buried-prevent-climate-change-study-says (Accessed 18 March 2019).

Center for Responsive Politics. (n.d.) *Energy/Natural Resources* [online]. Available at: www.opensecrets.org/industries/indus.php?Ind=E (Accessed 18 March 2019).

Clark, P. (2015) 'Mark Carney warns investors face "huge" climate change losses', *Financial Times*, 29 September [online]. Available at: www.ft.com/content/622de3da-66e6-11e5-97d0-1456a776a4f5 (Accessed 18 March 2019).

Climate Action Tracker. (2017) 'Equitable emissions reductions under the Paris Agreement', 19 September [online]. Available at: http://climateactiontracker.org/assets/publications/briefing_papers/EquiteUpdate2017/ CAT_EquityUpdateBriefing2017.pdf (Accessed 18 March 2019).

Coady, D., I. Parry, L. Sears, and B. Shang. (2017) 'How large are global fossil fuel subsidies?' *World Development* 91, pp. 11–27.

Corporate Accountability. (2017) *Polluting Paris: How Big Polluters are Undermining Global Climate Policy.*

Gaworecki, M. (2016) 'How much are fossil fuel interests spending to sway your vote for congress?' *Desmog*, 31 October [online]. Available at: www.desmogblog.com/2016/10/31/how-much-money-fossil-fuels- interests-have-spent-sway-your-vote-congress (Accessed 18 March 2019).

Genus, Fossil Fuel Divestment Report, September 2016.

Go 100% Renewable Energy. Available at: www.go100percent.org/cms/

Hussein, Y. (2015) 'University of Calgary will not divest from fossil fuels', *Financial Post*, 12 February [online]. Available at: http://business.financialpost.com/commod

ities/energy/university-of-calgary-will-not-divest-from-fossil-fuels (Accessed 18 March 2019).

Inside Climate News. (2015) 'Exxon: The Road Not Taken' [online]. Available at: insideclimatenews.org/content/exxon-the-road-not-taken

Intergovernmental Panel on Climate Change. (2018) 'Mitigation pathways compatible with 1.5°C in the context of sustainable development,' in Masson-Delmotte, V., P. Zhai, H.-O. Pörtner, D. Roberts, J. Skea, P.R. Shukla, A. Pirani, W. Moufouma-Okia, C. Péan, R. Pidcock, S. Connors, J.B.R. Matthews, Y. Chen, X. Zhou, M.I. Gomis, E. Lonnoy, T. Maycock, M. Tignor, and T. Waterfield (eds.) *Global Warming of 1.5°C. An IPCC Special Report on the Impacts of Global Warming of 1.5°C Above Pre-Industrial Levels and Related Global Greenhouse Gas Emission Pathways, in the Context of Strengthening the Global Response to the Threat of Climate Change, Sustainable Development, and Efforts to Eradicate Poverty*, in print, pp. 93–174.

Jennings, K., D. Grandoni, and S. Rust. (2015) 'How Exxon went from leader to skeptic on climate change research', *LA Times*, 23 October [online]. Available at: graphics.latimes.com/exxon-research/ (Accessed 18 March 2019).

Jerving, S., K. Jennings, M.M. Hirsch, and S. Rust. (2015) 'What Exxon knew about the Earth's melting Arctic', *LA Times*, 9 October [online]. Available at: graphics.latimes.com/exxon-arctic/ (Accessed 18 March 2019).

McGlade, C. and P. Ekins. (2015) 'The geographical distribution of fossil fuels unused when limiting global warming to 2°C', *Nature* 517, pp. 187–190.

McKibben, B. (2012) 'Global warming's terrifying new math', *Rolling Stone*, 19 July [online]. Available at: www.rollingstone.com/politics/politics-news/global-warmings-terrifying-new-math-188550/ (Accessed 24 March 2019).

McKibben, B. (2018) 'At last, divestment is hitting the fossil fuel industry where it hurts', *Guardian*, 16 December [online]. Available at: www.theguardian.com/commentisfree/2018/dec/16/divestment-fossil-fuel-industry-trillions-dollars-investments-carbon (Accessed 18 March 2019).

Meinshausen, M. et al. (2009) 'Greenhouse gas emission targets for limiting global warming to 2°C', *Nature* 458, pp. 1158–1162.

Morgan, J. (2017) 'University fossil fuel divestment total tips £80 billion globally', *Times Higher Education*, 17 August [online]. Available at: www.timeshighereducation.com/news/university-fossil-fuel-divestment-total-tips-ps80-billion-globally (Accessed 18 March 2019).

Oil Change International, "Dirty Energy Money," Oil Change International, http://dirtyenergymoney.org/ (Accessed 18 March 2019).

Oil Change International. (2016) *The Sky's the Limit: Why the Paris Climate Goals Require a Managed Decline of Fossil Fuel Production*. Washington, D.C.

Patsky, M. (2017) 'Fossil free investing and portfolio performance', 10 May [online]. Available at: https://divestmentguide.org/how-to/fossil-free-investing-and-portfolio-performance/ (Accessed 18 March 2019).

Peters, G. (2017) 'How much carbon dioxide can we emit?', CICERO, 16 March [online]. Available at: www.cicero.uio.no/no/posts/klima/how-much-carbon-dioxide-can-we-emit# (Accessed 18 March 2019).

Rainforest Action Network. (2017) *Banking on Climate Change: Fossil Fuel Finance Report Card 2017*.

Saad, A. (2017) 'Divestment debacle at York U casts school in a bad light', *Ricochet*, 19 April [online]. Available at: https://ricochet.media/en/1773/divestment-debacle-at-york-u-casts-school-in-a-bad-light (Accessed 18 March 2019).

Saad, A. (2018) 'Pathways of harm: the consequences of Trump's withdrawal from the Paris climate agreement', *Environmental Justice* 11(1). Available at: http://doi.org/10.1089/env.2017.0033

Simard, A. (2017) 'Laval makes history with fossil fuel divestment: how did they do it?' *Ricochet*, 20 February [online]. Available at: https://ricochet.media/en/1688/laval-makes-history-with-fossil-fuel-divestment-how-did-they-do-it (Accessed 18 March 2019).

Sparks, R. (2017). 'Teachers urge $175 billion pension fund to flex muscle on climate change', *National Observer*, 6 April [online]. Available at: www.nationalobserver.com/2017/04/06/news/teachers-urge-175-billion-pension-fund-send-climate-change-message (Accessed 18 March 2019).

Steffen, W. et al. (2018) 'Trajectories of the earth system in the anthropocene', *Proceedings of the National Academy of Sciences* 115, pp. 8252–8259.

Tutu, D. (2010). 'Divesting from injustice', *Huffington Post*, 13 June [online]. Available at: www.huffingtonpost.com/desmond-tutu/divesting-from-injustice_b_534994.html (Accessed 18 March 2019).

United Nations Environment Programme. (2018) Emissions Gap Report 2018.

Victor, D.G. et al. (2017) 'Prove Paris was more than Paper Promises', *Nature* 548, pp. 25–27.

World Bank. (2015) *State and Trends of Carbon Pricing 2015* (Washington) [online]. Available at: wwwwds.worldbank.org/external/default/WDSContentServer/WDSP/IB/2015/09/21/090224b0830f0f31/2_0/Rendered/PDF/State0and0trends0of0carbon0pricing02015.pdf (Accessed 18 March 2019).

11

I EAT, THEREFORE I'M EVIL

The dilemmas of applying climate justice to food choice

Caitlin Bradley Morgan

Introduction: The basics of agriculture, food, and climate

It is well known that the causes and effects of climate change are distributed unequally across the world. The top 10% of greenhouse gas (GHG) emitters are responsible for 45% of all emissions, and the bottom 50% for only 13% of emissions (Piketty and Chancel 2015). Moreover, nations and groups of people who have historically emitted less carbon often experience greater effects of climate change. Less developed countries, for instance, experience extreme weather events at higher rates than industrialized countries that have done much more to cause those events (Kreft et al. 2015). This is climate injustice.

There are many arenas for addressing climate injustice, and while the most immediately obvious might be sectors like transportation—where fossil fuels are visibly consumed and emitted—food and agriculture represent a huge portion of global climate concerns. Our industrial, globalized food system is inextricably linked to climate change, especially in North American diets. Agriculture accounts for over a quarter of all greenhouse gas emissions worldwide (Vermeulen et al. 2012), and meat production alone causes 15% of global emissions (Gerber and Food and Agriculture Organization of the United Nations 2013). Agriculture helps drive climate change and is also affected by it, in ways that are sometimes hard to predict. Between 1964 and 2007, droughts decreased cereal production by 10%, and more recent droughts have resulted in proportionally greater damage to food production—a reinforcing feedback loop in which food production is vulnerable to the climate variations that it contributes to.

Developing countries are more likely to suffer the damage than developed ones (Lesk et al. 2016). Not only are poor people more likely to feel climate effects, but some of the most significant effects will be felt by poor *farmers*. Subsistence farmers globally tend to live in the tropics, in places where climate adaptation is economically, geographically, or politically challenging (Morton 2007; Vermeulen et al. 2012).

The idea of "climate justice" has emerged as a framework for understanding inequities in the causes and effects of our changing climate broadly. This discourse, like environmental justice, centres on ideas of social justice, democracy, and sustainability. It centres on "local impacts and experience, inequitable vulnerabilities, the importance of community voice, and demands for community sovereignty and functioning" (Schlosberg and Collins 2014, p. 359).

The worse the climate crisis becomes, the more we hear about the need to dramatically increase production to feed the world, using the same carbon-intensive methods that, rather than scaling back emissions, will heighten them (Tomlinson 2013). Increasing—or improving—food production is a "supply-side" solution in the food system, but those of us on the receiving ends—consumers—may feel compelled to do something about our diets' contributions to climate change. We may feel compelled to eat in a way that seems more climate-just. This chapter examines the intersections between climate justice, food systems, and individual food choice, acknowledging and exploring how applying systemic global problems to personal action is wickedly complicated and often contradictory.

Applying climate justice to personal food choice in affluent countries

Personal food choice has emerged as an important avenue for people concerned with the ecological and social effects of their daily lives. Choosing ethically produced food is a way to do one's part to minimize injustice when we cannot simply exit the system. But how do we start to make better choices about what is a fairer way to eat?

Project Drawdown, a highly visible initiative that has aggregated and ranked climate solutions, lists seven sectors that can be immediately and effectively reformed to address climate change. Food is one of them. Of the food sector solutions, four are "demand-side" solutions, things that distributors and consumers can affect. These solutions are clean cook stoves (in places where people still use firewood as cooking fuel), composting, a plant-rich diet, and reduced food waste. Across seven sectors and 80 large-scale solutions, reduced food waste is ranked third overall, and adopting plant-rich diets is ranked fourth.

For food waste, the Project recommends systems-level action, arguing that for higher-income countries, change is needed at both the retail and consumer levels, supported by national policy (Project Drawdown 2017b). Plant-rich diets, on the other hand, are actionable on the individual level. The Project recognizes that dietary change "is not simple because eating is profoundly personal and cultural" but supports making plant-based options visible, plentiful, and desirable and ending subsidies that make animal protein distortedly cheap for consumers. It cites the Zen master Thich Nhat Hanh's belief that transitioning to a plant-based diet may be the most effective way individuals can stop climate change (Project Drawdown 2017a).

The devil is in the details. It is widely reported that beef and dairy are big climate offenders (Majot and Kuyek, 2017; Carrington, 2014), for example, and that vegetable-based foods represent far less fossil fuel use per kilogram of edible food (González et al. 2011; Carlsson-Kanyama and González 2009). But food transported on planes, including vegetables, can result in much higher emissions than ones transported by boat (Carlsson-Kanyama and González 2009). Vegetables will not necessarily represent much lower emissions than animal products if the vegetables are grown in greenhouses (González et al. 2011). And it is hard to know whether emissions estimates for one location would hold for consumers in another part of the world. While top-level trends may result in lower food-related GHG emissions, it is hard to know if individual choices are truly contributing to those trends.

Systems thinking and the complications of tackling justice through personal choice

If the climate justice motivation in one's food choice is related only to emissions, it is possible to make real efforts toward eating a more climate-friendly diet. These choices get more complicated when we examine food as a system, rather than a straight line from producer to consumer.

As a concept, the "food system" moves past the linearity of the "food chain," and past the narrow economic perspective of the "food economy," broadening the understanding to include interconnections between biological, economic, political, social, and cultural aspects of food (Tansey and Worsley 1999). Environmental effects are not the only effects of a system—economies, people, and cultures are bound up in system functioning. Thinking in systems also means acknowledging the logic of how a particular system operates.

In a capitalist food system, the system's purpose is capital accumulation and "efficient" distribution of food resources based on individual ability to pay. The global marketplace, in which traded commodity foods damage the climate and go to the highest bidder, ensures that poor people will go hungry even when there is enough food to go around (Farley et al. 2015). As renowned systems thinker Donella Meadows (2008) has pointed out:

> A free market does allow producers and consumers, who have the best information about production opportunities and consumption choices, to make fairly uninhibited and locally rational decisions. But those decisions can't, by themselves, correct the overall system's tendency to create monopolies and undesirable side effects (externalities), to discriminate against the poor, or to overshoot its sustainable carrying capacity.
>
> *(p. 109)*

This is the problem with framing climate justice around consumer choice. There is much to be done in agriculture, before food ever gets to consumers. Project

Drawdown lists thirteen supply-side climate solutions in the food sector, compared with the four demand-side solutions. Despite much conversation about "food miles," the great distances that most food travels from farm to plate, studies show that transportation only accounts for an average of 11% of foods' GHG emissions (Weber and Matthews 2008). Instead, as of 2012, agricultural production practices constituted up to 86% of all food systems' GHG emissions (Vermeulen et al. 2012).

Corporations, not individuals, run the system. Three companies own 40% of the international beverage market; ten companies account for 90% of agrochemicals, which are widely used in global agriculture; and three seed companies own 70% of genetically modified seeds (Pol and Luis 2013). Agricultural subsidies—which invariably go to industrial-sized farms—and low fuel prices mean that highly processed foods from distant places are more affordable than whole foods grown nearby. Consumers don't get to decide whether a producer employs good environmental practices, pays their workers well, sources ethical ingredients, tries to minimize packaging or fossil fuel inputs, or does any number of other things that climate activists might want to support.

Not only do individuals not get to decide on these issues, we often can't tell how they play out in the supply chain. We get information about the good practices a company wants to advertise, and nothing about the things they would rather hide. It takes time and research to learn how to read food labels, which are essentially in code, and are sometimes unregulated; words like "natural" or "sustainable" have no official designation in the United States and can mean anything, or nothing.

There is no food label for climate justice, and there is no label that can comprehensively capture food impact. Carbon emissions could be measured, but other impacts are not easily accounted for—impacts like

> water use, land use, nitrogen released, energy required to cultivate, process, package, and transport foods, as well as emissions and materials associated with packaging and food lost in production … chemicals (including pesticides and hormones) used in production … [with] varied impacts on the environment.
>
> *(Baker 2018)*

One of the closest approximations of such labelling, the Rainforest Alliance seal, has come under fire for labour exploitation in its certified production chains (Yeung 2016).

In these ways, consumer "choice" is shaped by other actors in the food chain (Silbey and Parker 2013).

Applying systems thinking: Food and climate justice considerations beyond emissions

If one moves past a conception of climate justice as a linear idea from GHG-emitter to climate-affected, allowing it to encompass other aspects of food justice, then the idea of eating a climate-just diet becomes even more complicated, and more untenable. Eating to reduce emissions-related injustice may overlook other environmental or social injustices.

The problem is, what "just" food is depends on which aspect of it you're considering, and even within the perspective of climate justice, particular foods may have both beneficial and damaging impacts. This premise, that we can and should consider the social and environmental justice components of our own diets, is essentially a "vote with your fork" approach to food choice and climate justice. This approach to reform has its own ethical considerations, bypassing democratic control by allowing people with greater wealth to have greater power in changing food system demands.

Quinoa is a classic case of the global food system's paradoxes for seeking justice. Quinoa came into fashion as a health food in the United States in the 1990s (Albert 2017). It is a complete vegetable protein, which makes it better environmentally than animal proteins in general (González et al. 2011). It supports individual health and nutrition, which has its own, positive climate implications; in 2013, 10% of greenhouse gas emissions in the United States came from the healthcare industry (Eckelman and Sherman 2016). Quinoa farmers, some of the poorest people in Peru, have experienced a 46% increase in welfare—measured by the overall value of goods and services consumed by a household—thanks to high quinoa prices internationally (Kasterine 2016).

But North American quinoa purchases have also transformed South American economies negatively. In 2012, violence erupted in conflicts over superior quinoa-growing land in Bolivia (Friedman-Rudovsky 2012). Changes in agricultural practices have degraded soil quality and changed traditional practices that protected soil health (FAO 2014). Biodiversity decreased as heritage varieties of quinoa and potatoes were abandoned in favour of more consistent, marketable varietals (Hellin and Higman 2005). Western industrialized nations' extraction of quinoa violated community food sovereignty by making it harder for people to afford and access culturally appropriate foods. Hundreds of articles between 2013 and 2016 debated the ethics of quinoa consumption (Albert 2017). If food has fewer emissions but other negative environmental and social issues, does it contribute to climate justice or injustice?

The quinoa story illustrates essential problems with striving for justice in a complex and global food system: lack of transparency and lack of control over outcomes. Years go by before investigative journalists uncover problems that consumers have been unknowingly contributing to at the checkout counter. Recent cases of food-system exposés include horrendous working conditions and slavery in Florida tomato fields (Paul 2015); avocado farms in Mexico leading to deforestation of pine forests (Nelson 2016); and the Thai shrimp industry's use of slave labour (Hodal and Lawrence 2014). Shrimp production is doubly damaging, as deforestation of mangroves can push shrimp's greenhouse gas footprint past that of beef, a notoriously carbon-intensive food (Boone Kauffman et al. 2017). From a systems perspective, the sheer amount that we *don't* know makes it difficult, if not impossible, to make informed choices about trade-offs between environmental and social justice impacts.

And yet, some of us cannot stop trying to maximize our goodness—or at least minimize our badness—through personal food choice. When we feel complicit in an unjust system, some of us will attempt to participate as little as possible. The problem is, when thinking about climate justice in a holistic sense, it's hard—maybe impossible—to eat a truly blameless diet.

When we back out of the tangled decisions in day-to-day life and examine them from a theoretical standpoint, the problem becomes clear: however much we care about our choices, we cannot fully know what they represent. Ecofeminist scholar Mary Mellor (1997) has written extensively about the connection between social oppression and ecological destruction. She argues that we cannot escape the fact of immanence, that our human lives are embodied lives, embedded in physical worlds.

> Awareness of immanence makes the concrete relations of any product virtually infinite. Who grew/extracted the raw materials? Who made the components? Who made the transport that brought it here? Who drove it? What energy was involved? How do all those people live? What do they consume to support their work? What emissions or elements will the object and the processes that created it break down into? Where will they go and with what effects? ... the life history of a product destroys the neoliberal notion of the independent consumer and the autonomy of economics processes.
>
> *(p. 195)*

Can local food reduce food-related climate injustices?

The local food movement sprang in part from worries about these kinds of unseen impacts, as well as from aspirations like eating fresher food and supporting farmers and local resiliency. In Vermont, where I live, it is relatively easy to get high-quality local food, if you have disposable income and reliable transportation. But there are problems even with eating regionally. If you drive to buy your local food, you might burn more fossil fuels than if it were shipped from a much greater distance (Coley et al. 2009). Bananas shipped to North America are still one of the lowest-carbon foods, thanks to how easy it is to grow them (Berners-Lee 2011). Transportation inputs may pale in comparison to geographic particularities, such as trying to grow food in an arid environment rather than shipping it in (Born and Purcell 2006). Perversely, eating locally can contribute to already outsized North American carbon footprints.

Consider the example of buying dairy products in Vermont, a case that further complicates choices, taking climate justice concerns beyond carbon emissions or land degradation. Dairy is the bedrock of Vermont agriculture and state economy, bringing $2.2 billion annually to the small state and accounting for over 70% of agricultural sales ("Vermont's Dairy Economy" n.d.). Animal husbandry has been embedded in Vermont's cultural history since European colonization. While most US dairy production has moved west, cow farmers here have held on against the odds and currently produce over 60% of milk consumed in New England (Vermont Farm to Plate n.d.). The dairy industry, especially on smaller farms, helps maintain the bucolic pastures and cornfields that residents and tourists expect (Wilson 2017). In these ways, dairy is a crucial part of the current culture and economy of Vermont.

Dairy has negative consequences, too—all the contradictions of eating quinoa, but the problems are local. Cows in general have a substantial carbon footprint (Desjardins et al. 2012). Lake Champlain, which stretches nearly the length of Vermont, suffers from increasing phosphorous pollution (US EPA 2016), half of which comes from agricultural and development runoff (Environmental Studies Spring Seminar 2011) and results in E. coli and algal blooms. Like climate change, the pollution is a failing of intergenerational justice, a case of leaving a mess for future generations to clean up or endure, because the system was created before questions of pollution, accumulation, and environmental legacy were mainstream considerations. Furthermore, Vermont's dairy industry employs an estimated 1,500 migrant farm workers (VT Migrant Farmworker Solidarity Project n.d.). Some farmers deal fairly with these unprotected workers, but many do not, and migrant life on the farm can be isolating, depressing, unsafe, and food insecure ("About Migrant Justice" 2010).

Even if we ignore the cultural and economic reasons to buy local milk products, it is not easy to simply exit the system. Buying dairy products from other places likely results in similar, if not worse, problems, in someone else's backyard (Gardiner 2015).

Milk replacements, whether almond, soy, or coconut, aren't panaceas, either. Almonds require over one gallon of water *per nut* and are often grown in drought-prone California (Park and Lurie n.d.). Ninety per cent of soy grown in the United States is genetically modified (Friedlander n.d.), which has its own issues and controversies, and in other places, soy production has displaced small farmers and contributed to clearcutting of the Amazon rainforest (Monahan 2005). The Caribbean is reportedly struggling to keep up with new global demand for coconuts for coconut milk, while locals go without the formerly cheap staple (Fieser 2016). Even if you give up all milk and cheese substitutes, you face the issue of where to get your fats. In Malaysia and Indonesia, people clear cut the rainforest (crucial carbon sinks for the atmosphere) to plant palm oil trees for vegan butter and many other products (The Union of Concerned Scientists n.d.).

To summarize so far: both local and global food have climate repercussions; local food is often set up as an alternative and as environmentally and socially preferable, but it is not inherently better on either count. The specifics of a particular food matter, and also complicate decisions. How can we balance these competing claims on our morality, knowing that we know so little? Perhaps the only way forward is to do the best we can with what we do know.

Working through systems contradictions and personal complicity toward climate-just food

For me, it is a daily struggle to balance all these things: guilt over my outsized carbon footprint, desire to support the things I believe in, the reminder that I cannot change the system through my own actions, and constant lack of transparency that might allow me to make better choices. I weigh my limited knowledge against my powerlessness and, often, my desire to eat certain foods. As social and political theorist Alexis Shotwell argues, it is not possible or even productive for us to achieve moral purity. Rather, we should embrace complexity and complicity as foundational aspects of living in a compromised system, working toward collective and imperfect change (Beck 2017).

For the most part, I try to eat locally. As I've discussed, local food is not necessarily more sustainable or more fair than non-local food, but for me, eating locally allows me to know more about what I am buying. I'm lucky to live in a town with many farms nearby, some of them within city limits, and as a member of a CSA share (community-supported agriculture, where members pay a fixed rate up front to mitigate farmer vulnerability), I can bike to the farm pick-up and get all my vegetables for the week—although I often drive. The farm is organic, the farm manager is salaried, and although I wish that the farmers themselves made a better living, I at least know that I'm not contributing to synthetic chemical use or forced labour, which is more than I can say for buying vegetables at the grocery store. Also, all the vegetables themselves are whole and unwrapped, which means no carbon was used to process or package them for me; packaging might be a minor contributor to carbon emissions, but avoiding it avoids adding more plastic bags to ocean trash piles (Rice 2018). The processing of vegetables comes from the labour of my own or my spouse's hands. I have had to get much better at cooking from scratch.

I also believe that as fossil fuels become less plentiful and production and distribution chains are increasingly vulnerable to weather and political and economic upheaval (Crownshaw et al. 2018), we will have to re-localize food systems and re-learn how to farm without as many carbon inputs. Supporting local farms is a way for me to support that transition now, and I hope our community will be more prepared because of such choices. It also offers me the opportunity to patronize farms that engage in regenerative practices like the ones listed in Project Drawdown's (2017) supply-side solutions.

I also buy plenty of things shipped from all over the world. Not much fruit, although I buy citrus in the winter—otherwise, it's local apples from cold storage and frozen berries from my own freezer. I try to buy dry goods in bulk to minimize packaging. When I buy things like tea or coffee or sugar, I look for fair trade and single-origin options. I am allergic to beef and dairy, so I get a free pass to feel good about avoiding cow products, except butter, which I can and do eat in excess. I buy local, non-organic butter from a conventional dairy because my brother works on one of the company's farms, and because, as mentioned above, palm oil has been a huge driver of rainforest deforestation. I try to never eat shrimp because of the association with slavery and mangrove deforestation; although in writing this, I looked up the Monterey Bay Aquarium Seafood Watch recommendations, and they list nine "best choice" options, all of which are raised in ponds or indoors ("Monterey Bay Aquarium" n.d.). Like so many folks, I tried being a vegetarian in college, and I spent that period hungry and angry. Now, I eat meat, but in small amounts, and not every day. I try to purchase meat that is raised on small farms, not by large meat producers, whose carbon footprint is massive: the top three meat companies were estimated as emitting more GHG in 2016 than the country of France, an amount comparable to oil companies like Exxon and Shell (Majot and Kuyek 2017).

I admit I have tried dumpster diving for food thrown away behind grocery stores, diverting food from compost or landfills and getting a free meal. It is stunning to witness first-hand how much food goes to waste, bins and bins full of mostly good food at just one store. In the summer, I sometimes volunteer for local gleaning programs, where we harvest leftover vegetables and donate them to the local food bank. At home, we compost food scraps and add the soil to our garden beds. In these ways, I try to reduce food waste. I promise myself I will be better about turning vegetable scraps into vegetable stock so I can avoid buying it and unnecessarily using carbon to package and ship it. I take small actions when I can, because I can, even though they are small.

In the summer, we grow tomatoes in the garden, and sometimes a few other things like squash or herbs. We process and freeze fruits and vegetables for winter, filling a five-cubic-feet external freezer. I wonder sometimes whether this is the best option: local food stored in plastic bags and kept cold from electricity. It feels like the best I can do.

I fail in these efforts constantly. Every day, I consume things that I know are problematic in one way or another—and in ways of which I am completely unaware. In those moments, I remind myself that some effort is better than no effort, and that I cannot change the monolithic, global, industrial food system worth trillions of dollars by choosing correctly between local, sprayed onions and organic ones shipped from California.

And so I also support systems-level change whenever I can, recognizing that it is probably a better use of my time than agonizing over my shopping cart. One of the primary barriers to systems change in the United States is the Farm Bill, which is "crucial to practically everything in our food system: what crops get subsidized, how much foods cost, how land is used and whether low-income Americans have enough to eat" (Nestle 2016). Because I live in a very liberal state, our congressional delegation already supports most of the priorities I would want to see reflected in the Bill, like reducing large agribusiness subsidies and supporting the Supplemental Nutrition Assistance Program (SNAP). Corporate lobbying still strongly affects Farm Bill provisioning (Nestle 2019), which speaks to the need for broader political reform before we can see environmental and social justice goals truly integrated into the food system.

From these impossible personal dilemmas and from the awareness of how large systems operate, I have learned to see my own and others' food choices with as much compassion as possible. Personal desire, food culture, and food access are critical avenues through which most people make food choices, which I have only touched upon briefly here. Most people will not eat for climate justice alone, and asking them to would risk erasing many of the other culturally rich reasons that people eat the way they do. For people whose food access is tenuous, it may be unjust to further burden their choices with climate considerations. Low-income cooks already have many pressures on their food budgets (Bowen et al. 2014). There is an antiregulatory thought trend that shifts the burden from policy to people's individual bodies, that we "should and can teach people how and what to eat, as if you could 'change the world one meal at a time' without attention to policy" (Guthman 2007, p. 78). We should resist this trend whenever we can, and fight for better policy.

Even folks who have the privilege and inclination to make values-based food choices may possess a variety of different values. A good friend of mine works with migrant labourers on dairy farms, and partly as a result of seeing their living conditions, she has almost entirely given up eating dairy products. Although I cannot eat dairy, I see some value in having a dairy industry in Vermont, as I outline above. Both of us are thinking about the social justice impacts of our diets and are arriving at different conclusions. I am better off embracing these differences, honouring their motivations, and having productive and critical discussions than I am writing off a friend's choices and deciding she is wrong. I believe we will only create change with open hearts.

Conclusion

Food and agriculture are inextricably tied to climate change, in both causes and effects. There are ways to eat a lower-emissions diet, although there is so little transparency in food chains that such efforts are not always easy to achieve. Once we expand the idea of a climate-just diet to include social and environmental effects beyond GHG emissions, it becomes increasingly hard to make unmistakably good choices. Any "answers" to these wicked problems lie in systems change, not individual choice; but until those changes occur, we can engage in thoughtful, personal, imperfect negotiations in daily life and community.

May we all eat with reverence, joy, humility, and as little harm as possible.

References

About Migrant Justice (2010, January 6). [Text]. Retrieved June 23, 2017, from https://migrantjustice.net/about

Albert, V. (2017). King Quinoa: The Development of the Modern Export Market and its Implications for the Andean People. *Graduate Association for Food Studies*, *4*(1). Retrieved from https://gradfoodstudies.org/2017/03/01/king-quinoa/

Baker, K. (2018, February 27). Do we need sustainability labels on food products? Retrieved March 5, 2019, from www.planetforward.org/idea/do-we-need-sustainability-labels-on-food-products

Beck, J. (2017, January 20). The Folly of "Purity Politics." *The Atlantic*. Retrieved from www.theatlantic.com/health/archive/2017/01/purity-politics/513704/

Berners-Lee, M. (2011). *How Bad Are Bananas?: The Carbon Footprint of Everything*. London: Greystone Books.

Boone Kauffman, J., Arifanti, V. B., Hernández Trejo, H., del Carmen Jesús García, M., Norfolk, J., Cifuentes, M., Hadriyanto, D., & Murdiyarso, D. (2017). The jumbo carbon footprint of a shrimp: Carbon losses from mangrove deforestation. *Frontiers in Ecology and the Environment, 15*(4), 183–188. https://doi.org/10.1002/fee.1482

Born, B., & Purcell, M. (2006). Avoiding the local trap: Scale and food systems in planning research. *Journal of Planning Education and Research, 26*(2), 195–207. https://doi.org/10.1177/0739456X06291389

Bowen, S., Elliott, S., & Brenton, J. (2014). The joy of cooking? *Contexts, 13*(3), 20–25. https://doi.org/10.1177/1536504214545755

Carlsson-Kanyama, A., & González, A. D. (2009). Potential contributions of food consumption patterns to climate change. *The American Journal of Clinical Nutrition, 89*(5), 1704S–1709S. https://doi.org/10.3945/ajcn.2009.26736AA

Carrington, D. (2014, July 21). Giving up beef will reduce carbon footprint more than cars, says expert. *The Guardian*. Retrieved from www.theguardian.com/environment/2014/jul/21/giving-up-beef-reduce-carbon-footprint-more-than-cars

Coley, D., Howard, M., & Winter, M. (2009). Local food, food miles and carbon emissions: A comparison of farm shop and mass distribution approaches. *Food Policy, 34*(2), 150–155. https://doi.org/10.1016/j.foodpol.2008.11.001

Crownshaw, T., Morgan, C., Adams, A., Sers, M., Britto dos Santos, N., Damiano, A., Gilbert, L., Yahya Haage, G., & Horen Greenford, D. (2018). Over the horizon: Exploring the conditions of a post-growth world. *The Anthropocene Review*, 2053019618820350. https://doi.org/10.1177/2053019618820350

Desjardins, R., Worth, D., Vergé, X., Maxime, D., Dyer, J., & Cerkowniak, D. (2012). Carbon footprint of beef cattle. *Sustainability, 4*(12), 3279–3301. https://doi.org/10.3390/su4123279

Eckelman, M. J., & Sherman, J. (2016). Environmental impacts of the U.S. health care system and effects on public health. *PLOS ONE, 11*(6), e0157014. https://doi.org/10.1371/journal.pone.0157014

Environmental Studies Spring Seminar. (2011). *Phosphorus Loading in Lake Champlain: A Geographic, Environmental, Civic and Economic Investigation into its Causes, Effects, and Prospects for the Future*. Middlebury, VT: Middlebury College. Retrieved from www.middlebury.edu/media/view/276855/original/final_compiled_small.pdf

Farley, J., Schmitt, A., Burke, M., & Farr, M. (2015). Extending market allocation to ecosystem services: Moral and practical implications on a full and unequal planet. *Ecological Economics, 117*, 244–252. https://doi.org/10.1016/j.ecolecon.2014.06.021

Fieser, E. The Caribbean's Running Out of Coconuts. (2016, September 9). *Bloomberg.Com*. Retrieved from www.bloomberg.com/news/articles/2016-09-09/just-when-the-world-craves-coconuts-the-caribbean-s-running-out

Food and Agriculture Organization of the United Nations. Check out this infographic on the impact of the quinoa boom on Bolivian family and small-scale farmers. (2014, April 25). Retrieved June 23, 2017, from www.fao.org/family-farming-2014/news/news/details-press-room/en/c/223319/

Friedlander, J. (n.d.). Soy versus dairy: What's the footprint of milk? Retrieved June 23, 2017, from http://theconversation.com/soy-versus-dairy-whats-the-footprint-of-milk-8498

Friedman-Rudovsky, J. (2012, April 3). Quinoa: The Dark Side of an Andean Superfood. *Time*. Retrieved from http://content.time.com/time/world/article/0,8599,2110890,00.html

Gardiner, B. (2015, May 1). How Growth in Dairy Is Affecting the Environment. *The New York Times*. Retrieved from www.nytimes.com/2015/05/04/business/energy-environment/how-growth-in-dairy-is-affecting-the-environment.html

Gerber, P. J., & Food and Agriculture Organization of the United Nations (Eds.). (2013). *Tackling Climate Change Through Livestock: A Global Assessment of Emissions and Mitigation Opportunities*. Rome, Italy: Food and Agriculture Organization of the United Nations.

González, A. D., Frostell, B., & Carlsson-Kanyama, A. (2011). Protein efficiency per unit energy and per unit greenhouse gas emissions: Potential contribution of diet choices to climate change mitigation. *Food Policy*, *36*(5), 562–570. https://doi.org/10.1016/j.foodpol.2011.07.003

Guthman, J. (2007). Can't stomach it: How Michael Pollan et al. made me want to eat cheetos. *Gastronomica: The Journal of Critical Food Studies*, 7(3), 75–79. https://doi.org/10.1525/gfc.2007.7.3.75

Hellin, J., & Higman, S. (2005). Crop diversity and livelihood security in the Andes. *Development in Practice*, *15*(2), 165–174.

Hodal, K., & Lawrence, C. K. F. (2014, June 10). Revealed: Asian slave labour producing prawns for supermarkets in US, UK. The Guardian. Retrieved from www.theguardian.com/global-development/2014/jun/10/supermarket-prawns-thailand-produced-slave-labour

Kasterine, A. (2016, July 17). Quinoa isn't a threat to food security. It's improving Peruvian farmers' lives. *The Guardian*. Retrieved from www.theguardian.com/sustainable-business/2016/jul/17/quinoa-threat-food-security-improving-peruvian-farmers-lives-superfood

Kreft, S., Eckstein, D., Dorsch, L., & Fischer, L. (2015). *Global Climate Risk Index 2016: Who Suffers Most From Extreme Weather Events? Weather-related Loss Events in 2014 and 1995 to 2014* [Briefing Paper]. Berlin: GermanWatch.

Lesk, C., Rowhani, P., & Ramankutty, N. (2016). Influence of extreme weather disasters on global crop production. *Nature*, *529*(7584), 84–87. https://doi.org/10.1038/nature16467

Majot, J., & Kuyek, D. (2017, November 7). Big Meat and Big Dairy's climate emissions put Exxon Mobil to shame. *The Guardian*. Retrieved from www.theguardian.com/commentisfree/2017/nov/07/big-meat-big-dairy-carbon-emmissions-exxon-mobil

Meadows, D. H. (2008). *Thinking in Systems: A Primer* (D. Wright, Ed.). White River Junction, VT: Chelsea Green Publishing.

Mellor, M. (1997). *Feminism and Ecology: An Introduction*. New York, NY: NYU Press.

Monahan, J. (2005, June 6). Soybean fever transforms Paraguay. *BBC News*. Retrieved from http://news.bbc.co.uk/2/hi/business/4603729.stm

Monterey Bay Aquarium. Shrimp Recommendations. (n.d.). Monterey Bay Aquarium Seafood Watch. Retrieved February 26, 2019, from www.seafoodwatch.org/seafood-recommendations/groups/shrimp

Morton, J. F. (2007). The impact of climate change on smallholder and subsistence agriculture. *Proceedings of the National Academy of Sciences*, *104*(50), 19680–19685. https://doi.org/10.1073/pnas.0701855104

Nestle, M. (2016, March 17). The farm bill drove me insane. *Politico*. Retrieved from http://politi.co/1ppK0bp

Nestle, M. (2019). Forward. In D. Imhoff & C. Badaracco (Eds.), *The Farm Bill: A Citizen's Guide*. Washington, DC: Island Press.

Nelson, Kate. Our love of avocados is causing a huge problem for Mexico. (2016, August 10). Retrieved June 23, 2017, from www.independent.co.uk/news/uk/home-news/avocados-deforestation-demand-price-rise-destroy-mexico-pine-forests-a7182571.html

Park, A., & Lurie, J. (n.d.). It takes how much water to grow an almond?! Retrieved June 23, 2017, from www.motherjones.com/environment/2014/02/wheres-californias-water-going/

Paul, E. T. (2015, February 25). You need to know: The slavery conditions on tomato farms. Retrieved from www.huffingtonpost.com/eve-turow/you-need-to-know-the-slavery-conditions-on-tomato-farms_b_6735842.html

Piketty, T., & Chancel, L. (2015). Carbon and inequality: From Kyoto to Paris. Paris: Paris School of Economics. Retrieved from www.ledevoir.com/documents/pdf/chancelpiketty2015.pdf

Pol, V., & Luis, J. (2013). *Food as a Commons: Reframing the Narrative of the Food System* (SSRN Scholarly Paper No. ID 2255447). Rochester, NY: Social Science Research Network. Retrieved from https://papers.ssrn.com/abstract=2255447

Project Drawdown. Food. (2017, June 19). Project Drawdown. Retrieved February 21, 2019 from www.drawdown.org/solutions/food

Project Drawdown. Plant-Rich Diet. (2017a, February 7). Retrieved February 28, 2019, from www.drawdown.org/solutions/food/plant-rich-diet

Project Drawdown. Reduced Food Waste. (2017b, February 7). Project Drawdown. Retrieved February 28, 2019 from www.drawdown.org/solutions/food/reduced-food-waste

Rice, D. (2018, March 22). World's largest collection of ocean garbage is twice the size of Texas. *USA TODAY*. Retrieved from www.usatoday.com/story/tech/science/2018/03/22/great-pacific-garbage-patch-grows/446405002/

Schlosberg, D., & Collins, L. B. (2014). From environmental to climate justice: Climate change and the discourse of environmental justice. *Wiley Interdisciplinary Reviews: Climate Change*, *5*(3), 359–374. https://doi.org/10.1002/wcc.275

Silbey, S. S., & Parker, C. (2013). Voting with your fork? Industrial free-range eggs and the regulatory construction of consumer choice. *The ANNALS of the American Academy of Political and Social Science*, *649*(1), 52–73. https://doi.org/10.1177/0002716213487303

Tansey, G., & Worsley, T. (1999). *The Food System: A Guide* (Reprinted). London, UK: Earthscan.

The Union of Concerned Scientists. What's Driving Deforestation: Palm Oil. (n.d.). Retrieved June 23, 2017, from www.ucsusa.org/global-warming/stop-deforestation/drivers-of-deforestation-2016-palm-oil

Tomlinson, I. (2013). Doubling food production to feed the 9 billion: A critical perspective on a key discourse of food security in the UK. *Journal of Rural Studies*, *29*, 81–90. https://doi.org/10.1016/j.jrurstud.2011.09.001

US EPA. (2016, June 17). EPA releases final phosphorus limits for Vermont segments of Lake Champlain [speeches, testimony and transcripts]. Retrieved June 23, 2017, from www.epa.gov/newsreleases/epa-releases-final-phosphorus-limits-vermont-segments-lake-champlain

Vermeulen, S. J., Campbell, B. M., & Ingram, J. S. I. (2012). Climate change and food systems. *Annual Review of Environment and Resources*, *37*(1), 195–222. https://doi.org/10.1146/annurev-environ-020411-130608

Vermont's Dairy Economy. (n.d.). Retrieved February 27, 2017, from Vermont Dairy website: http://vermontdairy.com/economic-impact/vermonts-dairy-economy/

Vermont Farm to Plate. Dairy Sustainability | Getting to 2020. (n.d.). Vermont Farm to Plate. Retrieved March 5, 2019, from www.vtfarmtoplate.com/getting-to-2020/8-dairy-sustainability

VT Migrant Farmworker Solidarity Project. (n.d.). *Myths and Realities.* Burlington, VT. Retrieved from https://migrantjustice.net/sites/default/files/mythsandrealities.pdf

Weber, C. L., & Matthews, H. S. (2008). Food-miles and the relative climate impacts of food choices in the United States. *Environmental Science & Technology, 42*(10), 3508–3513. https://doi.org/10.1021/es702969f

Wilson, J. (2017, Spring). How Artisanal Cheese Is Helping Save Vermont's Historic Landscape. Retrieved June 23, 2017, from National Trust for Historic Preservation website: https://savingplaces.org/stories/how-artisanal-cheese-is-helping-save-vermonts-historic-landscape?utm_medium=email&utm_source=NTHP_newsletter_040617&utm_campaign=NTHP_eNewsletter-FY17_Apr6

Yeung, B. (2016, November 28). Beware the little green frog logo on your sustainable food. *Reveal.* Retrieved from https://www.revealnews.org/blog/beware-the-little-green-frog-logo-on-your-sustainable-food/

12

FREE FOOD FOR JUSTICE

Sam Bliss

"You are what you eat" is not a metaphor. Your body makes itself from the food you eat, plus the water you drink and the air you breathe. Human survival depends on regular food intake.

People eat other organisms—plants, animals, fungi, microorganisms. Climate change is disrupting the conditions in which these organisms and the ecosystems they inhabit have coevolved. Scientists warn that the food supply, and thus food security, is in danger (Wheeler and von Braun 2013).

Here in the Great Lakes region, one cannot see global warming's full impacts on food systems, because international markets protect the world's wealthy people and places from shortages. Money attracts food. In many other regions, the effects are quite visible; markets generally push problems on the poor. In the first half of this chapter, I focus on the Middle East, where climate change's impacts on the food system are already quite visible, and are exacerbated by markets. I argue that market mechanisms put up inherent barriers to achieving climate justice in food systems. Therefore, climate justice activists should focus not just on making markets more just or sustainable, but on building non-market food systems of solidarity, charity, and community self-sufficiency. In the second half of the chapter, I review some examples from Vermont, where I participate in and study non-market food systems. I finish by questioning whether anything we can do locally will address global food injustice in the context of climate change.

Price spikes cause conflicts

In the winter that spanned 2006 and 2007, it stopped raining in the Middle East's "fertile crescent" region. A three-year drought ensued across much of Syria and parts of surrounding countries. It was the region's worst on record (Trigo et al. 2010). Human-caused climate change had more than doubled the likelihood of a drought of that length and severity (Kelley et al. 2015). Temperatures have been rising in the fertile crescent for over a century, together with the carbon dioxide concentration in Earth's atmosphere.

Syria had already been facing water scarcity. Thirst, undernourishment, and heat killed more than half of the country's livestock during the drought. Wheat production fell by half in 2008 and did not fully recover in subsequent years. The country began importing wheat again after a decade of self-sufficiency, just as wheat's price in the international commodity market was rising sharply (Ali 2010). Small farms failed to produce enough food or money to feed rural families. Hundreds of thousands of people migrated from agricultural areas to the edges of cities (Solh 2010; IRIN 2009). Over a million Iraqi refugees had arrived to Syria's urban fringes in the four years before the drought. Disconnected from the land, these new city dwellers needed cash to buy food in the marketplace.

Food was getting more expensive everywhere. The prices of most commodities were rising sharply. By June 2008, global financial markets had crashed completely. The bubble had popped, as bubbles do. At that point, life was expensive and it was hard to get money. The price of oil had doubled, yet there were fewer jobs and banks had stopped making loans.

The agricultural systems that produce staple food for trade happen to run on oil and money. The oil price hike and global financial crash disrupted international food markets. The price of commodity wheat had more than tripled, from less than $3 to over $10 a bushel (Tadesse et al. 2014). Other grain prices had risen substantially, too.

People took to the streets. The crisis led to food riots in about 30 countries (UNCTAD 2008; Berazneva and Lee 2013; Bello and Baviera 2009; Lagi et al. 2011). Statistical studies have confirmed what a skim of world history makes one suspect: rising food prices bring not just hunger but bloodshed (Raleigh et al. 2015; Holt-Gimenez et al. 2012; Walton and Seddon 1994). Commodities like oil and grains got cheap again in the months following the crash, as tends to happen after economic crises, but the damage was done. The number of undernourished people worldwide had gone up by about 7%; more than 40 million had entered the ranks of the chronically food insecure (Zaman and Tiwari 2010). Seeds of discontent were planted.

Markets allocate food according to purchasing power

When food prices increase suddenly, poor people go hungry while the wealthy continue eating like nothing happened (Regmi and Meade 2013, Seale et al. 2003; Mendes et al. 2018). This friction can lead to social strife and violent conflicts often erupt.

This can only happen because foods are market goods—private property that can be bought and sold. Markets allocate food to those who can pay for it, not those who need it. That is a big part of why one in ten people do not get enough to eat in a world that produces enough plant calories to feed the whole global population plus a few billion more (Holt-Giménez et al. 2012). It is why more than 40% of those calories go to livestock, biofuel refineries, and other industrial processes, rather than feeding humans directly (Cassidy et al. 2013). It is a big reason why at least another 25% of edible calories become food waste (Kummu et al. 2012). Some people pay for food they will throw away while others cannot pay for food to save the lives of their starving children (Farley et al. 2015).

In places where food is not a market good, people distribute it according to cultural norms of reciprocity and redistribution (Polanyi 1944). Some societies have strict sharing rules, especially for wild-harvested meat sources (Berking 1999). Where people have discretion over distributing food, they share with kin, or as part of reciprocal gift exchanges, or to signal unobservable things like their intention to collaborate or their fitness as a mate (see, for example, Nolin 2012). The processes of urbanization and development continuously bring more people into societies where market food systems dominate, but even in cities, complementary non-market systems often keep poor people from dying of hunger.

Climate change decreases food supply

Just as the rains were returning to the fertile crescent after three years of drought, extreme weather disrupted food production in regions around the world. The supply of wheat fell again in 2010. Summer drought and extreme heat across the Eurasian steppe diminished the wheat harvest by 30% in Russia and nearly 20% in Ukraine. Cold weather and excessive rain, on the other hand, led to production decreases in Canada and Australia (Sternberg 2012).

Then drought hit China's eastern wheat-growing region, threatening the winter crop. China both consumes and produces more wheat than any other country in the world. Most years, it is self-sufficient. It seemed 2011 would be different; China would need to buy wheat on the international market. This would increase demand—and thus the market price—considerably. Thanks to the inclement weather around the world, the price of commodity wheat was already on the upswing. Having dropped below $3 again briefly in 2009, wheat was dancing around $7 per bushel in November 2010 when reports of low precipitation levels in the Chinese winter wheat belt began to be broadcast (Yu 2011; Sternberg 2012).

Periodic food supply troubles will continue as climate change unsettles weather patterns and the ecosystems where communities produce food for themselves and the world's ever-growing urban populations (Parry et al. 2005; Porter et al. 2014). Gradually shifting conditions will stress global food systems. Extreme events could destabilize them (Wheeler and von Braun 2013; Thornton et al. 2014).

Populations that already suffer from food insecurity will likely feel the worst of climate change's effects on farming, fishing, and foraging. Models project that rising temperatures will negatively affect crop production in the tropical and subtropical regions where most of the world's poor people live (Porter et al. 2014).

At higher latitudes, where more food-secure countries lie, agriculture might become *more* productive as growing seasons lengthen.

The unfairness of climate change's effect on food systems becomes even more apparent when one thinks about responsibility for the disruptions. Every stage of modern food systems releases heat-trapping greenhouse gases: livestock digestive systems, fertilizer factories, over-fertilized fields, refrigerators, stovetops, landfills filled with food, and of course petroleum-powered tractors, trucks, trains, and ships. Peasant farmers, on the other hand, sometimes slow down climate change by storing carbon in the soil. The fossil-fueled industrial food systems—and economies, in general—of rich nations like the United States and Canada are culpable for changing the climate in ways that threaten food production in places whose populations have done little to cause the problem. This double inequality of responsibility and vulnerability is climate injustice.

Markets exacerbate climate injustice

The fact that food is a market good makes the injustice much worse. Markets coerce farmers into replacing human labour with machines, substituting carbon emissions for jobs. They reward simplification and uniformity, which makes the global food system less diverse, and thus less resilient against climate change and less able to adapt to it. Markets impoverish peasant farmers, who cannot compete with (often subsidized) industrial agriculture. They often direct the produce of smallholders in the Global South to faraway people who have purchasing power, away from their own communities. Markets systematically punish their poorest participants.

Those might feel like overstatements. Popular wisdom says that markets provide opportunities for enterprising poor people to help themselves, to get their hands on some wealth by selling their labour or things they make. Economists tell society that markets motivate efficient production, allocation, and consumption decisions. Free trade advocates preach that global markets will protect the world from shortages. With food, at least, markets do not work that way.

In rich countries, higher commodity prices hardly curb consumption. US wheat consumption *increased* slightly when its price tripled, because Americans do not notice when global food commodities fluctuate. In 2008, when food prices spiked, an average loaf of bread in the United States cost $1.37 and contained 12 cents of commodity wheat. In 2016, it contained 6 cents of commodity wheat and still cost exactly $1.37 (Wech 2017). Americans and Canadians spend less than 10% of their incomes on food, on average. When the global food supply is disrupted, markets send no discernible signal to the world's wealthy to consume less.

The poor, on the other hand, must choose to cut consumption of food or of other things when prices rise. Egyptians spend a third to a half of their income on food, depending on prices. Within any country, lower-income families spend much greater portions of their income on food. In Syria, that portion can be well over half. For the lowest fifth of income earners in the United States, this share is over 30%. US households with incomes less than 185% of the poverty line—families of four that took home less than $56,000 in 2017, for example—are nearly three times more likely than the average household to report very low food security (Coleman-Jensen et al. 2014). The poor and hungry get poorer and hungrier when food supplies fail.

Financial markets make food commodity markets even more unjust. Demand for food does not adjust quickly to price increases because the poor cannot immediately reorganize their lives to buy less food and the rich hardly change their consumption at all. Supply does not respond quickly either, since it takes at least a growing season to provide more of a given crop. Since neither demand nor supply can react to mitigate food price increases, small disruptions to food production can cause wild price escalations. Speculators can purchase paper food commodities when prices begin to rise, planning to resell them at a higher price later. This speculative demand further increases food prices (Tadesse et al. 2014). This is what happened in 2008 (Lagi et al. 2015). To Wall Street and Bay Street bankers it is simply a gambling game. But food price spikes cause real hunger for real people.

At the end of 2010, with China set to purchase an enormous amount of wheat on the international market, speculative demand surged again. Early in 2011, wheat was above $8 a bushel once more. Other food prices were soaring too, in part thanks to speculation (Lagi et al. 2015). Social unrest simmered across North Africa and the Middle East, where people rely on imported commodity wheat for staple sustenance (Zurayk 2011). These repeated food price spikes were one major factor behind the Arab Spring, a revolutionary eruption of demonstrations, violent upheaval, and civil wars (Lagi et al. 2011).

Subsidies do not prevent shortage

Government intervention cannot easily alleviate the hunger caused by international food price shocks. In countries like Syria and Egypt, authoritarian governments sell highly subsidized bread, in part to quell opposition. These regimes buy market food and then give it away nearly for free. The idea is to protect poor people from food insecurity—since hungry people overthrow dictators—but this does not work when prices rise dramatically. More and more people start relying on subsidized bread as everything else, especially bread in private bakeries, becomes increasingly expensive. The cash-strapped governments turn up their wheat imports precisely when commodity wheat starts getting costly. This additional demand drives commodity prices up further. Since the subsidized public bakeries do not raise prices to ration bread, long queues form. When their bread runs out, large groups of hungry, angry people are left standing in line in the streets.

In January 2011, collectively enraged citizens turned against their rulers. Long-time autocrats resigned and fled amidst popular uprising in Tunisia, Egypt, and Libya. But overthrowing dictators did not bring political freedoms or peace. Many countries in the region are still trying to gain some semblance of stability. People still rely on government-subsidized bread.

Protests in Syria began in March 2011. Armed clashes soon escalated into a bloody, multi-sided war between the government and several opposition groups. The conflict is entering its ninth year as I write this. The prolonged drought and subsequent food shortages played a substantial role in triggering Syria's civil war (Gleick 2014; Kelley et al. 2015). For many scholars, this is some of the first evidence of climate change creating violent conflict over resources. Academics and military generals agree that climate extremes increase the potential for large-scale violence, partly via shocks to food systems (Barnett and Adger 2007; Raleigh et al. 2015; US Department of Defense 2014).

Once war breaks out, food insecurity worsens. In Syria, the domestic price of wheat flour, the main staple, has more than tripled since 2011. More than one in three Syrians are food insecure (FAO and WFP 2017). Civil war has created a general state of insecurity there.

Freeing food from markets feeds people for free

Consumers buy food knowing only its price and the information on its packaging. It is often impossible to know if it comes from high-emitting industrial operations that pay farmworkers too little or take too much from the land (Clapp 2015). Our purchases may divert resources from populations who need them much more than we do.

Egyptians call bread *aish*, which means "life" (Zurayk 2011). They became reliant on imported wheat when farmers—with the encouragement of the Egyptian government, the International Monetary Fund, and the World Bank—shifted from growing staple grains to growing fruits and vegetables for export to higher-income Europeans who were willing and able to pay high prices to have fresh produce all winter. The same can be said about many countries that export coffee, cacao, tobacco, palm oil, and tropical fruits to the Global North, and then must use the money to buy their staple sustenance. With global markets, it pays better to grow dessert for richer countries than to grow your own dinner, so to speak. So poorer countries become extremely vulnerable to the whims of markets for both the commodities they produce and those they import (Bren d'Amour et al. 2016).

It is not clear what individuals in wealthy countries can do about this, especially when many working people struggle to afford basics like food in the Great Lakes region, as anywhere. Ten percent of Vermonters live in food insecure households, according to the organization Hunger Free Vermont. Even more Vermont residents, about 13%, receive SNAP benefits—basically, special purpose money to buy market food (from the commodity system that cannot feed the world reliably). Markets are failing Vermont food producers, too. About a third of the state's dairy farms have closed in the past decade as industrial-sized dairies elsewhere have driven prices down. The farms that have survived have had to become much larger to compete; the number of operations with over 700 cows doubled from 2013 to 2018 (Heintz 2018).

But Vermont is also a hotbed for making food markets work in more just and sustainable ways, both through policy and by creating alternative food networks—fair trade, organic, farmers markets, and such (Maye and Kirwan 2010). Strong local food systems with ecological production practices can reduce our diets' harmful effects on distant others and the climate. Fairer terms of trade for producers in the Global South help too. Yet, these modified, ethical markets can ensure food security for the poor and prioritize social and environmental goals more or less to the extent that they subordinate market mechanisms to other, non-market rules and values (Bliss 2019). The less they resemble unregulated, financialized commodity markets—the type Karl Polanyi (1944) called "disembedded" from social institutions—the better they can support the struggle for climate justice.

Thus, those curious about how to achieve climate-just food systems should take non-market food seriously (Bliss 2019). However, researchers and policymakers focus nearly all their attention on commercial farming, fishing, and food processing and distribution. North American culture often considers free food only for charity. Food systems without markets seem quaint, like an unrealistic step backwards on the imaginary one-way path of development. But non-market food systems might become critical when flows of oil and money falter as a result of the ecological-economic effects of climate breakdown. Because of markets' inherent inequities and unsustainabilities, I believe non-market food systems must play a large role in any future global food regime that is to approach some semblance of climate justice. To explore this, I study, participate in, encourage, and support non-market food systems in Vermont while learning about non-market food systems in other places and times, and imagining desirable ones that might come to be.

People everywhere and always produce food that is not for sale—whether as a way of life, a hobby, or a coping mechanism against unreliable markets. Communities share, gift, and redistribute food even in the heart of capitalism. I have participated in some self-organized efforts to foster just, carbon-saving non-market food systems. The Beacon Food Forest is a three-acre communal permaculture garden and gathering space in a Seattle city park, where everyone is free to participate, or even harvest without participating (Bliss 2015). The *Xarxa d'Aliments*—"nourishment network," roughly—is a collective that recycles edible-but-not-sellable food from a network of about 15 shops in Barcelona's fast-gentrifying Gràcia neighbourhood (Bliss and Rubio Herranz 2016). Once a week, members collect food that collaborating businesses would otherwise throw in the garbage and bring it to an assembly where they divvy it up, discuss issues related to local politics, make group decisions by consensus, and then take home their free groceries. Both collectives occasionally make big, free meals to share with their communities.

Non-market food production and sharing is prevalent in Vermont. Vermonters hunt, fish, forage, glean, garden, and dumpster dive. Many have chickens or bees in their backyards. Some tap maple trees with no intent to sell the syrup. Others clean and butcher roadkill for their freezers and friends. They share food at potluck parties and community meals. They barter without set prices, reciprocate without keeping track, and share without self-interest. Food banks, food shelves, food pantries, and other charities provide free food to low-income families. Nobody had really thought of all these activities as part of some greater category called "non-market food systems" before I came along with my research project, but they are all happening all at once, freeing food from the market. In what follows, I report some examples from the preliminary stages of a mixed methods research project on Vermont's non-market food landscape.

Caring for the commons can free food from markets

Gardening is the unofficial state pastime in Vermont. Every March, Vermonters start seedlings by sunny windows and under fluorescent lights while several feet of snow still cover the ground. Folks in towns like Burlington frequently have community garden plots if there's no space to grow food at home. A colleague of mine gardens in buckets hanging from the windows of her second-story apartment. Neighbours share veggies from their yards. People preserve their harvests, canning homegrown tomatoes for the winter or fermenting cabbage for kimchi. Many of my friends get a large portion of their vegetables and fruits outside the market.

In Burlington's Old North End, resettled refugees from Somalia, Bhutan, Bosnia, the Congo, and elsewhere farm the thin strip of earth between the sidewalk and the street as if their food security depended on it. Maybe it does. Maybe subsistence farming is what they know. New Farms for New Americans is a program that makes sure these migrants have access to subsistence-sized plots in community gardens. At a minimum, people in an unfamiliar place can grow, cook, and eat familiar foods (AALV n.d.; Bose and Laramee 2011).

Huertas is a project that supports Latinx dairy workers in planting gardens of veggies, herbs, and other culturally important crops that are hard to find in Vermont (Mares 2017). Many migrant farmworkers have trouble accessing fresh food because they do not have documentation and cannot leave the farms where they live and work without worrying about being harassed or detained by border police. Having a kitchen garden at home helps.

Producing non-market food is difficult because so many of the things one needs to produce it are market goods—seeds, water, feed, fertilizer, fuel, land, our labour. Even some agricultural knowledge is considered private intellectual property. There's a lot of work to be done to free these things from the market, too: seed saving, rainwater harvesting, community composting, knowledge sharing, and reclaiming common land for grazing and growing. Non-market food systems rely on commons—resources that groups of people manage collectively, in common (Vivero Pol et al. 2019).

Some edible organisms remain legally owned in common in the United States and Canada. It is difficult to exclude people from collecting wild foods like fruits, roots, mushrooms, leaves, and nuts. Government agencies regulate hunting and fishing but have not assigned private property rights over wild animals. Ecosystems produce some food for free, provided that people interact with them in ways that are sustainable and defend them against extraction and pollution. This requires collective governance.

Constructing non-market food systems requires harvesting from, caring for, propagating, and creating community around food commons. At least one Vermont town has a public orchard. Many have communally managed gardens. Even community gardens that are separated into individually managed plots need commons governance to apportion water, fight pests, divvy up space, and plan and implement projects. The idea of food as a commons includes alternative markets and supportive states: cooperative farms, community-supported agriculture, farm-to-school programs, farmer-to-farmer agroecology education, and local, direct sales (Vivero Pol et al. 2019). But even in these cases it is self-governance and self-organization beyond the market and the state that make food commons.

Food markets make non-market food "waste"

The commodity food system, paradoxically, produces a lot of free food. One supermarket in Burlington throws hundreds of pounds of food into accessible, sanitary containers every day: organic veggies, exotic fruits, prepared dishes, bakery items. This grocery store is not an outlier. Dumpster diving is among the best ways to access free food in urban areas.

Gleaning works well in rural places. Market farms leave a lot of edible crop unharvested, every time bringing it to market would not turn a profit. Each fall, Vermonters make non-market apple butter, cider, vinegar, sauce, and pies from apples that have fallen under neglected trees and in orchards after harvest. Like everywhere in North America, food banks and other charities in Vermont move massive amounts of would-be food waste from commercial farms and the industrial distribution system to people in need.

In a world where more than a quarter of all calories produced go to waste, there is almost no limit to how many people "recycled" or rescued food can feed. The *Xarxa d'Aliments* collective in Barcelona contributes substantially to the diets of 15 to 20 regular participants, plus a smattering of flatmates, friends, and family, from the would-be waste of just a handful of small food sellers in 1 section of 1 neighbourhood, 1 day per week. Similar groups could feed tens of thousands if they recycled every evening from every fruit stand and supermarket in the city.

Some activists and I have revived Burlington's long-dormant Food Not Bombs chapter, to turn that would-be waste into free meals for anyone at City Hall Park (see Routledge and Heynen 2010). Eating food that would otherwise go to waste instead of buying food in the marketplace avoids degrading distant lands or diverting nourishment away from humans.

Freeing food is an emancipatory act

The idea is to create communities of mutual aid. When prices shoot up quickly, poor people go hungry. When they fall slowly, over time—thanks to the globalization of markets and the competition they entail—small farmers suffer and peasants are impoverished. Changing market conditions create winners and losers. The poor rarely win and the wealthy do not lose much. States like Syria and Egypt offer subsidized bread to make people dependent on dictatorships for protection from the whims of markets that have no morals. But what if people organize collectively to feed each other for free?

Climate change will wreak havoc on the world's food production and distribution systems. What if communities could ensure that no one goes hungry unless everyone does? A desirable food system must prioritize necessity over luxury, sufficiency over efficiency. Markets do the opposite.

But does organizing non-market food systems in the Great Lakes region address the effects of wheat price spikes in the Middle East? Not unless someone starts producing non-market wheat, or people replace some wheat with other non-market foods in their diet. But maybe that is fine for climate justice. Maybe there is no direct intervention from places like the Great Lakes that can help the Middle East. Food aid, commodity trading, agricultural development, and other interventions from the rich world have historically come with the twin undesirable consequences of injustice and unsustainability for the global majority (George 1977; Kaufman 2012; Chappell 2018). Working toward climate justice might mean removing wealthy Northern hands from the Global South.

This is not to argue that each world region should be food self-sufficient and disconnected. But, given that markets serve the rich at the direct expense of the poor—because of their rules, not necessarily malicious intent—disentangling the highly unequal world marketplace somewhat might not be a bad idea. Non-market food systems preclude most possibilities of exploiting unseen others.

Transforming food systems means changing society's stories about food. Narrative matters. Non-market food systems may feel naïve or idealistic. Yet new, perhaps naïvely utopian (Wright 2013), ways of thinking about food must necessarily accompany a new food system that feeds all humans, regardless of purchasing power, without destroying the ecosystems and the climate that produce the organisms we eat. Those organisms become us. Down the road, that matter cycles back around and becomes other organisms, including other humans. Food never really *could* be private property.

References

AALV (n.d.) *New Farms for New Americans* [online]. Available from www.aalv-vt.org/farms [15 February 2018].

Ali, M. (2010) *Years of Drought: A Report on the Effects of Drought on the Syrian Peninsula*. Middle East Office: Heinrich – Böll – Stiftung.

Barnett, J. and Adger, W.N. (2007) 'Climate Change, Human Security and Violent Conflict'. *Political Geography* 26 (6), 639–655.

Bello, W. and Baviera, M. (2009) 'Food Wars'. *Monthly Review* 61 (3), 17–31.

Berazneva, J. and Lee, D.R. (2013) 'Explaining the African Food Riots of 2007–2008: An Empirical Analysis'. *Food Policy* 39, 28–39.

Berking, H. (1999) *Sociology of Giving*. trans. by Camiller, P. London, UK: Sage.

Bliss, S. (2015) 'These Urban Farmers Want to Feed the Whole Neighborhood — for Free'. Grist [online] 13 March. Available from https://grist.org/food/these-urban-farmers-want-to-feed-the-whole-neighborhood-for-free/ [17 February 2018].

Bliss, S. (2019) 'The Case for Studying Non-Market Food Systems'. *Sustainability* 11 (11), 3224.

Bliss, S. and Rubio Herranz, M. (2016) 'Malbaratem Un Món d'aliments'. *La Metxa* 9, 11–13.

Bose, P. and Laramee, A. (2011) 'Taste of Home: Migration, Food and Belonging in a Changing Vermont'. *Opportunities for Agriculture Working Paper Series* 4, 1–8.

Bren d'Amour, C., Wenz, L., Kalkuhl, M., Steckel, J.C., and Creutzig, F. (2016) 'Teleconnected Food Supply Shocks'. *Environmental Research Letters* 11 (3), 035007.

Cassidy, E.S., West, P.C., Gerber, J.S., and Foley, J.A. (2013) 'Redefining Agricultural Yields: From Tonnes to People Nourished per Hectare'. *Environmental Research Letters* 8 (3), 034015.

Chappell, M.J. (2018) *Beginning to End Hunger* [online]. Berkeley, CA: University of California Press. Available from www.ucpress.edu/book/9780520293090/beginning-to-end-hunger [17 September 2018].

Clapp, J. (2015) 'Distant Agricultural Landscapes'. *Sustainability Science* 10 (2), 305–316.

Coleman-Jensen, A., Gregory, C., and Singh, A. (2014) *Household Food Security in the United States in 2013* [online] ID 2504067. ERR No. 173. Washington, D.C.: US Department of Agriculture, Economic Research Service. Available from https://papers.ssrn.com/abstract=2504067 [17 June 2017].

FAO and WFP (2017) *FAO/WFP Crop and Food Security Assessment Mission to the Syrian Arab Republic*. Rome, Italy: Food and Agriculture Organization of the United Nations and World Food Program.

Farley, J., Schmitt Filho, A., Burke, M., and Farr, M. (2015) 'Extending Market Allocation to Ecosystem Services: Moral and Practical Implications on a Full and Unequal Planet'. *Ecological Economics* 117, 244–252.

George, S. (1977) *How the Other Half Dies: The Real Reasons for World Hunger*. Montclair, NJ: Rowman & Littlefield.

Gleick, P.H. (2014) 'Water, Drought, Climate Change, and Conflict in Syria'. *Weather, Climate, and Society* 6 (3), 331–340.

Heintz, P. (2018) 'Selling the Herd: A Milk Price Crisis Is Devastating Vermont's Dairy Farms'. Seven Days [online] 11 April. Available from www.sevendaysvt.com/vermont/selling-the-herd-a-milk-price-crisis-is-devastating-vermonts-dairy-farms/Content?oid=14631009 [15 March 2019].

Holt-Gimenez, E., Patel, R., and Shattuck, A. (2012) *Food Rebellions: Crisis and the Hunger for Justice*. Oakland, CA: Food First Books.

Holt-Giménez, E., Shattuck, A., Altieri, M., Herren, H., and Gliessman, S. (2012) 'We Already Grow Enough Food for 10 Billion People … and Still Can't End Hunger'. *Journal of Sustainable Agriculture* 36 (6), 595–598.

IRIN (2009) *Drought Response Faces Funding Shortfall* [online]. Available from www.irinnews.org/news/2009/11/24/drought-response-faces-funding-shortfall [13 February 2018].

Kaufman, F. (2012) *Bet the Farm: How Food Stopped Being Food*. New York, NY: Wiley.

Kelley, C.P., Mohtadi, S., Cane, M.A., Seager, R., and Kushnir, Y. (2015) 'Climate Change in the Fertile Crescent and Implications of the Recent Syrian Drought'. *Proceedings of the National Academy of Sciences* 112 (11), 3241–3246.

Kummu, M., de Moel, H., Porkka, M., Siebert, S., Varis, O., and Ward, P.J. (2012) 'Lost Food, Wasted Resources: Global Food Supply Chain Losses and Their Impacts on Freshwater, Cropland, and Fertiliser Use'. *Science of The Total Environment* 438, 477–489.

Lagi, M., Bar-Yam, Yavni, Bertrand, K.Z., and Bar-Yam, Yaneer (2015) 'Accurate Market Price Formation Model with Both Supply-Demand and Trend-Following for Global Food Prices Providing Policy Recommendations'. *Proceedings of the National Academy of Sciences* 112 (45), E6119–E6128.

Lagi, M., Bertrand, K.Z., and Bar-Yam, Y. (2011) *The Food Crises and Political Instability in North Africa and the Middle East* [online]. Working Paper available via Social Science Research Network. Cambridge, MA: New England Complex Systems Institute. Available from https://papers.ssrn.com/abstract=1910031 [17 June 2017].

Mares, T.M. (2017) 'Cultivating Comida: Cacao Fields and Dairy Cows: The Interdependencies between Mexican Workers and the US Food System'. *Journal of Agriculture, Food Systems, and Community Development* 7 (4), 9–12.

Maye, D. and Kirwan, J. (2010) 'Alternative Food Networks'. *Sociology of Agriculture and Food* 20, 383–389.

Mendes, R., Farley, J., and Corta, A.C. (2018) *International Food Consumption Patterns: Food Allocation and Demand Curve Distortion in an Unequal Market Economy.* Working Paper.

Nolin, D.A. (2012) 'Food-Sharing Networks in Lamalera, Indonesia: Status, Sharing, and Signaling'. *Evolution and Human Behavior* 33 (4), 334–345.

Parry, M., Rosenzweig, C., and Livermore, M. (2005) 'Climate Change, Global Food Supply and Risk of Hunger'. *Philosophical Transactions of the Royal Society B: Biological Sciences* 360 (1463), 2125–2138.

Polanyi, K. (1944) *The Great Transformation: The Political and Economic Origins of Our Time.* Beacon Press.

Porter, J.R., Xie, L., Challinor, A.J., Cochrane, K., Howden, S.M., Iqbal, M.M., Lobell, D.B., and Travasso, M.I. (2014) 'Food Security and Food Production Systems'. in *2014: Impacts, Adaptation, and Vulnerability. Part A: Global and Sectoral Aspects. Contribution of Working Group II to the Fifth Assessment Report of the Intergovernmental Panel on Climate Change.* ed. by IPCC. Cambridge, UK: Cambridge University Press, 485–533.

Raleigh, C., Choi, H.J., and Kniveton, D. (2015) 'The Devil Is in the Details: An Investigation of the Relationships between Conflict, Food Price and Climate across Africa'. *Global Environmental Change* 32, 187–199.

Regmi, A. and Meade, B. (2013) 'Demand Side Drivers of Global Food Security'. *Global Food Security* 2 (3), 166–171.

Routledge, P. and Heynen, N. (2010) 'Cooking up Non-Violent Civil-Disobedient Direct Action for the Hungry: "Food Not Bombs" and the Resurgence of Radical Democracy in the US'. *Urban Studies* 47 (6), 1225–1240.

Seale, J., Regmi, A., and Bernstein, J. (2003) *International Evidence on Food Consumption Patterns* [online] 33580. Technical Bulletins. United States Department of Agriculture, Economic Research Service. Available from http://econpapers.repec.org/paper/agsuerstb/33580.htm [7 May 2017].

Solh, M. (2010) 'Tackling the Drought in Syria'. Nature Middle East [online]. Available from www.natureasia.com/en/nmiddleeast/article/10.1038/nmiddleeast.2010.206 [13 February 2018].

Sternberg, T. (2012) 'Chinese Drought, Bread and the Arab Spring'. *Applied Geography* 34, 519–524.

Tadesse, G., Algieri, B., Kalkuhl, M., and von Braun, J. (2014) 'Drivers and Triggers of International Food Price Spikes and Volatility'. *Food Policy* 47, 117–128.

Thornton, P.K., Ericksen, P.J., Herrero, M., and Challinor, A.J. (2014) 'Climate Variability and Vulnerability to Climate Change: A Review'. *Global Change Biology* 20 (11), 3313–3328.

Trigo, R.M., Gouveia, C.M., and Barriopedro, D. (2010) 'The Intense 2007–2009 Drought in the Fertile Crescent: Impacts and Associated Atmospheric Circulation'. *Agricultural and Forest Meteorology* 150 (9), 1245–1257.

UNCTAD (2008) *The Least Developed Countries Report 2008: Growth, Poverty and the Terms of Development Partnership*. New York and Geneva: United Nations Conference on Trade and Development.

US Department of Defense (2014) *Quadrennial Defense Review 2014* [online]. Available from http://archive.defense.gov/pubs/2014_Quadrennial_Defense_Review.pdf

Vivero Pol, J.L., Ferrando, T., de Schutter, O., and Mattei, U. (eds.) (2019) *Routledge Handbook of Food as a Commons*. London, UK: Routledge.

Walton, J.K. and Seddon, D. (1994) *Free Markets and Food Riots: The Politics of Global Adjustment*. Cambridge, MA: John Wiley & Sons.

Wech, M. (2017) 'What's the Value of Wheat in a Loaf of Bread?' *Abilene Reporter-News* [online] 19 March. Available from www.reporternews.com/story/money/industries/agriculture/2017/03/19/whats-value-wheat-loaf-bread/99329496/ [22 June 2017].

Wheeler, T. and von Braun, J. (2013) 'Climate Change Impacts on Global Food Security'. *Science* 341 (6145), 508–513.

Wright, E.O. (2013) 'Transforming Capitalism through Real Utopias'. *American Sociological Review* 78 (1), 1–25.

Yu, C. (2011) 'China's Water Crisis Needs More than Words'. *Nature News* 470 (7334), 307–307.

Zaman, H. and Tiwari, S. (2010) *The Impact of Economic Shocks on Global Undernourishment* [online] WPS5215. The World Bank. Available from http://documents.worldbank.org/curated/en/509661468163742397/The-impact-of-economic-shocks-on-global-undernourishment [5 June 2017].

Zurayk, R. (2011) 'Use Your Loaf: Why Food Prices Were Crucial in the Arab Spring'. *The Guardian* [online] 16 July. Available from www.theguardian.com/lifeandstyle/2011/jul/17/bread-food-arab-spring [22 June 2017].

13

BUILDING SOCIAL CAPITAL TO INCREASE DISASTER RESILIENCE

Stephen M. Clare

Introduction

As climate change fuels more frequent and more extreme weather events around the world, researchers, policymakers, and activists are working to make communities more resilient to natural disasters. Often these efforts focus on investments in infrastructure, like improved drainage systems or higher sea walls (Gill et al. 2007). Such projects are important, but they can feel beyond the influence of individuals concerned about climate change and activists working at the local level. An alternative way to invest in community resilience is to build social capital. Community-building initiatives that foster connections between people, build trust, facilitate knowledge-sharing, or spread a common vision build networks that people can draw on when disaster strikes. The experiences of communities that have been struck by natural disasters around the world, from Chicago to Tobago, show how social capital-building initiatives can save lives and mitigate damage when disasters occur. This can be particularly important for people who are especially vulnerable, so it represents climate justice in action.

Social drivers of natural disasters

In July of 1995, Chicago suffered an extreme heat wave. Heavy rains and the arrival of a hot summer air mass brought a combination of heat and humidity that effectively sent temperatures soaring up to 52°C (Healy 2005). The effects were dramatic. The city's water pressure plummeted in certain neighbourhoods as people opened their blocks' fire hydrants to cool themselves off (Gladwell 2002). The electricity demand from thousands of air conditioners caused rolling blackouts. Roads warped and buckled and cars stalled.

The effects were also deadly. Emergency services were overwhelmed by the volume of calls from people stricken by heat stress and dehydration. Hospitals filled with people seeking help. As the week wore on, morgues filled too. By the time the heat relented and Chicago began to recover, reports were that 739 lives had been lost due to the heat (Semenza et al. 1996).

The Chicago heat wave is an example of the incidents climate modellers are forecasting when they predict increases in the frequency of "extreme weather events" due to climate change. Other examples from the Great Lakes watershed include the January 1998 ice storm in eastern Ontario and southern Quebec which killed 35 and left nearly 5 million people without power, some for weeks; a Toronto thunderstorm on August 19, 2005 that washed out arterial road Finch Avenue and thousands of basements, resulting in $600 million in insurance payouts and a public infrastructure bill of $47 million (Wells 2012); the December 2013 ice storm in Ontario, Quebec and the Maritimes that left 400,000 without power over the Christmas holiday; Hurricane Irene, which devastated Vermont, New York, and the eastern seaboard of the United States in August 2011; and another thunderstorm-caused flood in Toronto on July 9, 2013 which was termed Ontario's most costly natural disaster yet. This last event led to more than $850 million in insurance payouts and cost the city more than $60 million (Mills 2013). While the degree to which any one of these events can be attributed to climate change is difficult to determine, researchers anticipate an overall increase in the number and severity of all such events as the planet warms (Meehl and Tabaldi 2004; Luber and McGeehin 2008). A 1998 study found a 20% increase in the frequency of heat waves between 1949 and 1995 (Gaffen and Ross 1998).

Although such events are termed "natural" disasters, social factors significantly shape their impacts. For example, deaths caused by the Chicago heat wave were not randomly distributed: the death rate among Chicagoans who lived alone was twice as high. In contrast, the mortality rate was about 30% lower for people who "participated in group activities" like church services and 70% lower for people who had local friends (Semenza et al. 1996). Neighbourhood effects also strongly affected mortality rates during the heat wave. While people in the neighbouring communities of North and South Lawndale were equally likely to be elderly, live alone, or be poor, the per capita death rate during the heat wave was ten times higher in North Lawndale (Klinenberg 2002). Researchers attribute this disparity to North Lawndale's higher crime rate, which made vulnerable seniors afraid to leave their homes to seek help. The safer community of South Lawndale also housed a busy commercial centre and a strong community of churches that offered services and events. During the heat wave, people could rely on these everyday centres of social interaction for support.

These findings suggest that at the root of many environmental problems are social factors shaped by people just "going about their everyday lives" (Rudd 2000, p. 131). Simply having friends in the city significantly lowered a given individual's mortality risk during the Chicago heat wave. On the one hand, this is disheartening because it means that many deaths could have been prevented if social networks were stronger. On the other hand, it means that strengthening social bonds can make a community more resilient to extreme weather events.

Social capital and the value of community

One way to gauge the strength of a community is to measure its social capital. Social capital can be defined most simply as the goodwill created by positive social interactions (Adler and Kwon 2002). Despite the use of the term "capital," measures of social capital focus on "what groups *do* rather than what people *own*" (Bowles and Gintis 2002, p. 420). Measuring social capital allows researchers to identify the intangible social factors which proved so important in Chicago. Learning how to strengthen social capital is an important research topic as scholars, policymakers, and community advocates make plans to cope with the challenges posed by climate change-fueled disasters.

Social capital is built through increased connectedness, which can take many different forms. Pretty and Ward (2001) write that "connectedness manifests itself in different types of groups at the local level, from guilds and mutual aid societies, to sports clubs and credit groups, to forest and pest management groups, and to literary societies and mother and toddler groups" (p. 211). In Chicago, connections made in stores, churches, and sports teams built the social capital that people later drew upon to survive the heat wave.

Social capital can be further divided into three subcategories: bonding, bridging, and linking social capital (Kawachihi et al. 2004). Bonding social capital measures the strength of relationships between people *within* social groups in terms of trust and cooperation. For example, neighbourhoods in Chicago which had more church groups had more bonding social capital. Similarly, individuals with strong friendships have more bonding social capital. Bridging social capital refers to relationships between people who are "unlike" each other but whose interactions are not characterized by significant power differences. Thinking again of Chicago, one could consider people who shared air conditioning with their neighbours. Connecting people from different social circles, perhaps through community events or classes, builds bridging social capital. This kind of social capital is important because it facilitates the transfer of knowledge and skills between different kinds of people (Schuller et al. 2000). Lastly, linking social capital measures the strength of relationships in which a power differential between participants must be accounted for, such as between a boss and an employee in an organization or between an official and a constituent in a community. The failure of some government officials in Chicago to recognize the vulnerability of some communities may be attributable to a lack of linking social capital.

The different roles of each type of social capital are explored in a study on the recovery of families in New Orleans following Hurricane Katrina (Hawkins and Maurer 2010). That research found that bonding social capital was an important factor in supporting people during and immediately following the disaster. People drew on bonding social capital when they met as families or friends to plan for the hurricane and pool their knowledge and financial resources. Meanwhile, bridging and linking social capital contributed to long-term recovery at both the individual and the community level. For example, several community organizations formed in the wake of the disaster, some connecting neighbourhoods of diverse socioeconomic statuses, and played key roles in obtaining and distributing resources during the recovery.

Building social capital as a way of increasing resilience to extreme weather events or natural disasters does not supplant investments in more conventional kinds of capital such as disaster relief funds, disaster preparedness plans, or storm surge infrastructure. Nor should it distract from the need to mitigate climate change to decrease the frequency and severity of such disasters in the first place. But the fact that trust, close relationships, and generosity can promote the effective use of knowledge and resources is significant (Ishihara and Pascual 2009). Measuring social capital focuses attention on how communities can act pre-emptively to increase their resilience to disasters.

What next?

Adaptive capacity describes a community's ability to respond to shocks, including heat waves and other extreme weather events. A community's potential for collective action partly determines its adaptive capacity, and the ability to act collectively is in turn shaped by social capital (Adger 2003). This is why high-quality social interactions help communities respond to disasters (Pelling and High 2005). Three dimensions of social capital help make communities more adaptive: structural, relational, and cognitive factors (Nahapiet and Ghoshal 1998):

THREE DIMENSIONS OF SOCIAL CAPITAL

1. **Structural**: a relationship network that connects people and helps individuals to find people for assistance or cooperation
2. **Relational**: the sense of trust that individuals have toward each other along with connections
3. **Cognitive**: the bonding force, such as shared understanding, interest, or problems, that holds the group together

So actions that connect people, deepen existing connections by building trust, or spread shared values and interests that can hold groups together build the social capital that make communities more adaptive.

There are many examples of this process in action. Collaborative governance models have been used to build a community's linking social capital. On the island of Tobago, a productive collaboration between the national government and local coastal communities allowed for the development of sustainable fishing industries. The management of Buccoo Reef Marine Park, a local marine protected area, gradually progressed from state management to co-management with substantial local involvement as various parties worked together and built trust (Adger 2003). This "mutual learning relationship" shows the importance of the relational and cognitive aspects of social capital. Communities and governments had a shared interest in solving the problem of unsustainable fishing. However, to successfully work together, the government had to trust that communities would follow their own rules about fishing allowances. Communities had to trust that the government would negotiate in good faith and devolve some regulatory and management authority. The social capital required to sustain this trust had to be built over time but grew in a positive feedback loop as each party consistently met the expectations of the other.

In Denmark, social capital has been built through agricultural cooperatives (Chloupkova et al. 2003). These organizations give members a common cause to rally around, cultivating the cognitive dimension of social capital. Sharing knowledge and resources has increased trust between farmers and made them more resilient to the economic and environmental shocks with which agricultural producers must contend. Trustworthy, community-based cooperatives also allow farmers to enhance their bridging social capital, which encourages customers to support them by paying higher prices for produce. The pooled interests of the farmers also strengthen their linking social capital with governments, facilitating knowledge-sharing and democratic participation to improve policymaking.

Local associations are another way to build social capital. For example, the community of Crest Street in Durham, North Carolina, united around a neighbourhood council to maintain community cohesion in response to the perceived threat of a highway development project (Rohe 2004). In the 1990s, the city government proposed relocating the Crest Street community to make way for a proposed highway expansion. In response, residents drew on their bonding social capital to form a democratic neighbourhood council and coordinate political action. Allying with influential municipal actors and organizations strengthened bridging social capital. This coordinated social action initiated a positive feedback loop that led to lasting changes in the community. Not only was the proposed highway rerouted, but improvements to the neighbourhood's streets, parks, and houses were made as part of the project. After fulfilling its original mandate, the Community Council turned its attention to other local challenges.

Another study, which focused on a town in rural Australia, linked a high degree of community involvement to engagement by residents in "purposive community activities" (Falk and Kilpatrick 2000). These findings suggest that social capital is built through "repeated high-quality interactions between community members," where quality interactions are characterized by mutual benefits and knowledge exchange.

Social capital-building initiatives in Ontario and Quebec include the privately funded work of Social Capital Partners; the Toronto Social Capital Project, linked to the city's Vital Signs Program; a Rural Ontario Institute project on Newcomer Engagement and Social Capital in Rural Communities; the Festival of Neighbourhoods in Kitchener, Ontario; the Quebec government's support for the social economy through locally managed funding mechanisms following the 1998 creation of the Chantier de l'économie sociale (Social Economy Workshop); and many local academic-community partnerships aimed at building social capital. Watershed organizations can play a key role in developing social capital and building networks (Floress et al. 2011).

Theory of change: Strategies for local engagement

Community involvement builds trust between community members, and increased trust leads individuals to involve themselves more deeply with community groups (Brehm and Rahn 1997). This is helpful for organizers because it means that hosting an event that gets people more involved in the community can trigger a positive feedback loop. Engagement builds trust, which encourages engagement, which builds yet more trust, and so on. This theory of social capital and the examples of community engagement described above give organizers something to focus on: ensuring community activities are purposeful, positive, and informative for participants.

Organizers should also focus on building a "shared social vision" in their communities (Rudd 2000, p. 132). This enhances the cognitive dimension of social capital: giving people something around which to rally. Communities which have a shared vision are more likely to undertake effective action because they have a tangible, common goal to pursue.

This theory reflects the anecdotal experience of many activists and organizers. Russell and Moore (2011) write that "we all need common strategic frameworks to move together" (p. 15). They see the purpose of climate justice work as developing visions of alternative economies, building equitable and democratic communities, and increasing people-power. These activities are united by "care" as "the thread that runs through this work" (Russell and Moore 2011, p. 46).

What the social capital literature offers organizers interested in combating climate change is not a panacea solution or a substitute for focused political action, but encouragement, reassurance, and hope. Even in the face of an overwhelming problem like climate change, simple local efforts like building trust or encouraging community engagement can be highly effective actions. The effects of global climate change will manifest most dramatically as extreme weather events at the local level, and while the sudden descent of a heat wave or flash flood may be beyond an individual's control, the way one's community responds to that event is not. Whether people have friends to rely on, safe places to go, and the knowledge to protect themselves and each other are the result of community connections—connections that we can all start building right now.

References

Adger, W.N. (2003) 'Social capital, collective action, and adaptation to climate change', *Economic Geography*, 79(4), 387–404.

Adler, P.S. and Kwon, S. (2002) 'Social capital: Prospects for a new concept', *Academy of Management Review*, 27(1), 17–40.

Bowles, S. and Gintis, H. (2002) 'Social capital and community governance', *The Economic Journal*, 112(483), 419–436.

Brehm, J. and Rahn, W. (1997) 'Individual-level evidence for the causes and consequences of social capital', *American Journal of Political Science*, 41(3), 999–1023.

Chloupkova, J., Svendsen, G.L.H., and Svendsen, G.T. (2003) 'Building and destroying social capital: The case of cooperative movements in Denmark and Poland', *Agriculture and Human Values*, 20(3), 241–252.

Falk, I. and Kilpatrick, S. (2000) 'What is social capital? A study of interaction in a rural community', *Sociologia Ruralis*, 40(1), 87–110.

Floress, K., Prokopy, L.S., and Allred, S.B. (2011) 'It's who you know: Social capital, social networks, and watershed groups', *Society and Natural Resources*, 24(9), 871–886.

Gaffen, D.J. and Ross, R.J. (1998) 'Increased summertime heat stress in the US', *Nature*, 396(6711), 529–530.

Gill, S.E., Handley, J.F., Ennos, A.R., and Pauleit, S. (2007) 'Adapting cities for climate change: The role of the green infrastructure', *Built Environment*, 33(1), 115–133.

Gladwell, M. (2002) 'Political Heat' *The New Yorker*, 12 August. Available at www.newyorker.com/magazine/2002/08/12/political-heat (Accessed: 23 June 2017).

Hawkins, R.L. and Maurer, K. (2010) 'Bonding, bridging and linking: How social capital operated in New Orleans following Hurricane Katrina', *British Journal of Social Work*, 40(6), 1777–1793.

Healy, K. (2005) 'Review: Heat wave: A social autopsy of disaster in Chicago', *Imprints*, 8(3), 283–289.

Ishihara, H. and Pascual, U. (2009) 'Social capital in community level environmental governance: A critique', *Ecological Economics*, 68(5), 1549–1562.

Kawachi, I., Kim, D., Coutts, A., and Subramanian, S.V. (2004) 'Commentary: Reconciling the three accounts of social capital', *International Journal of Epidemiology*, 33(4), 682–690.

Klinenberg, E. (2002) *Heat Wave: A social autopsy of disaster in Chicago*. Chicago, IL: University of Chicago Press.

Luber, G. and McGeehin, M. (2008) 'Climate change and extreme heat events', *American Journal of Preventive Medicine*, 35(5), 429–435.

Meehl, G.A. and Tebaldi, C. (2004) 'More intense, more frequent, and longer lasting heat waves in the 21st century', *Science*, 305(5686), 994–997.

Mills, C. (2013) 'Toronto's July flood listed as Ontario's most costly natural disaster' *Toronto Star*, 14 August.

Nahapiet, J. and Ghoshal, S. (1998) 'Social capital, intellectual capital, and the organizational advantage', *Academy of Management Review*, 23(2), 242–266.

Pelling, M. and High, C. (2005) 'Understanding adaptation: What can social capital offer assessments of adaptive capacity?' *Global Environmental Change*, 15(4), 308–319.

Pretty, J. and Ward, H. (2001) 'Social capital and the environment', *World Development*, 29(2), 209–227.

Rohe, W.M. (2004) 'Building social capital through community development', *Journal of the American Planning Association*, 70(2), 158–164.

Rudd, M.A. (2000) 'Live long and prosper: Collective action, social capital and social vision', *Ecological Economics*, 34(1), 131–144.

Russell, J.K. and Moore, H. (2011) *Organizing Cools the Planet: Tools and Reflections on Navigating the Climate Crisis.* Oakland, CA: PM Press.

Schuller, T., Baron, S., and Field, J. (2000) 'Social capital: A review and critique', in Schuller, T., Baron, S., and Field, J. (eds) *Social Capital: Critical Perspectives.* New York, NY: Oxford University Press, pp. 17–32.

Semenza, J.C., Rubin, C.H., Falter, K.H., Selanikio, J.D., Flanders, W.D., Howe, H.L., and Wilhelm, J.L. (1996) 'Heat-related deaths during the July 1995 heat wave in Chicago', *New England Journal of Medicine* 335(2), 84–90.

Wells, J. (2012) 'Climate change: How Toronto is adapting to our scary new reality' *Toronto Star,* 19 August.

14

CULTIVATING COMMUNITY RESILIENCE

Kelly Hamshaw and JoEllen Calderara

Introduction

As climate change intensifies, climate justice theory and past empirical studies indicate that local communities will have an increasingly important role in responding to and recovering from more frequent and more severe extreme weather events (Wise et al. 2014; Twigger-Ross et al. 2015; Burchell et al. 2017). Communities can take a variety of steps to improve resilience at the local scale in the face of climate change—such as implementing hazard mitigation strategies like reinforcing critical facilities, raising homes, or buying out properties with repetitive losses. A growing list of scholars also recognize the importance of investing in a community's social infrastructure as well as its physical infrastructure (Adger 2000; Magis 2010; Patterson, Weil and Patel, 2010; Berkes and Ross 2013; Aldrich and Mayer 2015). Magis (2010, p. 402) defines community resilience as "the existence, development, and engagement of community resources by community members to thrive in an environment characterized by change, uncertainty, unpredictability, and surprise."

This chapter provides a case study of a community-based response to Tropical Storm Irene, an extreme weather event. It summarizes the lessons learned, drawn from the experiences of the Central Vermont Long Term Recovery Committee (initially convened as the Barre Flood Center) about cultivating community resilience in the days, weeks, months, and years during the response and recovery phases. The authors were heavily involved in this effort. According to the Irene Recovery Report,

> On August 28, 2011, Vermont was forever changed. Tropical Storm Irene brought personal loss and damage unlike anything we have experienced in more than a generation. The rising waters took lives and the incredible

> damage to homes, property, land and our natural environment is difficult to comprehend.
>
> *(Lunderville 2012, p. 9)*

Recognizing the importance of understanding how communities can build resilience at the local scale, Ross and Berkes (2014) reviewed methods used in a selection of existing community resilience studies. They conclude their review by calling for researchers to continue efforts to gain deeper understanding of community resilience through a variety of methodological approaches, including participatory methods (2014, p. 787). Individuals in any community should find these lessons to be applicable to their own unique place and anticipated impacts. These reflections also seek to honour the contributions of all Vermonters to their communities in the aftermath while highlighting the importance of cultivating community connections pre-disaster in a world that is expected to continue to experience more frequent, severe, and unpredictable events due to climate change (Galford et al. 2014).

Case study background

Tropical Storm Irene ravaged two thirds of the state of Vermont in the northeastern United States on August 28, 2011. Six Vermonters lost their lives, including a father and son in Rutland County who died while checking on the municipal water system. Thirteen rural communities were completely isolated as streams, creeks, and rivers severed transportation networks. More than 3,500 households were damaged or destroyed by floodwaters—totalling over $21 million in Federal Emergency Management Agency (FEMA) total housing assistance funds under Disaster Declaration DR-4022 (FEMA 2017). There is no shortage of dramatic stories, such as the evacuation of Vermont Emergency Management's recently renovated office in the midst of the storm while farmers raced to ensure their dairy cows got milked despite extended electricity outages. The forceful waters splintered iconic cultural treasures, such as historic covered bridges and village centres. Perhaps as dramatic was the outpouring of volunteers, donations, and community spirit in the immediate aftermath in communities across the state once August 29 dawned with bright sunshine and blue skies. The slogan, *Vermont Strong*, soon began appearing on t-shirts and the front pages of Vermont's local newspapers as state agencies, municipal governments, homeowners, and business owners took stock of the damages and catalogued immediate needs.

In stark contrast to its coastal counterparts, Vermont had little experience responding to such an extreme and widespread flooding event. Irene served as an unprecedented focusing event for the state's climate change planning, adaptation, and mitigation efforts. It heightened the urgency of such efforts, leading to a high profile planning process that engaged over 45 organizations in conversations over an 18-month planning process (Institute for Sustainable Communities

2014). Important recommendations from this process included supporting "local resilience networks that bring together planning boards, conservation commissions, emergency managers, social service providers and other leaders to develop a shared vision for resilience within a community," (Institute for Sustainable Communities 2014). Community leadership was highlighted as a critical factor in towns and villages across Vermont as well as the strong sense of duty that spurred many volunteers to action.

The Central Vermont Long Term Recovery Committee (CVLTRC)

Serving 18 municipalities in the central region of Vermont, the CVLTRC sprung from a hyper-localized effort to an extreme flash flooding event in May 2011. The Barre Flood Center was a response catalysed by the mayor's office, community volunteers, and area social service agencies to address the gap in support and resources while the state was applying for individual disaster assistance to FEMA. The group drew from community volunteers and a variety of social service agencies within the region including the local chapter of the United Way, the regional office of the Vermont Agency for Human Services, a workforce development organization, the local community action agency, and local volunteers. Damages ranged from flooded basements and cars to homes deemed uninhabitable due to landslides or destroyed by floodwaters. The mayor essentially procured an empty storefront in the downtown centre and deputized a committee of volunteers to operate it as a drop-in centre for residents affected by flooding. Local leaders, mostly serving in an unpaid role, successfully sought financial contributions, donated clothing, appliances, and home goods as well as coordinated volunteers to help with clean-up reconstruction efforts.

Several weeks later, the Barre Flood Center, led by community volunteers and professionals from area social services agencies, transitioned into the Central Vermont Long Term Recovery Committee (CVLTRC) when FEMA granted individual disaster assistance funds to the State of Vermont approximately six weeks after the storm event. The primary purpose of these funds was to return affected households to a safe, sanitary, and functional living situation with funds up to the federally determined maximum award of $30,200. This transition brought increased resources in terms of organization, technical assistance, and a case management pilot program supported by FEMA's Voluntary Agency Liaison (VAL) Program. The VAL Program was specifically designed to foster coordination and collaboration amongst local agencies to close the inevitable resource gap between the individual assistance received by households and community resources to help people find their "new normal." At the time, FEMA officials estimated between 5–15% of the affected population would need additional assistance beyond what insurance payouts and FEMA individual assistance dollars would cover (K. Ash 2012, pers. comm., 19 March). These households tended to be those most vulnerable prior to the storm and were already dependent upon the social safety net for various types of support within the community—whether accessing the services of the local community action agency for services to meet their basic needs including food and transportation or the provision of mental health services. The VAL Program supported the CVLTRC as it defined its mission and scope, built volunteer capacity, and conducted outreach to affected households. It was fortunate that this partnership was already developing before Tropical Storm Irene's arrival just a few months later.

By the time Irene's floodwaters were forcing Vermonters from their homes in August, the CVLTRC volunteers were tracking the damage reports and preparing to stand up operations at the downtown storefront. Drawing on their experiences and connections forged in the aftermath of the Memorial Day flash floods, the CVLTRC was well-positioned to begin the urgent work of response and recovery. The chair of the CVLTRC, while listening to emergency calls for technical swift water rescues on the emergency scanner, began making calls to her fellow committee members to start organizing for much more widespread devastation than they had previously experienced. All told, the CVLTRC handled more than 250 cases over nearly 2 years—connecting people to local resources, Vermont Disaster Relief Fund dollars, and FEMA assistance. The CVLTRC conducted outreach to affected communities, organized donations, coordinated volunteers, and eventually provided case management to impacted households through the support of a pilot program from FEMA that hired social workers to work with the LTRCs on some of the most complicated cases. While some cases were relatively straightforward, such as seeking financial assistance from the Vermont Disaster Relief Fund to replace a flooded furnace, many cases were much more complicated, such as those seeking a stable housing situation while a federal buyout was processed.

CVLTRC served as a model for other communities around Vermont as they had essentially a jumpstart on getting mobilized due to the flash flooding events in May. Eventually, nine more Long-Term Recovery Committees formed across Vermont—either organically by volunteers or with the facilitation of FEMA VALs who would help convene stakeholders and provide training for those new to the recovery process (Bishop 2012). A weekly coordinating conference call helped to facilitate learning and troubleshooting across the different LTRCs starting in the early weeks of September. Trust and community connectedness were identified as key ingredients for successful LTRCs (Bishop 2012).

What can communities do?

Tropical Storm Irene was an opportunity for communities to develop, test, and improve their resilience at the grassroots. The prevailing question being asked across the state during the recovery phase was, "how do WE make OUR communities stronger and resilient?" Lessons from the CVLTRC experience offer these five recommendations:

1. Build connections "before the storm"

   As Cosmer and Milman (2016) found in interviews with LTRCs across the state of Vermont, social capital (relationships, networks, and trust) played an important role in the recovery process. Investing in social capital prior to disaster events enables such groups to convene quickly and more effectively. Not only does social capital between stakeholders serving on the LTRCs matter, social capital between the LTRC and the wider community are critical. In the case of CVLTRC, existing connections prior to the May flooding made it possible for the mayor to make a handful of phone calls that led to the hand-off of keys to an office space on Main Street which created an environment that organically fostered the eventual CVLTRC at a time of critical community need.
2. Recognize and leverage local assets

   Every community has particular sets of gifts, skills, and strengths that it draws upon when a disaster strikes. There are countless stories of neighbours with chainsaws or tractors lending a hand to neighbours in need. Consider the important role that anyone can fulfil for a flood survivor who just needs to share their story to someone who can simply listen. Of course, there are those within the community who may have more financial assets or political capital who could be important allies during the recovery process. Local volunteers should thoughtfully consider their assets from a holistic perspective. They are best positioned to think creatively about assets within the community that can be leveraged for recovery.
3. Understand and address community vulnerabilities

   Tropical Storm Irene powerfully demonstrated that disasters amplify pre-existing vulnerabilities within communities. For example, there is a severe shortage of affordable housing in Vermont. When Irene struck, over 15% of the FEMA registrants were owners of mobile homes, yet mobile homes represent just 7% of the state's total housing stock (Baker et al. 2012). Mobile homeowners, in comparison to stick-built homes, were more likely to be lower income and be older in age. These socioeconomic vulnerabilities were further compounded by the structural vulnerabilities of the housing type—essentially that flooded mobile homes tend to not be cost-effective to repair. The CVLTRC's area was home to a large mobile home park that saw 70 of 82 homes flooded—many beyond repair. Recognizing that many homeowners were elderly and/or had limited financial means, the CVLTRC Chair focused efforts on ensuring that FEMA individual assistance awards were maximized, and directed volunteer hours and fundraising towards supporting mobile homeowners.
4. Embrace local community spirit

   The recovery process can be a powerful healing process when approached with compassion and persistence. By drawing on local assets and energy of community volunteers, CVLTRC's work reflected the independent and self-reliant nature of this region of Vermont. Volunteers and affected residents took great pride in being able to come together for a common goal in the face of an otherwise seemingly daunting task. Some emergency management practitioners strongly caution against spontaneous volunteerism over concerns regarding training and liability. The CVLTRC was able to skillfully coordinate volunteers in ways that enabled volunteers to participate and contribute without creating chaos. For example, a class from the state's flagship university partnered with the CVLTRC to assist homeowners looking for assistance with "mucking and gutting" their homes, distributed large clothing donations at several community facilities in the area, and replanted garden beds for elderly homeowners.

5. Understand local needs within the community

   Operating at the local level, the CVLTRC was well-positioned to know what the current needs were within their community. When Irene struck, many people fled their homes with just the shoes on their feet as extra shoes were certainly not their primary concern. At the time, in August, many folks were wearing sandals and were storing their winter boots in their basements. As September turned into October and the temperatures dropped, the CVLTRC volunteers realized that people desperately needed more seasonally appropriate footwear. Understanding how unique footwear is for each person, the CVLTRC knew they needed to find a way to fund the purchase of shoes, but in such a way that would allow the client the ability to select their own shoes. The CVLTRC secured a small grant from the Vermont Community Foundation for $5,000, allowing them to spend $50 on 100 local clients. To maximize the benefit, CVLTRC partnered with a beloved local clothing store. Working together, the business was willing to give an additional 25% discount for each shoe voucher, including shoes already on sale, thus allowing a person the choice to replace work boots, winter boots, school shoes, and work shoes. This is just one example of a partnership program that helped fill a small need that had been totally overlooked but made a huge impact on the recovery of the clients. These sorts of needs will vary drastically by community so it requires community members to be observant, thoughtful, and creative.

Conclusion: Theory of change

Communities across the Great Lakes watershed and beyond face increased frequency and severity of extreme weather events. Whether a particular community should expect increased flooding, droughts, or wildfires due to climate changes, the lessons offered in this chapter illustrate tangible ways that anyone can pursue to increase resilience within their own community starting at the grassroots. While global climate change trends can be discouraging to the average citizen, focusing efforts on the local level on issues that directly impact the well-being of your loved ones and community at large is one way that people can realize the power of their own agency. Disasters can serve as catalysts—whether inspiring community volunteerism or the hosting of real conversations of how climate change is impacting livelihoods in real time. Having been on the frontlines during the Irene response and recovery efforts, we draw upon those first-hand experiences to encourage others to make the most of such catalysts. These are moments where change is most possible, to forge new connections and change local policies to increase resilience to climate change with a particular focus on equity within and across communities by building back stronger and better for all.

References

Adger, W 2000, 'Social and ecological resilience: are they related?,' *Progress in Human Geography*, vol. 24, no. 3, pp. 347–364.

Aldrich, D & Meyer, M 2015, 'Social capital and community resilience,' *American Behavioral Scientist*, vol. 59, no. 2, pp. 254–269.

Baker, D, Hamshaw, S, & Hamshaw, K 2012, 'Rapid flood exposure assessment of Vermont mobile home parks following Tropical Storm Irene,' *Natural Hazards Review*, vol. 15, pp. 27–37.

Berkes, F & Ross, H 2013, 'Community resilience: toward an integrated approach,' *Society & Natural Resources*, vol. 26, no. 1, pp. 5–20.

Bishop, D 2012, '*Lessons Learned from Vermont's Long-Term Recovery Experience*,' Vermont Disaster Relief Fund, Montpelier, France.

Burchell, K, Fagan-Watson, B, King, M, & Watson, T 2017, *Urban heat: developing the role of community groups in local climate resilience*, viewed 25 June 2017, https://westminsterresearch.westminster.ac.uk/item/q179v/urban-heat-developing-the-role-of-community-groups-in-local-climate-resilience.

Consoer, M & Milman, A 2016, 'The dynamic process of social capital during recovery from Tropical Storm Irene in Vermont,' *Natural Hazards*, vol. 84, pp. 155–174.

FEMA 2017, *Vermont Tropical Storm Irene (DR-4022)*, viewed 25 June 2017, www.fema.gov/disaster/4022?page=1.

Galford, G, Hoogenboom, A, Carlson, S, Ford, S, Nash, J, Palchak, E, Pears, S, Underwood, K, & Baker, D 2014, '*Considering Vermont's Future in a Changing Climate: The First Vermont Climate Assessment*,' Gund Institute for Ecological Economics, Burlington, VT.

Institute for Sustainable Communities 2014, *Vermont's roadmap to resilience*, viewed 25 June 2017, https://resilientvt.org/project-documents/.

Lunderville, NF 2012, *Irene Recovery Report: A Stronger Future*. State of Vermont, Agency of Administration, Office of the Secretary, Montpelier, Vermont, January. http://54.172.27.91/EM/Irene_Recovery_Report_201201.pdf, Accessed 29 March 2019.

Magis, K 2010, 'Community resilience: an indicator of social sustainability,' *Society and Natural Resources*, vol. 23, no. 5, pp. 401–416.

Patterson, O, Weil, F, & Patel, K 2010, 'The role of community in disaster response: conceptual models,' *Population Research and Policy Review*, vol. 29, no. 2, pp. 127–141.

Ross, H & Berkes, F 2014, 'Research approaches for understanding, enhancing, and monitoring community resilience,' *Society and Natural Resources*, vol. 27, pp. 787–804.

Twigger-Ross, C, Brooks, K, Papadopoulou, L, Orr, P, Sadauskis, R, Coke, A, Simcock, N, Stirling, A, & Walker, G 2015, *Community Resilience to Climate Change: An Evidence*

Review, viewed 25 June 2017 www.jrf.org.uk/report/community-resilience-climate-change.

Wise, R, Fazey, I, Stafford Smith, M, Park, SE, Eakin, H, Archer Van Garderen, E, & Campbell, B 2014, 'Reconceptualising adaptation to climate change as part of pathways of change and response,' *Global Environmental Change,* vol. 28, pp. 325–336.

15

AFTER THE FLOOD

Coming together for Toronto

Laura Gilbert and Claire-Hélène Heese-Boutin

This chapter is a call to action to you, the reader—a call to participate at home and collectively to help lessen the impact of flooding in our communities. Periodic flooding due to climate change is increasing nearly everywhere. Planning for it advances climate justice.

We have grounded our call in Toronto. We hope that Toronto's story may also inspire those of you who call another place home. Climate change predictions tell an ugly story of intensified storm events and increased precipitation for the City of Toronto, which will only increase the risk of floods. An examination of historical examples of flooding in the city reveals an all too common story, where the people and communities who already face social and economic injustices are disproportionately affected by flooding. This story of disproportionate impacts continues to this day and is repeated in communities across Canada and the world.

Climate change: It's about to get a whole lot worse

The City of Toronto is vulnerable to flooding due to the size of its population and its geographic features (Nirupama et al. 2014). Located in the southeastern part of the province of Ontario, the city borders Lake Ontario, one of the Great Lakes. Toronto's eastern lakefront was once the largest wetland in North America (TRCA 2017). Until a large storm on April 13, 1858, it was even possible to walk to Hiawatha Island, now known as the Toronto Islands. Toronto's hydrological cycle is affected by air masses from as far away as the Gulf of Mexico, the Atlantic Ocean, and the Arctic (Nirupama et al. 2014).

Toronto's paved urban landscape increases rainfall runoff, exacerbating flooding (Nirupama et al. 2014). When hard impermeable surfaces on the ground, such as roads and buildings, do not allow water to penetrate the soil, water flows on the surface instead, picking up debris and pollutants (such as oils and chemical wastes) until it reaches either a sewer or a permeable surface (Nirupama et al. 2014). The city is continuously adapting to deal with this extra runoff (D'Andrea 2017). What has changed, and will continue to change, is the amount of rainfall and the intensity of storms. Environment Canada's Coupled Global Climate Model (CGCM) forecasts that yearly precipitation will increase by about 10% compared to 1975–1995 average precipitation, while runoff will increase significantly and the region around the Great Lakes will become more vulnerable to flooding (Natural Resources Canada 2017). Toronto's urban infrastructure, such as the storm sewer system, is not equipped to handle sudden large volumes of water and risks overflowing. This is especially problematic for older infrastructure which combines sanitary (household sewerage) and storm water in the same pipes. This means that when they overflow, human waste is released untreated into waterways, and that can cause even more damage to infrastructure through contaminants found in human waste (City of Toronto).

A short history of flooding—Who is most affected and why?

Some Torontonians remember Hurricane Hazel, which hit Toronto in October 1954. The storm was severe, but it was the subsequent flooding that caused the most damage. Over 7,000 homes and 81 lives were swept away by the rising waters of the Humber and Don Rivers, which flow into Lake Ontario (Gifford 2004). Due to the storm, the city established new 100-year floodplain lines within which no new residential or commercial developments could occur. Existing residential and commercial areas within the floodplains were converted to parks (Environment and Climate Change Canada 2013). The deep web of ravines and river systems that make Toronto one of the greenest cities in the world were preserved in part due to the legacy of Hurricane Hazel. After Hurricane Hazel, the city also established the Toronto and Region Conservation Authority (TRCA), which was tasked with flood control, floodplain management, and water conservation issues (Environment and Climate Change Canada 2013).

However, despite the post-Hazel efforts, new drivers of flooding have emerged. A growing population has exacerbated the risks of flooding by increasing paved impermeable surfaces without investing in sufficient expansion and maintenance of sewage systems. The ecological risks are also higher for poor and racialized communities who are more likely to live along the city's flood plains. This is evident when we look at maps of flood vulnerability and flood risk relative to the City of Toronto's map of marginalized communities, termed Neighbourhood Improvement Areas, which is available online for consultation (City of Toronto 2014).

The northwest quadrant of the city is located within the Humber River floodplain and has a higher level of all three variables: flood risk, flood vulnerability, and the marginality captured in the designation of Neighbourhood Improvement Areas. It is also near the municipal border with York Region, where rapid development of suburban malls and parking lots drains runoff into Toronto's old sewage system and the Humber watershed.

Community attention to the causes and risks of flooding, and the development of municipal strategies to address it, have the potential to significantly help those living in the most vulnerable areas of the city who are likely to be the most vulnerable in socio-economic terms as well.

Since the 2000s, urban development plus increasing precipitation has resulted in severe floods. Between Hurricane Hazel and the year 2000, there was only one other major storm event, in 1976. The 2000s have already witnessed two major storms, in 2005 and 2013 (Nirupama et al. 2014), as well as record-breaking rainfall through the spring of 2017 (Fearon 2017). Significant strides have been made in flood management, and Torontonians generally do not fear death or the total destruction of their homes (Mann and Wolfe 2016). Today, flooding represents a predominantly financial burden. The costs of the 2005 flood were quantified at $500 million in insurable damages (Armenakis and Nirumpama 2014), and in 2013 insurable damages exceeded $850 million (Armenakis and Nirumpama 2014).

In Canada, homeowner and tenant insurance policies rarely cover flooding (Lamond and Penning-Roswell 2014). Some older homeowner insurance policies do include flood protection, but this is becoming increasingly rare or can only be bought as additional stand-alone insurance (Sandink et al. 2016). The extra flood insurance has variable costs based on risk determined by proximity to floodplains, or whether you have a driveway that slopes down toward the house. Depending on your postal code, it may not be possible to get any stand-alone flood coverage. Often, ineligible postal codes are within the areas of high risk and vulnerability. This leaves homeowners responsible for covering the costs of damages (with government aid only in extreme disaster situations).

What is Toronto doing about it?

The City of Toronto and the TRCA have both acknowledged risks of flooding and are working to improve the situation for all citizens. Policies, guidelines, and protocols have been created in certain high-risk areas to address problems of flooding and pollution. Unfortunately, the boundaries of watersheds do not align with political boundaries, which makes it harder to manage water at the watershed level. Similarly, groundwater doesn't follow the same boundaries as surface water, making it hard to coordinate protection efforts.

The City of Toronto's Wet Weather Flow Management Master Plan (WWFMMP), created in 2003, is taking action to address this situation (City of Toronto, 2003). It consists of the Wet Weather Flow Management Policy and the Wet Weather Flow Management Guidelines. These two documents are meant to direct and inform development of the WWFMMP. The first part of the master plan is set to be completed over the course of 25 years (starting in 2003). It targets four watersheds (Mimico and Etobicoke Creeks, Humber River, Don River, Rouge River and Highland Creek, all of which flow into Lake Ontario) and the combined sewer systems (City of Toronto 2006). The plan aims to reduce runoff not only for flood protection purposes but also to protect water quality in Lake Ontario (City of Toronto 2003). For example, high density residential areas are allowed up to 49% runoff (as a percentage of annual precipitation), while agricultural land is allowed as little as 2% depending on the type of soil (City of Toronto 2006). A variety of techniques will be used to reach this goal, including increasing green spaces and intercepting rainfall through catchment systems (City of Toronto 2006). Another important goal of the WWFMMP is to replace combined sewer systems with separate storm and sanitary sewers. Toronto still has about 23% of its sewer system as combined sewers, mostly in older neighbourhoods (City of Toronto 2017a).

The WWFMMP is an expensive endeavour that promises results that will mitigate the cost of flood damages to homes or apartments. It is a preventive strategy that allows the owner/renter to pay a fraction of the cost of flood insurance to reduce the likelihood of damage occurring in the first place. To meet the cost of replacing and improving its sewer infrastructure, the City of Toronto has raised water rates by 8% in both 2015 and 2016, with a further increase of 5% in both 2017 and 2018 (City of Toronto 2017b). For an average Toronto household, this represents an increase of $46 in 2017 (from $914 in 2016 to $960 in 2017) (City of Toronto 2017b). The Canadian federal government in early 2019 announced more than $400 million in new support for climate change induced flood-protection infrastructure in Toronto and several other nearby cities; Infrastructure Minister François-Philippe Champagne stated that cities need to adapt to the "sad and new and complex reality of climate change" (Rider 2019 n.p.)

To learn more about the WWFMMP, a copy can be accessed on the City of Toronto's website (City of Toronto 2003). An entire section of the city's website is devoted to water. It is user-friendly and provides information on a variety of flood-related issues, including a featured section on basement flooding. Here, homeowners can find information on a subsidy program that provides up to $3,400 for flood protection devices, as well as videos and text resources to help residents prevent flood damage.

In case flooding does occur, the TRCA has put in place a system to alert residents on their website by announcing Flood Forecasting around Lake Ontario and providing updates on shoreline erosion (Toronto and Region Conservation Authority 2017). Some of the TRCA's more important work involves working with people and businesses in the area to help address specific problems and concerns (Mann and Wolfe 2016). They have a wide network of resources and volunteers which fosters community engagement and are always looking for volunteers of all ages (Toronto and Region Conservation Authority 2017). They also do community outreach through workshops and events for educators, citizens, corporations, and volunteer groups. The TRCA prides itself in helping citizens build a relationship with nature by organizing outdoor activities to teach them about their community and the problems it faces. These events create opportunities to meet others that share an interest in the well-being of water resources in the community.

What can you do to protect yourself and your community?

For many people, a responsible relationship with water begins with respect and acknowledgement of the life-giving sacredness of water. We can take a minute to reflect on all the ways that water is essential to our lives, and take another to marvel at its power and destructive potential. This exercise is both spiritual and practical. With a heightened awareness of the value of water, it becomes easier to adjust our behaviours towards a responsible relationship. Perhaps each time we wash our hands or drink a glass of water, we could try to express gratitude for the cleansing and refreshing power of water, as well as for the vast infrastructure that brings clean drinkable water through pipes to us, on demand, and imagine what it takes to build and maintain this system. Many people in Canada and around the world do not benefit from this privilege, and must labour and toil to obtain, purify, and distribute their water.

When we consider climate change as an exacerbating factor for flooding, and the impact this may have on our lives, we can also think about the impact it will have on the marginalized residents of Toronto who live in the most flood-vulnerable areas of the city. Toronto is a wealthy city in a wealthy nation, and we can adapt and implement policies like the WWFMMP. How will less fortunate communities be able to respond? If we began this reflection with gratitude for our privileges, we should end with considering the responsibilities linked to our capacity. These can be as simple as respecting the need to increase the price of water, or utilizing the City of Toronto and TRCA's resources to help protect homes so that public aid and emergency resources are directed to the most in need.

We may even go further and participate in the broader fight for climate justice. In Toronto, the Toronto Climate Action Network and the LEAP Manifesto organizations are dedicated to citizen involvement towards the fight against climate change (The LEAP Manifesto 2018). For a more regional viewpoint, Great Lakes Commons is a grassroots organization whose mandate is to create a "thriving, living commons," with goals that involve restoring our relationship to the Great Lakes, fostering a spirit of responsibility, and stewardship and governance by all communities adjoining the Great Lakes in order to protect these waters. The Great Lakes Commons Charter Declaration is a great place to start getting informed and involved; it outlines the importance of the Great Lakes region and our dependence on this ecosystem for our survival (Great Lakes Commons 2017; see also Chapter 17 in this book).

Conclusion

As changes in climate become more severe and frequent, flooding will continue to cause severe damages to people and infrastructure, particularly in areas of rapid urbanization. Toronto is taking action to mitigate current and future risks. Local residents can help by making sure we know the risks to our homes and communities, and how to protect them, allowing for emergency response to assist the most vulnerable neighbourhoods. It is also necessary for citizens to reduce water use to increase conservation and decrease the stress water use puts on our ageing water infrastructure system, particularly in times of floods. The city of Toronto has a variety of initiatives, including much-needed infrastructure improvements and resources for households including subsidies for flood prevention devices. Community members who are already vulnerable are the ones most affected by flooding. It is up to all of us, as community members, to support one another in the fight against climate change, and towards our common future.

References

Armenakis, C. & Nirumpama, N. (2014) 'Flood Risk Mapping for the City of Toronto' 4th International Conference on Building Resilience, Salford Quays, United Kingdom. *Procedia Economics and Finance*, 320–326.

City of Toronto. (2003) *Wet Weather Flow Management Policy* [Online]. Available at: www1.toronto.ca/city_of_toronto/toronto_water/files/pdf/wwfmmp_policy.pdf [Accessed].

City of Toronto. (2006) *Wet Weather Flow Management Guidelines* [Online]. Available at: www1.toronto.ca/city_of_toronto/toronto_water/files/pdf/wwfm_guidelines_2006-11.pdf [Accessed].

City of Toronto. (2014) *Neighbourhood Improvement Area Profiles* [Online]. Available at: www.toronto.ca/city-government/data-research-maps/neighbourhoods-communities/nia-profiles/ [Accessed February 2018].

City of Toronto. (2017a) *About the Sewers on Your Street* [Online]. Available at: www1.toronto.ca/wps/portal/contentonly?vgnextoid=fc8807ceb6f8e310VgnVCM10000071d60f89RCRD&vgnextchannel=094cfe4eda8ae310VgnVCM10000071d60f89RCRD [Accessed June 2017].

City of Toronto. (2017b) *2017 Wastewater and Wastewater Consumption Rates and Service Fees* [Online]. Available at: www.toronto.ca/legdocs/mmis/2016/ex/bgrd/backgroundfile-98484.pdf [Accessed February 2018].

City of Toronto. (2018) *Combined Sewer Overflows* [Online]. Available at: www.toronto.ca/services-payments/water-environment/managing-rain-melted-snow/what-is-stormwater-where-does-it-go/combined-sewer-overflows/ [Accessed February 2018].

D'Andrea, M. (2017) City of Toronto's Climate Change Adaptation Strategy to Address Urban Flooding. Congrès INFRA 2017. Montreal, QC.

Environment and Climate Change Canada. (2013) Hurricane Hazel—Storm Information [Online]. Available at: www.ec.gc.ca/ouragans-hurricanes/default.asp?lang=En&n=5C4829A9-1 [Accessed February 2018].

Fearon, E. (2017) Toronto Rainfall Breaks Record Set in 1953. *The Star* [Online]. Available at: www.thestar.com/news/gta/2017/05/25/toronto-rainfall-breaks-record-set-in-1953.html [Accessed May 2017].

First Story Toronto: Exploring the Indigenous History of Toronto Walking Tour. (2017) [Online] Available at: firststoryblog.wordpress.com.

Gifford, J. (2004) *Hurricane Hazel: Canada's Storm of the Century.* Toronto, ON, Dundurn Press.

Great Lakes Commons. (2017) Charter Declaration [Online]. Available at: www.greatlakescommons.org/charter-declaration [Accessed November 2017].

Lamond, J. & Penning-Roswell, E. (2014) 'The Robustness of Floor Insurance Regimes Given Changing Risk Resulting from Climate Change' *Climate Risk Management*, 2, 1–10.

Mann, C. & Wolfe, S.E. (2016) 'Risk Perception and Terror Management Theory: Assessing Public Responses to Urban Flooding in Toronto, Canada' *Water Resource Management*, 30, 2651–2670.

Natural Resources Canada. (2017) *Climate Warming—Global Annual Precipitation Scenario: 2100* [Online]. Available at: www.nrcan.gc.ca/earth-sciences/geography/atlas-canada/selected-thematic-maps/16888 [Accessed June 2017].

Nirupama, N., Armenakis, C. & Montpetit, M. (2014) 'Is Flooding in Toronto a Concern?' *Natural Hazards*, 72, 1259–1264.

Rider, A. (2019) 'Flood mitigation efforts in full swing." *Toronto Star*, March 27, p. GT1.

Sandink, D., Kovacs, P., Oulahen, G. & Shrubsole, D. (2016) 'Public Relief and Insurance for Residential Flood Losses in Canada: Current Status and Commentary' *Canadian Water Resources Journal/Revue Canadienne des Ressources Hydriques*, 41, 220–237.

The Leap Manifesto. (2018) Sign the Manifesto [Online]. Available at: https://leapmanifesto.org/en/sign-the-manifesto/ [Accessed February 2018].

Toronto and Region Conservation Authority (TRCA). (2017) *Flood Contingency Plan—TRCA* [Online]. Available at: https://trca.ca/wp-content/uploads/2016/02/TRCA-2017-Flood-Contingency-Plan-for-web-.pdf [Accessed February 2018].

Toronto and Region Conservation Authority (TRCA). (2018) *Get Involved* [Online]. Available at: https://trca.ca/get-involved [Accessed February 2018].

PART III

Education, consciousness-raising, and collective visions

The chapters in this section focus directly on education and cultural change processes. Whether climate justice is understood as a moral imperative, a visionary hope, or a movement, research and education are crucial and cultural change is fundamental. The individual actions of researchers, organizers, teachers, and artists—and their collective activism—are no less important than in the cases discussed in the previous sections; here the focus is on how information and awareness about climate justice spread among people, with sensitivity and beauty, in expressing and modelling the change processes required for a speedy and fair energy transition to counter the worst impacts of climate change.

When artists such as Matisse, Picasso, Rousseau, and Gaugin "normalized" African masks and inhabited tropical landscapes in their art, Europeans became aware, perhaps subliminally, of the extent to which their own privileges and lifestyles were interrelated with imperialist and colonial economic processes and violence. It is impossible to separate art and culture from politics. The global atmospheric and ecological linkages among all humans, in their complexity, require more-than-verbal expression. In communicating these linkages, teachers and artists help everyone produce new visions of the world's future.

16

AAMJIWNAANG TOXIC TOURS AND CLIMATE JUSTICE

Lindsay Gray (with poem by Alice Damiano)

Aamjiwnaang First Nation is located near where the St. Clair River leaves Lake Huron on its way to Lake Erie, in southwestern Ontario.[1] We live in Treaty #29 territory, in the centre of Canada's Chemical Valley—a site of petrochemical-industrial injustice where 63 refineries and chemical plants surround a small reservation of 850 people living within a 25 km radius. The highest-polluting facilities are those within 5 km of the Aamjiwnaang community. Chemical Valley holds 40% of Canada's petrochemical industry, which is a large burden across a fence line from a small First Nations community. Imperial Oil, the first company to set up there, started operation in the 1880s after the first commercial well in North America began operating in nearby Oil Springs, Ontario, in 1858. Pollution has been accumulating and contaminating Aamjiwnaang for over 150 years.

As Anishinaabe people, we hold a huge connection to the land and are the first caretakers of the land. The pollution has taken a toll on our ability to carry on our teachings and our traditional way of life. Community members carry many stories of their experience of living next to so many refineries. Pollution has contaminated the air, water, and land, which poses a problem for a people whose teachings are focused on respecting and taking care of Mother Earth. Aamjiwnaang has changed greatly due to the growth of the petroleum industry.

Spills from the refineries happen almost every week. In the past 2 years (2015–2017), there have been more than 500 spills or leaks from the Chemical Valley. Spills and leaks affect our daily life because they can happen any day or

1 For a map of the First Nations reserves in Ontario, see https://files.ontario.ca/pictures/firstnations_map.jpg .

time. They can make the air harmful to breathe, soil unfit for planting medicines or food, and they have impacted Talfourd Creek, which runs through the reserve, to the point where there are signs warning of the dangers of contamination. People have gotten sick from being exposed to toxic chemicals such as benzene and other hydrocarbons. Because of dangerous spills, people in Aamjiwnaang have been evacuated from their homes until it was safe to be back in the area.

Aamjiwnaang community members face health problems such as respiratory diseases, cancer, and pollution contaminating their bodies. Scientific studies have found mercury in some community members' hair, and polychlorinated biphenyls (PCBs) (now banned in Canada) have been found in people's bloodstreams. The pollution has cut the birth ratio in half: two girls are now being born for every boy. Women in Aamjiwnaang face a 39% chance of having a miscarriage. Mutation-causing chemicals are being passed down from generation to generation—things our grandparents were exposed to have been passed on to their descendants. Our homelands and bodies have been trespassed by pollution.

It is important for all people to learn about Aamjiwnaang and the impacts of Chemical Valley, to create awareness of the ongoing colonization of Indigenous people and their homelands. Indigenous people are not the only ones affected. Aamjiwnaang is sacred land, even the land that was taken over by industry. The medicines and people that call this place home are sacred.

Community members who have gotten fed up with the industries' continuing destruction of the land, and failure to provide even basic information when spills are happening so that people can take action to protect themselves, have come up with different ways to create awareness. One of the most successful of the Indigenous-led organizing strategies through community events in Aamjiwnaang is the Toxic Tours.

Toxic Tours started out as a grassroots-led, community-based initiative. It began with community members giving small tours to each other and sometimes small groups of visitors, in their own cars, driving through the Chemical Valley and sharing information and stories from the community. Aamjiwnaang and Sarnia against Pipelines is a grassroots direct-action group that took on the responsibility of growing the tours into something bigger. The first larger-scale Toxic Tour took place in 2013 and was a 10 km walk through the main roads of Chemical Valley and through Aamjiwnaang. We noticed that each person who comes on the tour experiences it differently and in their own way; for many people it is a shock to witness what Chemical Valley really looks, smells, and tastes like (see poem below). By the fifth Toxic Tour in 2016, this had grown into a weekend-long event with a welcome feast at the Maawn Doosh Gumig Youth and Community Centre on the reserve, meals catered by local groups, and traditional teachings. Buses brought visitors from Montreal, Toronto, Hamilton, London, Kitchener/Waterloo, and Guelph, who were invited to camp on the beautiful community centre grounds. The event grew

FIGURE 16.1 Toxic Tour poster 2017. Poster credit: Aamjiwnaang and Sarnia Against Pipelines, http://muskratmagazine.com/aamjiwnaang-water-gathering-toxic-tour-2017-2/

FIGURE 16.2 One of the many refineries near Aamjiwnaang First Nation. Photo taken by Alice Damiano during the Toxic Tour in 2017. Used with permission.

further into the Aamjiwnaang Water Gathering and Toxic Tour 2017, incorporating traditional water teachings for visitors about responsibilities to the water (Garrick 2015) .

Toxic Tours provide information about the refineries—what they make, the chemical process information, and what to do when there is a spill—and their effects on our community. It's important to learn about Canada's Chemical Valley because it impacts Aamjiwnaang people's lives and it impacts the Great Lakes, a huge source of freshwater for millions of people. The products and by-products that come out of Chemical Valley are things people use every day. It's good for people to get their own experience of the area and see its impacts so they can really learn the cost of the petrochemical industry. It's also good for visitors to acknowledge and learn about local stories from elders or community members. Local people are encouraged to come to learn something new about the mysterious polluting neighbours behind the fences, and hundreds of people have come from all over Canada and the United States to attend. The tours share local stories with people from elsewhere, and also create a knowledge-sharing space for community members.

Toxic Tours help to educate everyone about the impacts of fossil fuel extraction across Canada—tar sands, fracking, coal mines, and oil wells, and their links via pipelines, refineries, air pollution, carbon emissions, and spills—which are not distributed equitably.

Toxic Tours also connect people to a larger picture. They focus on showing people that environmental racism is happening to First Nation communities, which is a continuation of the long history of shady land deals and ongoing colonization of First Nations peoples. It is important to bring to light the struggle Indigenous people have when their homelands become the site of environmental contamination. This poses a large problem not just for Aamjiwnaang but also for all the Great Lakes, and brings everyone into the struggle to protect our common freshwater source. We need to heal the water which is suffering from the accumulation of all the toxins, since the industries continue polluting the air and their spills and leaks into the St. Clair River are an ongoing threat.

Since the Toxic Tours began in 2013, the grave situation at Aamjiwnaang has received more and more attention. In October 2017, a joint investigative report by Global News, the *Toronto Star*, the *National Observer* and journalism schools at Ryerson University, Concordia University, the University of Regina, and the University of British Columbia, entitled *The Price of Oil*, highlighted environmental problems in Chemical Valley including more than 500 "incident reports" of spills and leaks in 2014–2015. It raised concerns about poor government enforcement of environmental regulations and repeated failures to inform and protect Aamjiwnaang residents. The Ontario Environmental Commissioner called the situation "outrageous" and "shameful." Ontario Minister of the Environment Chris Ballard promised a health study to examine the impacts of the pollution. In February 2018, the Ontario government launched a new real-time air quality website for Chemical Valley.

Following shutdowns of the Line 9 pipeline by activists in Sarnia and Montreal in December 2015, and the National Energy Board's approval of the reversal of Line 9 so it could carry tar sands bitumen to refineries and ports in New Brunswick, Quebec courts called for more examination of the effects on belugas of the Energy East port construction in Cacouna, Quebec. TransCanada Corporation then abandoned its plans to build the $12 billion terminal. The TransCanada Corporation also announced on October 5, 2017, that it would not be proceeding with its proposed Energy East and Eastern Mainline pipeline projects. As activists, voters, politicians, and courts learn more about the inequitable and dangerous impacts of fossil fuels, this awareness permeates public decision-making.

Climate justice organizers need to work in solidarity with Indigenous communities. Indigenous people uphold the responsibility to this land and are faced with broken treaties. Indigenous people have rights to these lands, which should be upheld and restored. Environmental racism needs to end, along with the pollution of the land. This is a good way to work towards reconciliation and act so that seven generations can live respectfully with the land.

FROM BLUE AND GREEN, TO GREY: THE TOXIC TOUR AS A TRIP THROUGH TIME AND INJUSTICES

Alice Damiano

Participating in a Toxic Tour in the Sarnia Chemical Valley means being plunged, for a day, into a context that is anything but easy. Historical injustices and recent ones pile up in the same place, mostly unsolved, adding every day a new little straw to a camel's back that is already broken.

Colonialism and pollution are already too deadly when they are alone … and in the Sarnia Chemical Valley, they act together. This combination of colonialism and pollution creates, in this area, a reality that some might label as *alternative* or *dystopic*, but that is, in fact, the reality where some children have to grow up, *in Canada*.

The personal stories shared by Lindsay and Vanessa Gray during the Toxic Tour show, in a way that is heartbreaking, what growing up in the Chemical Valley means. It means arguing with your teachers, because what they explain about the formation of clouds does not match your everyday experience that *clouds are created by chimneys*. It means being forbidden to swim in a pond, because it is too polluted. It means playing with a frog, seeing it fall in a polluted pond, and collecting it dead and shrunken. It means having one's everyday life regularly interrupted by spill news and safety concerns. It means often being sick, being surrounded by people who are often sick as well, and knowing that this will probably be passed to your offspring. And, for an Anishinaabe, it means experiencing all this on the same land where your ancestors lived, a land that was not divided by the Canada–US border, a land where people used to live with respect for the land and consideration for the following *seven generations*.

As a person with a Western background, I cannot speak for the Anishinaabe people, or for any people who have experienced colonization or environmental racism. Nevertheless, I can try to make a connection by walking across the Chemical Valley, imagining how it looked before all this happened, thinking about the principles life used to be based on, and trying to feel in my bones—even if I will never manage to do it fully—the intergenerational wounds and the sense of injustice described by Lindsay and Vanessa.

This is what I did at the end of the Toxic Tour, when I wrote the poem "The blue, the green and the grey" (below) and then shared it on my personal blog,[1] and on the *Economics for the Anthropocene* blog.[2] I quote it here, with the hope of making a connection with the reader, and perhaps helping the reader build an emotional connection with what happens in the Chemical Valley.

The blue, the green and the grey

I breathe blue and green
and feel the air on my face

while the wind pushes the clouds
in their playful race

the ponds are a sheer shelter
for fishes and frogs
while the crickets accompany
the birds in their songs

the river flows calmly
in the middle of one land
one home, one host
one respected older friend

Seven generations
lived here before me
seven more generations
are meant to be …

… … … … … … …
These lines
are a lie
… … … … … … …

… I'm breathing grey
feeling the chemicals in my mouth
while the chimneys shoot out
their rude, toxic clouds

the ponds are acidic
I can't swim, nor fishes can
and the frogs falling in
are skeletons in the end

the river flows resigned
dividing the land in two
the neighbor is now a stranger
and the sibling is too

Seven generations
lived here before me …
… but I am the parasite
threatening the tree.

Living once
and living for all
living to the fullest
living each goal

Living exploiting
just because I can
living proud of
the supremacy of the white man

How could I think
this egotism was right
how could I believe in it
and in its name even fight!

I stole my offspring's future
I wonder how I could dare
but I won't be anymore
the one that does not care

So, what am I loyal to?
Should it be my land?
The planet as a whole,
to the smallest grain of sand?

One day I'll find out the *what*,
the *how* and the *why*
I am supposed to do
in this time under this sky
meanwhile I search as hard as I can …
… until my heart will quit being sore
and its rotten side won't be here anymore.

—Alice Damiano

All the information and personal stories reported about the Sarnia Chemical Valley were stated orally by Lindsay and/or Vanessa Gray during a Toxic Tour in the Chemical Valley held in June 2017.

[1] Page of the *Alice in the Anthropocene* personal blog:
https://aliceintheanthropocene.wordpress.com/2017/06/08/the-blue-the-green-and-the-grey/

[2] Post on the *Economics for the Anthropocene* blog: https://e4a-net.org/2017/06/08/the-blue-the-green-and-the-grey-by-alice-damiano/

Further Reading

Studies of environmental contamination and the history of petrochemical pollution at Aamjiwnaang First Nation include the following:

Black, T. (2014), "Petro-chemical legacies and Tar Sands Frontiers: Chemical Valley versus Environmental Justice," in Black, D'Arcy, et al., *A Line in the Tar Sands* (Toronto, ON: Between the Lines) pp. 135–136.

County of Lambton (2010), Oil Heritage Conservation District Study. Retrieved from www.lambtononline.ca/home/residents/planninganddevelopment/Oil%20Heritage%20Conservation%20District%20Plan%20Documents/Conservation%20District%20Study%20Text.pdf

Cribb, R. et al. (2017), "In Sarnia's Chemical Valley, is 'toxic soup' making people sick?" Retrieved from www.thestar.com/news/world/2017/10/14/in-sarnias-chemical-valley-is-toxic-soup-making-people-sick.html

Cryderman, Diana, Lisa R. Letourneau, Fiona Miller, and Niladri Basu (2016), "An ecological and human biomonitoring investigation of mercury contamination at the Aamjiwnaang First Nation," *EcoHealth* vol. 13, pp. 784–795. www.semanticscholar.org/paper/An-Ecological-and-Human-Biomonitoring-Investigation-Cryderman-Letourneau/3cc02031022d2b82723a59a7bfdd618d5068926f

Garrick, R. (2015), "Sisters host 'Toxic Tours' of their home in Canada's Chemical Valley," *Anishinabek News*. Retrieved from http://anishinabeknews.ca/2015/01/07/sisters-host-toxic-tours-of-their-home-in-canadas-chemical-valley/

Jackson, D.D. (2010), "Shelter in place: a First Nation community in Canada's Chemical Valley," *Interdisciplinary Environmental Review* vol. 11, no. 4, pp. 249–262. Retrieved from http://deborahdavisjackson.org/documents/ShelterInPlace.pdf

Konsmo, E, and Pacheco, A.M.K. (2016), "Violence on the Land, Violence on our Bodies: Building an Indigenous Response to Environmental Violence," [online]. Available at: http://landbodydefense.org/uploads/files/VLVBReportToolkit_2017.pdf

MacKenzie, C.A., A. Lockridge, and M. Keith (2005), "Declining sex ratio in a First Nation community. University of Ottawa: Ontario, Canada," *Environmental Health Perspectives* vol. 113, no. 10, pp. 1295–1298. Retrieved from www.ncbi.nlm.nih.gov/pmc/articles/PMC1281269/

McGuire, P. (2013), "I Left My Lungs in Aamjiwnaang." *VICE Canada*. Retrieved from www.vice.com/en_ca/read/i-left-my-lungs-in-aamjiwnaang-000300-v20n8

McIntosh, E. (2018), "Sarnia Air Quality Site 'Misleading', Critics Say," *Toronto Star*, February 24.

Scott, D.S. (2008), "Confronting chronic pollution: a socio-legal analysis of risk and precaution," *Osgoode Hall Journal* vol. 46, p. 293.

Toledano, Michael (2013), "A Toxic Tour of Canada's Chemical Valley," *Vice*, March 22. Retrieved from www.vice.com/en_ca/article/vdy948/a-toxic-tour-of-canadas-chemical-valley

Weibe, S. (2010), Anatomy of place: ecological citizenship in Canada's Chemical Valley. Retrieved from https://oatd.org/oatd/record?record=handle%5C%3A10393%5C%2F26187

17

THE GREAT LAKES COMMONS

Working with water and adapting our movement to the Great Lakes

Paul Baines

Just as we've seen governments' failure to protect the climate from dangerous CO2 pollution, water protection is also in crisis. Both of these earthly elements—watersheds and the atmosphere—cross political borders, bear witness to generations of abuse, and highlight the brokenness of long-trusted regulatory institutions. Both are also the focus of strong and growing campaigns to protect the common, shared necessities of life.

Since 2012, the Great Lakes Commons initiative has been working towards the recovery and renewal of water protection. We work on community building, constructive visioning, collaborative storytelling, and a host of projects and platforms that respond to society's broken relationship with water. Our work has been shaped by (and is helping to shape) other social movements by collectively improvising and adapting human ways into more respectful and reciprocal relations with all life on Mother Earth.

To introduce a few of our water justice strategies and to share reflections for the climate justice movement, this chapter asks the following questions:

- What is the Great Lakes Commons initiative?
- How to define the problems and solutions on our own terms?
- How to create a community of people across a large territory?
- How to design critical platforms and resources that foster belonging and participation?
- How to question the types of metrics and values that are at the root of the problems, and then reimagine alternatives?
- How to decolonize our efforts by addressing the colonial norms within progressive social movements?

Great Lakes Commons (GLC) is an ambitious bioregional initiative to transform water governance in the Great Lakes so that the long-term needs of communities and the watershed take precedence over quarterly profits and settler-colonial state-making. Leadership of the Great Lakes Commons reflects a confluence of people across Nations, geography, and ancestry. We highlight governance principles that are rooted in commons and Indigenous-based traditions. GLC is based on the assertion that all people in the region have a stake in the future of these freshwater gifts and a role in shaping their perpetual care (see Figure 17.1). Ultimately, we want policy, law, education, and cultural expression that seed a social transformation rooted in reconnection and decolonization. By activating a spirit of responsibility, belonging, and leadership across communities and issues, our work fosters a shared and sacred water commons. We believe that *how* Great Lakes problems are defined ultimately shape the focus of our actions and imagination. The same can be true for climate justice. We can ask everyone: How are problems being defined *by* you and *for* you?

Most Great Lakes engagement is divided by specific water issues and political boundaries. Water pollution is separated from water privatization. Ontario's bottled water permits don't consider Michigan's permits. People on one side of Lake Huron's shoreline have no power to stop nuclear waste piled on the opposite side. Political agreements continuously ignore the inherent sovereignty of Indigenous nations and the quilt of Treaties that cover this watershed and most of Turtle Island. Clean water access and affordability are given to Nestle, Coke, Pepsi, and teams of polluting industries, while over 40 First Nations and cities

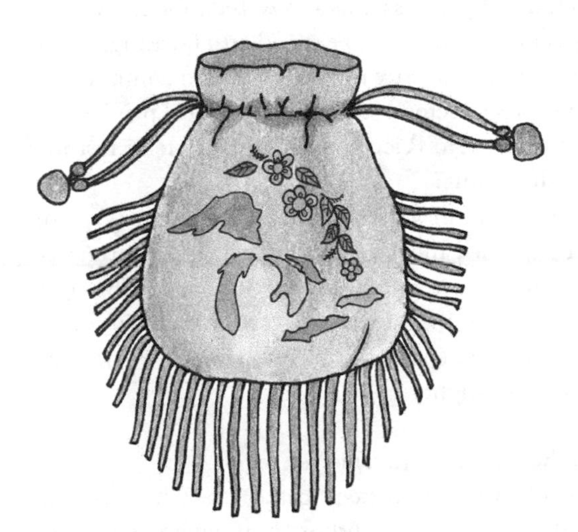

FIGURE 17.1 Great Lakes Pouch. The Great Lakes are a gift and responsibility for all who share them. Image by Lena Maude Wilson (lenamaude.com) for Great Lakes Commons. Reproduced with permission.

like Flint and Detroit go without clean water. Water is not a human right in the Great Lakes region. Water has been distorted into a commodity and "utility" that commands relations based on scarcity, mistrust, disconnection, competition, and secrecy.

For instance, simple and real-time data about sewage releases and E.coli measurements are still not shared with the public. Most Great Lakes research and policy respects political boundaries over ecological ones and obscures connections between food, energy, and land-use regimes. Water indicators are confined to measurements of quality and quantity, while many people are effectively illiterate when trying to understand the quality of their relationship with water. A water-resources outlook produces complex legal, chemical, and engineering responses to problems that are stripped from people's cultural and embodied experiences of water. Legal rights granted to corporations don't apply to the very communities and water bodies that are directly affected by these corporate decisions. A new water ethic is desperately needed, based on a different set of principles and relationships. Water holds magical properties and one of them is the ability to magnify what we can see. Looking at this broken water relationship presents a vivid picture of what the bigger problems are.

Climate activists can likely relate. While the atmosphere might not have the sensory or spiritual bond that many have for water, sustained political will to protect the commons (all that we share, inherit, and pass on) has largely disintegrated. The stories and structure of capitalism and its political kin only make CO2 reduction more impossible. In the summer of 2018, Ontario cancelled green energy investments and incentives, Canada spent $4.5 billion to finance a tar sands pipeline through sovereign Indigenous lands and waters, and the United States pulled out of the Paris Agreement and tried to expand coal burning. At the same time, fires, floods, and heat waves strain already marginalized communities who lack the financial wealth to protect themselves and the political privilege to be helped out. It took 11 months for Puerto Rico to have its electricity restored after Hurricane Maria. Where is this going?

GLC has focused on building a community, rather than a campaign. We want to position our work within larger social movements and we try to reflect on what roles we are playing—including unsettling traditional notions of "the commons." In our blog post called "Mapping the Movement," we introduce several frameworks for this mapping to help us navigate our position and roles within a larger water justice movement. We highlight:

- Working within, against, or beyond the current system
- Unpacking different layers of water injustices based on who gets the resources, who makes the rules, who has the authority, and who legitimizes authority through culture
- Seeing our levels of perception (the events, the patterns, the structures, the beliefs)
- Playing the roles of the acupuncturist, questioner, gardener, and broker

GLC was founded within a context of emergence. How are we animating the shift from "our water" to "we are water?" How are we questioning the structures and beliefs of water governance to address water sovereignty and authority? How can we support water justice movements struggling with access and affordability, while collaborating on water guardianship projects that imagine beyond the current system? How are we planting the seeds for the world we want? Rather than run a campaign, GLC decided to write a Great Lakes Commons Charter—a 500-word agreement naming the type of governance practice and ethic we desire.

The Great Lakes Commons Charter

Within the complexities of political systems and the split between economic and ecological issues, the Leap Manifesto has offered a vision change and the opportunity for unity. Similarly, GLC spent two years drafting an agreement to renew our relationship with these waters and their governance. This Great Lakes Commons Charter is a participatory effort to build community engagement, agreement, and shared responsibility for the waters.

Since the Charter's launch, hundreds of people have responded and grown a community affirming that these waters:

- Are the source of our well-being and identity
- Work as one interdependent whole connecting their health with ours
- Are a gift and eternal responsibility that every generation protects for the next one
- Have been shared and sustained by Indigenous nations throughout the ages
- Obligate nations into shared decision-making through honourable treaty

While the Leap Manifesto is mainly a policy document, the Charter stresses key principles. To grow the climate and water justice movements, we need bold statements like these about our collective future. Signing 50 petitions a year to stop government or corporate action is important, but naming our common values and desires feeds our heads and hearts and touches a wider circle of allies. We focus on collaborative tools and shared experiences to animate our Commons Charter agreement.

Commons Charter Toolkit

Several GLC strategies enliven how we work beyond the current system, how we ask questions, how we look at the deeper patterns, and how we shift water authority.

The Commons Charter asks us "What action will you take to bring this Declaration to life?" In 2017, we released a set of resources to help answer this question and to enable individuals and groups bringing transformative change into their communities. This Toolkit currently includes:

- The Commons Charter in 5 Great Lakes languages (French, Mohawk, Anishinaabemowin, Spanish, and English)
- Anishinabek, Haudenosaunee, Canadian, and American Great Lakes governance introductions
- Expressive ways-of-knowing practices and a water-caring workshop
- Conversation starters about the commons using a series of bottled-water questions
- Community organizing and visioning tools

We wanted to bridge the Charter's intention with a set of practices, information, and opportunities. The Charter and the Toolkit reinforce one another with shared vision, knowledge, and action. For instance, we ask ten questions about bottled water to unpack water ethics and authority. Each question is paired with a short response that helps navigate how to be a commoner and how to decolonize our relationship to place. What would the central questions for climate justice activism look like within the Great Lakes? Five of GLC's questions are:

- If water permits us all to be here, why are we not asking the water's permission when giving out government permits for our use and treatment of this gift?
- Since Canada and the United States are treaty nations, why are they not making water decisions with the Anishinabek nation and Haudenosaunee confederacy?
- What is the difference between water as a public good and water as a commons?
- Since the Great Lakes are one interconnected watershed, why do we have different and uncoordinated processes for water withdrawal permits divided by political boundaries?
- What is the permit process for bottled water companies using municipal tap water for their product, and how is this permit and monitoring data transparent, shared, and in the public's interest?

Water journey

We are waterbodies travelling through these watersheds. In 2016 we initiated over ten new partnerships with allies travelling the lakes by walking, biking, running, paddling, and performing. Through these networks we distributed 1,000 copies of an original 22-page Water Journey Guide that contained the Commons Charter and voices from our partners about transformative change. We expanded people's relationship with water by looking at micro-plastics, re-wilding the lands, oil pipelines, wampum treaties, the nibi (water) song, and sacred water walks.

This Water Journey guide invited reflection on our common intentions about:

- What ceremonies of gratitude can we offer?
- What harmful water issues can we learn about?
- What communities can we visit and highlight their struggles or successes?
- What protocols can we respect and share to show our responsibilities as great ancestors?
- What expressions of hope, grief, and beauty can make a Great Lakes Commons?
- What collective power can we harness crossing boundaries of place?

Online communities and platforms are a key element of our work, since our focus and supporters are spread across the basin. But it was important for us to collaborate with artists, healers, teachers, and researchers who were grounding their water intentions in a particular place. Shared experiences on or by the water fuel how we come to understand and feel what's needed in these places. This journey project linked a diversity of movement-building strategies within a larger and unifying Great Lakes Commons project that puts principles into practice.

Great Lakes Commons map

Examining how we define the water problem prompts us to question *who* is telling the story about Great Lakes protection. To foster a more collaborative story-making practice, GLC created a Great Lakes Commons map to host our expressions of worry and wisdom. We use an open-source online mapping platform to showcase our community's ideas through text, photo, and video. Each post is pinned to the Great Lakes map so our community can generate a greater sense of belonging to up/downstream neighbours and to this massive watershed.

Now with over 250 stories, our commons map connects people across borders and issues. Whether the post is about pipelines, plastics, bottled water, sacred water walks, community art projects, water privatization, student research, or a celebration of meaningful places, our map works when people take action to protect their commons. This participation and witnessing is called "commoning" since we can't rely on corporate self-regulation or the state protection of the public's interest. In fact, something only becomes a commons when we recognize it as such and take individual and collective action to protect its shared access and value. The mapping platform we use is called "Ushahidi," a pan-African Swahili word meaning "testimony" or "witness."

Water-commoning in the Great Lakes has so far designed a unity Charter, learning and action Toolkit, shared journey experiences, and a community story-telling platform. We work with other groups across the basin on water quality and quantity issues, and we also piloted a project asking the question "what does a good relationship with water look like?" The actions we take on water justice depend on the questions we ask and what gets counted and valued when doing water research, education, and advocacy. Part of defining the problem and solutions on our own terms is redesigning the foundations upon which decisions are being made.

Water Friendship

In 2017, we initiated the Water Friendship project to explore what patterns or indicators could be used to gauge a respectful and reciprocal relationship with water. Professionals have never known more about water's quality and quantity than now, yet the state of the lakes is collapsing. Knowledge about water as a "resource" has developed important metrics to guide policy, literacy, and advocacy, but new relational indicators are needed. For social policy we measure people's social isolation, fear of walking the streets at night, ratios of students to teachers, gendered violence, diet and life-expectancy links, and the list goes on. We have thousands of indicators that analyse two or more factors in order to make meaningful and impactful policy choices. Measuring just H2O (water quality), biodiversity, or water levels limits how our complex interactions safeguard water as a human right and water as the source of life.

The Water Friendship Project sets out to name these relationships, present their value and role in water protection, and offer concrete actions to guide water policy, curriculum, and advocacy. Additionally, uncovering lost connections to water invites more people into the movement from different cultures, professions, and locations. One of the major challenges facing the Great Lakes is their size. As the source of 20% of the world's surface freshwater, they are home to 40 million people and divided by over 1,000 laws, boundaries, cultures, agreements, jurisdictions, and protocols. Scaling our water justice movement implies looking at our message. It needs to be accessible and relatable. What relationships could be highlighted to track our climate impacts? Along with emission and mitigation planning, what key relationships will seed a system change for climate justice?

For the Water Friendship project, we summarized nine key relationships we felt were essential for water protection and connection.

Water is a:

- Source of life, livelihood, and being
- Form of life with its own purpose and right to be
- Source of wonder, sacredness, and mystery
- Guide for living in unity, cooperation, and peace

Water gives us:

- Our connections to ancestry, place, and our identity
- Our multi-sensory awareness of life

- Guidance for a respectful relationship with creation
- Power to heal ourselves from pain and conflict

Water invites:

- Our expressions of gratitude and humility

Our blog post on Water Friendship documents the entire project including the three communities who created this friendship-model and implemented local projects to animate our findings. With more capacity, we would like to develop this summary into a broader assessment framework for researchers and educators.

Commons Currency project

GLC's newest pilot project challenges our collective relationship with money. The Great Lakes Commons Currency project rethinks currency, value, and the impact money has on the world around us, especially water. For eight months, we created our own currency to reflect and reproduce values such as gratitude, reciprocity, reverence, and friendship. We printed and distributed 5,000 currency notes to our supporters and asked them to give them away.

When calculating the likelihood of the current economic system's protecting our bond with clean, accessible, and life-giving water, the situation seems impossible. There is no money to be made by protecting water as the source of life. Financing Great Lakes care today comes through either altruistic charity or legislated compensation. Water restoration costs are a fractional expense for a pollution-based economic system. Advocating for a friendlier version of the current system denies its core impulses and interests. How can our collective and radical imaginations connect our desire to link money's value with our values?

Money is not just a medium of exchange, but a disciplinary force on what we value, the story of a meaningful life, and our position within this story. We need a new story for money and a new currency can help us tell it. Right now our money commodifies time, ideas, muscle, relationships, and all of creation in order to create more money. But what if the value of money was based on caring for water? Money works as a magic system. It is created out of nothing and yet becomes the measure of everything in our society. Money's value is anchored to greed, commodification, competition, and scarcity. These are the human values behind the dollar value. Wished into creation by stimulus and debt, money performs the market's promise of happiness, while a global audience suspends their disbelief and ecological grief.

Imagine the experience of using a form of money (a currency) imbued with our collective effort and ethic of grateful reciprocity! We don't have to choose between protecting water as a shared and sacred commons and growing the current economy.

Let's imagine that the value of money is tied to the quality and availability of water to serve life in the Great Lakes basin. Since we are water, the water's benefit is our benefit. We know economics is a sub-system of ecology and our money system should reflect this, not subvert it.

Let's imagine that watersheds return as the de facto boundaries for trade and treaty- making. Canada and the United States are not nations, but settler-states enforced through divine proclamation, genocide, and broken promises on Indigenous people's homelands. The Anishinabek nation and Haudenosaunee Confederacy have been living with the Great Lakes since these waters were formed. A new currency could promote common peace and prosperity.

We gave out ten note packages to our Charter supporters and at water events. Each note represents the act of giving gratitude or requesting action. Each note carries the most precious value: acts of thanks and care for the Great Lakes. Rather than based on dollars, the value of these notes is our collective agreement and intention to reward people for their water protection through past actions (saying "thanks") or future actions (saying "please"). Because our current money systems only acknowledge economic utility and gain, our Great Lakes Commons currency needs a wildly different theory of value, such as past/future actions for water care.

We asked people to keep one note for themselves as our "thank you" for being a Great Lakes Commons supporter. Each person would then give the other nine notes away as a way to thank someone or an organization for protecting water or as an invitation for future action—a "please." To encourage the movement of this currency and the cultural critique of money, the notes expired at the end of 2017. We hosted a call-in forum afterwards to hear people's experiences of the project and some participants shared their stories on the commons map. We would like to relaunch this project in the future to expand its reach and influence about our economic system.

Returning to the various roles mapped out above, we hope it's now clearer how we routinely choose projects that question the current water governance system, while co-creating and affirming new relationships, experiences, roles, values, and measurements. This last section specifically locates water justice strategy and organizing within stolen territories. More than just a form of equitable inclusion, decolonizing our movement reveals new commitments and forms of accountability.

Unsettling the Commons

From the start, Great Lakes Commons has been seeding a transformative approach to current water governance. Using the histories and frameworks from both "commons" and "Indigenous" sources, we continue to map how these principles and practices enrich our connection and protection with these waters. But there's always also been a critical tension between these sources.

Craig Fortier's 2017 book *Unsettling the Commons: social movements within, against, and beyond settler colonialism* helped us name this tension more explicitly. By referencing Adam J. Barker's work (2015), this book names three ways "commons" can erase Indigenous ways of knowing and being:

- By evading complicity in producing and maintaining structures of colonization
- By naturalizing settler spaces and systems of governance
- By appropriating Indigenous territories and ways of being (Fortier 2017, p. 35)

Evading practices include the languages of "reclaiming" and "occupy." Fortier illustrates how progressive movements, such as "reclaim the streets" and "Occupy Wall Street," can evade the ways setter-colonialism claims space as its own. We see this in the water-justice movement too. Settler-activists want to "take back" the water from corporations, while not taking responsibility for their own theft of water access and authority from Indigenous nations.

Naturalizing settler spaces is also common in Great Lakes protection. These waters are often referred to as "bi-national" and the domain of both Canada and the United States, while erasing the Indigenous nations who not only preceded Canada and the United States, but continue to resist assimilation and recover their sovereignty. Naturalization could easily include creating a "commons." When looking at decision-making and rights, GLC believes everyone in the basin should participate. We focus on the inclusivity of a commons. However, not all members of the watershed should be equal. Through starvation, genocide, and betrayal by settler state agents, Indigenous peoples in Canada currently control only 0.2% of the land. Everyone knows Canadian and American sovereignty is contested based on this land theft, so how is it even possible to hold a commons-inclusive ethic within an oppressively exclusionary context? GLC encourages these critical conversations using commons-logics (non-ownership-based, bioregional, intergenerational, and participatory) to advance systems of Indigenous authority and futurity.

Naturalizing settler systems of governance also includes policy, education, and campaigns that appeal to "our" democratic and "public" institutions. However, Indigenous nations in the Great Lakes (and across Turtle Island) don't consider settler society and institutions as their own. In fact, it's these "public" claims and powers that have stripped away Indigenous authority from their own homelands.

When it comes to water quantity, the 2008 Great Lakes Compact, signed by eight US states, denies Indigenous authority outright. Ontario's permit-to-take water regulations are unilaterally controlled by the Provincial government—an arbitrary boundary of Canadian state-making that interferes with and yet ignores the ongoing assertions of multiple Indigenous nations that are enclosed by and extend past this Provincial settler-government.

In a recent letter to each political party in Ontario's 2018 election, environmental organizations listed three pages of concerns and political actions they wanted candidates to make commitments on. GLC asked for an edit that affirmed Indigenous authority for these lands and waters, but it was rejected. For some, limiting our desires to a better version of the status quo is more pragmatic and professional than acting as an ally to Indigenous nations, as a Treaty partner, and as a Canadian citizen whose own country has already recognized such Indigenous authority. When water protectors recuse themselves from issues of political authority, they are naturalizing the injustices of the day.

There is social, political, and legal power when asserting the need for "public" control of water and when using European-based legal traditions such as the Public Trust Doctrine that affirms the government's duty to protect the commons as a shared resource for generations to come. Compared to the shared benefits of privatization, robust public ownership has more democratic potential. While "public" water systems and institutions can also be failures (e.g. Flint's and Detroit's city water), they can protect long-term public investments and basic health standards. The challenge for water protectors must be to use their rights granted by the settler-state to minimize harm for the most-at-risk while also realigning their language, strategies, and goals to expose and replace the powers that give these rights.

Comparing the practices of "allies versus accomplices" (Accomplices not Allies 2014), Fortier and others highlight the need for settlers to unmake those relationships that maintain colonial power. Using the example of the Tar Sands Healing Walks, Fortier shows us an Indigenous/settler fight against fossil fuels that is led by Indigenous women and reconnects everyone to land and water through ceremony. This walk was not a protest. This was not a reclaiming of the commons. This was not asking for settler permission or recognition. Being an accomplice in decolonization should expose the normalized, violent, and ongoing erasure of Indigenous futures. GLC wants a commons framework to disrupt water-ownership and governance regimes that prioritize profit over people and this generation over all those to come. Through the use of various social movement mappings and decolonial alignments, our vision and projects above can stimulate water justice within, against, and beyond the current conditions.

References

Accomplices not Allies (2014). *Abolishing the Ally Industrial Complex. An Indigenous perspective & provocation.* http://www.indigenousaction.org/accomplices-not-allies-abolishing-the-ally-industrial-complex/ (Accessed 23 March 2019).

Barker, Adam J. (2015). *Settler: Identity and Colonialism in 21st Century Canada.* Black Point, NS/Winnipeg: Fernwood Publishing.

Fortier, Craig (2017). *Unsettling the Commons: Social Movements Against, Within, and Beyond Settler Colonialism.* Winnipeg, MB: ARP Books.

18

PLANTING SEEDS FOR GRASSROOTS ACTIVISM WITH YOUTH

Barbara Sniderman

As a parent, teacher of teens and social justice, and eco-activist, I work with young people growing up in a world that rewards consumption, self-promotion, and profit-making (Klein 2014). Anxiety and depression (Doherty and Clayton 2011) are commonplace, as are efforts to numb through gaming (Young 2009) and addictions (Leatherdale et al. 2008; Alexander 2000). Instances of violence and disaster are rampant in global and local news. Increasing evidence of climate disruption is linked to war, famine, violence, migration, displacement, energy scarcity, contamination, pollution, unfair resource distribution, polarized economies, and unjust laws and policies (Solnit 2014). All of this compounds tension and uncertainty. We live in times of frequent terrorist attacks (LaFree and Dugan 2007), the rise of populist leaders (Inglehart and Norris 2016), a worsening global economic market (Roberts and Parks 2006), climate migration and refugee crisis (White 2011), and unprecedented environmental degradation and destruction of ecosystems (Awâsis 2014).

Young people in North America may expect to change careers at least six times (Harris 2014), may accrue debilitating debt getting educated (Ehrenberg 2009), and many may live with their parents through their 30s (Picchi 2016). They may expect harsh competition for stable work (Cote and Bynner 2008), intern for too little pay (Perlin 2012), and commute or live in remote places to secure paid positions (Turcotte 2011). They are also the generation that may lack access to affordable housing, and access to local, healthy food, green spaces and nature (Sovacool et al. 2014).

It's a tricky business to try to introduce the imperative to advocate for environmental and social justice when the norm supports the idea that the efforts of a non-voting kid can achieve very little (Monbiot 2008). I have to be careful when introducing a concept, story, or topic of climate justice when working with kids: when I bring up an issue and an opportunity to explore it, a young person commonly thinks it's cool in the abstract, but when invited to watch a film about climate change, or read an article, or explore deeply the nuances of the issue, the response is a heart-wrenching emphatic no because who wants to hear how awful the world is? What follows is a brief overview of some ways to navigate these waters and try to spark a passion in young people to advocate for a future less dire.

Develop a psychology of care: Build deep connections to place at local and global scales to foster a love and ethic of care for all living things

The first step toward advocacy is fostering a connection to and love for the natural world and beings within it to establish direct and personal relationships with place (Curthoys 2007): If you know a place well, understand the way it works, learn the interconnectedness of systems attached to and supporting it, and feel a sense of belonging, wonder, gratitude or engagement as a result of this knowing, you will be more likely to advocate for it if—or when—it becomes vulnerable.

Get outside: With my son and students, getting outside acts as a re-set to energize and help to focus a positive outlook and enable openness for connection with each other, the land, and themselves. When I find young people distracted, spinning their wheels, disengaged, or engrossed in their digital worlds instead of the one they are taking up space in, I often shake things up by getting outside. The effect is instant. It gets them awake, engaged, primed for newness, curious, moving, and exploring. Wherever possible I try to find opportunities to foster a love of the natural world by first knowing local spaces, going on nature walks regularly to revisit nearby sites and mark shifts through the seasons (Krapfel 1999). We have a wetlands steps away from the school, which is located with a view of Lake Ontario that I have dubbed "mental hygiene" that is calming and provides direct access to daily learning by looking at or being outside. I live next to a ravine and two amazing parks, and bike to school daily even through the winter. I take my son regularly into natural spaces to hang out, explore or move our bodies in clean air. I try to model the healthy active activist lifestyle I hope will spark curiosity and rub off on the young people I see every day. I engage socially in the invisible curriculum of living life for this purpose. Cultivating a grounded sense of place encourages empathy and community that leads to pro-environment behaviour (Ernst 2014). In life and work I have found ways to foster an appreciation for the environment: Encourage finding a specific spot to tend, observe, cherish for its quiet, shade, beauty, and uniqueness. Engage in writing a nature journal that can use the written word, art, photography, poetry, drama or spoken word (Walker Leslie and Roth 2003). Watch and understand and talk about how leaves fall and grow, flowers bloom, weeds poke out of the sidewalk. Find natural things in urban spaces through parks, gardens, lawns, and playgrounds (Haluza-DeLay 2001). Watch and listen to the birds, squirrels, and raccoons. Look at the clouds and stars. Listen to sounds. Explore what kinds of bugs can be found at different times of year (Ernst 2014). Encourage using all of the senses to experience nature. My son regularly gets calmer when he can explore, move around, and engage in inquiry-based learning by following his curiosity when in nature. He can climb, run, get wet, touch things, photograph them, and study and learn in his own way. Doing this is calming: "We get to see all the animals and wildlife. You can actually breathe" (Kastner and Sniderman 2017).

Explore beyond your boundaries: Getting out of your comfort zone encourages one to visit new spaces and compare them with the ones you know to explore how and why they are different. I am a long-standing traveller. I learn through experiences, and take myself, my son, and students out of their comfort zone whenever possible: to see art, theatre, or performances; to ice skate or learn to ski or snowshoe; to go winter hiking, canoe tripping, or bike riding on trails. Whether in Ontario, Canada, the United States, or elsewhere, one can find natural areas in virtually any region with a quick consultation on the internet seeking out the natural parks. I try to regularly discover new places to understand and fall in love with, and connect to in order to make them part of my places to learn from and to advocate for: Hike the Bruce Trail (Bruce Trail Conservancy n.d.) or other natural pathways (Toronto and Region Conservation for the Living City 2016), swim in ponds, paddle and portage in Algonquin (Friends of Algonquin Park 2019), Killarney (Ontario Provincial Parks 2019a) or other natural parks (Ontario Provincial Parks 2019b) or kayak in the open waters of Georgian Bay (Parks Canada 2019). Canoe or kayak down rivers that have history (Canada's Historic Places 2017). Tell stories. Let a place tell you its story. Meet people who know places better than you and hear their telling of the place's story to compare to yours. Find and create magic that includes communities of more than human realms in the places you explore. Imagine what these beings would say if they spoke your language and had a voice. Learn to do this while minimizing your footprint on a place, and treading softly so as not to interrupt other beings you share space with. Learn past and create new histories. Construct narratives that can be shared with others about your local and new spaces and how they are each unique (Kiefer and Kemple 1999). With youth, this can be a real challenge. With anxieties already high, often people are reluctant to leave their comfort zone. I have found the use of anecdotal stories of travels, discovering new places, people, and experiences goes a long way to play up the adventure of engaging with new spaces. It is not often the being there that is the problem, but getting there that poses the largest barrier. Getting buy-in before the journey and also providing enough of an outline to let people know what the day will look like can go a long way. Having a strong and trusting relationship with the young person already is helpful, but using trips as a way of bonding is also effective. Our school each year starts with a school-wide community building trip to the Toronto Island, going across by ferry and setting up a barbecue, games, and sports, as well as art and cooking stations. It is our ice-breaker event that starts everyone on even ground, exploring the grounds near our school. My son and I annually embark on a canoe trip to a different place, often with different people to explore the world together. The act of discovery is multiple, learning how to be among each other, and learning how to navigate new waters and lands is figuratively and literally significant to us as we both grow and change in how we relate to one another. These trips regularly signify how we have shifted in how we are from year to year and reinvigorate our passion for being in the outdoors and love of the environment of all sorts of different natures.

Be curious: The extraordinary thing about being in nature is that it provides multiple opportunities to follow personal curiosities. Learning through inquiry within natural systems promotes wonder, and develops the practice of critical thinking about why things look and function the way they do. People young and old do this instinctively in the natural world. Activating this curiosity in young people is often easy, but can be finessed: Invent an encyclopaedia that comes from imagination rather than research (Henderson and Webber n.d.). Ban Google from the conversation of inquiry, as well as phones, unless they are used to document different ways of seeing. Make up names for plants and animals that reveal your own definitions and reasoning and explain why they evolved or adapted the way they did. Create a sound or water or plant or bug or bird map, or a map that tracks your journey symbolically or literally. The challenge is getting people to care enough to participate and overcome feelings of being silly exploring in nature. If a person is not used to being in natural spaces, it might be uncomfortable or uncool to take interest in something as mundane as a plant or a tree. Tactics to require a time-limited image-quest where students must capture and find images to bring to the teacher may entice people to engage. Activating specific questions before students go outside also helps. Providing a question at various points along a journey, or asking people to look at the experience from a specific point of view they will be asked to report on are also motivators. Collecting items (without ripping them off plants) to create a walking story stick that represents the journey can also be a way to entice people to explore and then articulate what they found and why it was interesting to them.

Find your people: Barriers to advocacy and to building a relationship with the environment often stem from lack of access to nature and also a lack of access to people who will spend time in nature who know and appreciate its importance. For the young, being social is more important than almost everything. Youth crave acceptance from one another more than access to learning in, for, and about the environment. It is easier to sit inside and connect via social media than to be vulnerable getting outside and unplugged. This makes it hard to get young people out the door. Once they are out, however, they tend to be grateful to be there. School clubs and teams are a way to start, but building intramurals or ways to meet with other young people beyond the school also makes it interesting. Connecting people with broader communities and providing opportunities to meet regional teams builds confidence and fosters positive identity traits that flow well into advocacy work. Beyond school, there are grassroots neighbourhood collectives popping up literally everywhere, accessible through the web and locally, to make it easy to find or build communities in natural spaces.

There are ways of making it cool to get outside, not just because being outside is cool, but because you can do cool things and find other cool people who like doing those things as well. Finding places to meet in nature is easy through community channels or asking around at local spaces, and also through more formal channels (Green Communities Canada 2019). In Toronto, Evergreen (Evergreen Brickworks 2019), eco schools (Toronto District School Board 2019), and City of Toronto (City of Toronto 2017a) are good places to start locally. There are social media groups to join to find people to go with on a hike, trip, paddle, bike, etc. Local outdoor shops often have boards at the store and online to connect people or sponsor events. Some places even offer bussing from start to end of hiking or biking trails. Many places also have outreach programming for kids with volunteers offering fun events or tours, camps, or after school programming. Educators can connect with specific school boards for access to field trip opportunities as well that connect young people with nature.

***Share spaces*:** A key element in advocacy is building a community of like-minded people interested and engaged in standing up for the same things. With climate justice and youth, this means building our communities by sharing our spaces, and sharing our connections to them. Schools are prime sources for this, with playgrounds that can offer opportunities for building agency in young people. One of the best ways to do this is to invite people and living things to share your spaces. Plant things that bring bees (David Suzuki Foundation 2014); plant food to share with neighbours and friends (Homesteading 2019); put up a bird feed centre (Beatty 2013); plant trees. Some cities offer incentives for this (LEAF 2019). Sign up for Doors Open in your city to allow tours to come through your spaces and wander through others to find the stories that connect to them (Ontario Heritage Trust 2017); get tours of familiar places and new ones to learn different perspectives (Tripadvisor Canada 2019). First Story Toronto is a tour and app that tells the Indigenous history embedded in specific locations around Toronto (First Story TO 2019). The growing number of dog parks, community gardens, and prevalence in recent years of planting pollinator gardens in public spaces is an indication that perhaps societally we are reaching a zeitgeist of understanding that this is important. Include young people in participating to build and be part of these things and have them promote their participation with the communities they are a part of.

Use art and metaphor to introduce difficult issues: Art can make learning about concepts of crises symbolic and less overwhelming

Advocating for climate justice is hard. It means you need to address the very real and scary issues of injustice that can be devastating and overwhelming. For the young person and teen, this can cause disengagement in very real and important issues that should not be ignored but do need to be approached in age-sensitive ways. Visual art, documentaries, sculptures, and storytelling are symbolic ways young people can learn about the concepts of crises of climate disruption without being too overwhelmed. Film festivals increasingly offer programming that teaches kids about issues that are important for climate justice. In Toronto, this includes Hotdocs International Documentary Festival (HotDocs 2019), Planet in Focus Film Festival (Planet In Focus 2019), Toronto International Film Festival (TIFF 2019), and Regent Park Film Festival (Regent Park Film Festival 2019), to name a few. The National Film Board of Canada offers online access to films (NFB 2019). These films range from literal and difficult to cartoon and metaphorical and easy to comprehend. There are also exhibits creating visual images that engage discourse. Ai Wei Wei's life vest sculpture used thousands of discarded life vests from the island of Lesbos to appeal to Europe to take action and take in refugees fleeing Syria (Sierzputowski 2016). Lorenzo Quinn's sculpture in Venice of hands supporting the Ca'Sagredo Hotel from out of the water makes a statement about rising waters due to climate change (Jobson 2017). Talking about where artworks are mounted, and why, allows for discussion without showing graphic images of death and destruction. There are many resources now available that are kid friendly, including videos and storybooks for the very young (Schwartz 1999). Kids can grasp the concept of fairness through art without having to be exposed to the traumatic graphic detail adults are more able to engage in. By developing an understanding of power through the symbolism of (un)fairness rather than the minutia of details surrounding climate justice issues, young people can clarify where they stand in a safe way using the language they are comfortable speaking with.

Be creative in articulating observations: Young people (and old) can also express their own personal responses to climate injustice through art. This can work toward fostering advocacy in two ways: one by providing a form to express oneself; and the other by using the arts as a forum for discussing issues of justice with the public and engaging the public through the exhibition of artistic works. Encourage young people to create symbolic representations of nature and justice narratives through art, dance, walks, stories, and films in response to what they know about the places they love and the things happening as a result of disruption (hooks 1994). Share these works in public places such as schools, cafes, libraries and other spaces that can help to build a sense of community. Arrange for and encourage mural making in parks and buildings where many people can see them. Society listens when young people create things with a strong message. Complex justice work by youth is heard more readily by adults, and helping these voices be heard empowers young people to speak out through creative means. My son drew a picture of himself hanging off a tree after a nature walk, and the view from the car window after a visit to an Indigenous burial site. He photographs birds and squirrels and dogs every time we see them. We make up songs about having rocks in our shoes and paddling in a canoe. These we post and put up in public places, and tell the story of where they come from. These stories are passed on to others who take interest in the narratives steeped in advocating for fairness and this also is passed along. When these creative works come from the young, they are listened to. When audiences are both a mix of young and old, the message spreads more widely.

Building communities and cultures: Make do-it-yourself projects that teach good lessons and practices

The next step toward fostering advocacy is to actively build communities and cultures of care to create and model justice work.

Create DIY bags, balms, bikes, and other non-toxic stuff: Each year, my school engages in building and solidifying our relationship with other stakeholders in our community. Housed in a building where a high school shares the space with a grade school, nursery school, and community centre, we are at an advantage to reach a wider cohort to participate in eco-justice work. To this end, we hold a summit every year where older high school students create workshops to teach younger kids to do justice work of their own and hopefully bring their learning to their families and communities at home. DIY projects make this fun, interesting, and easy to spread beyond the house or classroom. People learn better when they are asked to teach others, and older students learn better in this context. Younger students are hungry for mentors closer to their age, and connect well with learning when it comes from students rather than teachers. The building then becomes a community of reinforcing advocates for the environment. The challenge then becomes finding ways to keep this lasting throughout the year. Options include ideas to make your own reusable bags out of old t-shirts (Instructables 2019), make lip balm and body scrubs out of ingredients in your pantry (David Suzuki Foundation 2019), build and fix bikes and use them rather than drive (Flack 2014), and model and practice the concepts of reduce, reuse, recycle, and upcycle to underline the importance of being mindful of maintaining rather than depleting our natural resources (Green Planet 4Kids 2019). Make art out of natural things and about and within natural settings, and create posters, zines, blogs, films, websites, tweets, soundscapes, and silk screen t-shirts with messages that say something kids want adults to hear (Marino 1997). Teach about toxins in daily products, the negative effects of plastics, the concept of upcycling, and different ways of knowing by embedding them into the workshops making cool things out of natural products. Making advocacy fun while creating mentorship opportunities and connecting young people with common goals help to create a language of advocacy that can extend beyond the moments when they are built. My son annually makes designs for silk-screening images onto tee-shirts. Every year he makes a design that follows a passion for humour and characters and a second that connects to nature. We have a large wardrobe now of clothes with his personal impression of a wolf, eagle, fox, earth, tree, flower, and superhero advocate. People ask and we have a story to tell about how and why we wear these things close to our hearts.

Boycott buying bad items because they are toxic or wasteful or made by companies that are unfair to people and nature: In our globalized world, it is hard to impress upon young people forming their identities based on fads and social pressures that the clothes and makeup and items they use make a difference in terms of having an impact on global climate. It's hard to convince someone to avoid spending far less at a mall than they would at a local shop making original pieces. Luckily, there are growing numbers of options available to make ethical purchases based on justice work, and the costs are becoming more accessible. Bulk stores and places offering reduced packaging are becoming more the norm, as well as donating unused groceries rather than reducing them to waste. Navigating which companies are truly doing the good work, operating under fair trade models, and not capitalizing on marketing ploys to imply more than what is there is still tricky. Teaching young people how to make ethical decisions plays an important role in fostering young and confident and hopeful advocates. Model good behaviour and make it explicit so young people are able to make informed decisions. To lower your carbon footprint, favour veggies to meat (Gilpin 2015), reduce plastic (including gum) (Lourie and Smith 2013), travel by foot or bike or wind or muscle power whenever possible, use public transit, and use petrol-based vehicles like cars and planes minimally (Gertten 2015). Buy in bulk and carry your own reusable bags. Buy local. Grow or make your own. Explain why this is important and show how easy and how much fun it can be to make these ethical choices. In our rapidly changing world, it's impossible to provide a comprehensive list of ethical practices when making purchases. Modelling how to understand ethical literacy in purchasing is the important message as products come in and out of availability.

Find local groups that are doing good work: Give agency to young voices through participation, art, and activism

Be creative—engage in spaces without tampering with them: The golden rule is to imprint upon young people that it is always best to leave a space better than it was found. Whether it is on a canoe trip, hike, or walk in the park or schoolyard, treading with a conscientious aim at preserving a pristine space is essential in building young activists. Modelling this by not littering, picking up what others drop, and creating opportunities to make a space appreciated by the community and also by passers-by without disrupting its natural processes does well to become a place meaningful to young people that they will want to return to again and again and to advocate for. On a field trip once, students discovered a fallen bird. We stopped, gathered round, took photos, and found the local agency that might help it. The students were ultimately released from the site to go home for the day, while I followed up with the bird. Each one the next day returned asking what had become of the bird, and formed a bond with the agency who gave daily reports for the week on its progress. These students still recall that moment of advocacy, and retell the story when visiting the site, connecting with one another, and encountering me. One student even went on to volunteer at the agency. We have also done creative enhancements to natural spaces.

Rather than pick flowers or remove rocks from sites, create art opportunities to help others see the wonder of the natural world without altering or destroying it. Make rubbings of leaves (Firstpalette 2019). Make sculptures of things found on the ground like the work of Andy Goldsworthy (Artnet 2019). Draw, paint, sculpt or take pictures of things as they change and grow. Be a documenter of living things. Frame things you see to look at them closely, differently, and to help others appreciate them from new perspectives (Haluza-DeLay 2001). Make a communal fairy home in a local park. My son and I came upon a wooden house erected on an old tree stump years ago. We saw that daily there were little changes made and joined in to add to it, using stones to create a pathway, sticks to make a fence. Others responded in the same way, and it was continually and secretly updated by various members of the neighbourhood. It was a cherished space. Last year the stump was removed from the park. It made local news, and many were devastated to see it go. The spring revealed new pop up fairy spaces found in random parts of the park to keep the community participation alive. We still collect things to add to the magic (Kastner 2019).

Be outspoken and learn how to be heard: The most prevalent excuse for apathy in the young when it comes to justice work is that people feel nothing they do will make a difference. At the small high school I teach at, it is easy to initiate

FIGURE 18.1 The Fairy Tree. "Community-building through attention to nature: how does a Fairy Tree change over time?" Drawing by Max Kastner. Reproduced with permission.

school-wide programs without a lot of red tape. We are quite successful at helping students bring forth school-wide shifts for positive change. Making that leap then beyond the school boundaries becomes the new barrier, but not an impossible one to break through. Starting small and then expanding the boundaries of the work is helpful when encouraging young people to do what they believe is right. When you see something is not right—like building a parking lot where a park used to be, or cutting down trees, or putting up smokestacks on someone's sacred land, speak up. Be a model for young people and encourage them to work with you to tell an adult, write a letter, or tweet or FB or YouTube video a petition (Petition Online Canada 2019). Learn about what local government is doing or not doing to support your cause (City of Toronto 2017b), and how to engage to be heard. Events like the Little Turtles Water Walk with Nokomis Josephine Manadamin at the Naamwe Kendassing Child Care Centre of Whitefish River First Nation makes a difference to both the children who feel empowered and the adults who see the issue as important for future generations already passionate about the issues (Chiblow 2015). Make it communal, and then make it viral. Ask for what you want. Having young people participate in this process, and better yet leading it, can go a far distance to being heard and building a community that speaks to power to advocate for justice. An explicit task for an Indigenous studies course I taught to 14-year-olds revolved around creating a work of art that reflected learning and understanding power and its structures surrounding a reality in present day or the past between government policy, society and First Nations, Metis or Inuit peoples' experiences. The works were extraordinary, creatively outlining advocacy for water protection, resisting pipelines, reducing hunting rights, acknowledging the need to investigate missing and murdered Indigenous women in Canada, outlining the ongoing effects of residential schools and the 60s scoop, and many other things. On display, the culture of the school shifted, as others gathered around the works to read artist statements and ask questions and learn. These are lasting moments of learning when sparked by creative expression of an identified worthy causes.

Support the underdog: It can be daunting to stand up for something on your own. Building community and finding your people is essential in gaining traction. There are multitudes of opportunities to find others in the world, locally or beyond, standing up to injustice. Making this explicit to young people is important. Notice when you see things or people who are more vulnerable than you are. Do something to help (Centre for Social Justice 2012). It might be speaking as an ally (Centre for Social Innovation 2017), or it could be calling an animal shelter when a bird hits a window (FLAP Canada 2019). When something can't fight for itself, be there to help lend a voice (IPPF 2011). Kids know when something is not fair. Give them a voice to state it and be heard. Be an adult who is listening and helping to create confidence in the young person stating what should be done. Be a human megaphone, helping the young one's voice be heard by those who can help to engage in initiating change. Help to spread the word. Bringing the world into the classroom with speakers, films, or audio or video conferences with people does wonders when there is opportunity for live interaction. Bringing the young person to the world works in even more exciting and experiential ways.

Create shelters: Sometimes, the best way to stand up for something is to create a safe space for it. Make birdhouses (The Spruce 2019), let beaver dams stay intact, put water out for pets on sidewalks. Consider housing a shelter animal rather than a pure breed (Toronto Humane Society 2019). Share a meal with a refugee (Depanneur 2019), or sponsor someone in need through a community or school group (Lifeline Syria 2017). Be explicit about what you can do to help and include kids in brainstorming and participating in doing good work to help the vulnerable (Nabarro 2010). I took a recent workshop in creating seed bombs, wrapping bee pollinator seeds into a mixture of clay and earth. This allows the seeds to take root within the clay before a bird or other animal uses the seed for food. It is nurtured until rain washes away the protective clay shell, and by that time, roots will have begun to form and the plant is well under way to survive. My students and son have all been guerrilla planting pollinators in public spaces like grasslands and other properties where wild things grow. Of course they are careful not to introduce new species to areas, or to ingratiate on private spaces, but we are finding ways to provide shelter for animals, plants, and insects in need. These are adventures kids will not soon forget, and when they see a bee or a pollinator plant like the one they have contributed to creating, they will recognize their part in sustaining the system. During the height of the Syrian refugee crisis, I mobilized my local dance community—housed in a building with multiple dance communities—to allocate one week toward dancing for donations. The funds raised by collectively joining the communities into one action led to assisting three separate families to establish themselves in Canada with both financial and emotional supports for the next two years. My students and family were actively engaged in the process, learning that activism can be creative, throughout. A friend keeps bees and harvests honey, and works with an agency to teach others about this good work. While I have yet to gain permission to drive a couple of hours with a class to visit the farm where they are, I have been able to bring my son to a harvest and help to bottle the honey we use for the year. I take photos and bring them to school and show others as I feed them the golden deliciousness. I show them the honeycomb and teach them how it works. It sparks interest. They are open to planting pollinators so that when bees come, we can apply for grants nearby to build hives locally. We can partner with agencies that teach us how to do this.

Pushing the edges of systemic infrastructure: Initiate programming in schools, libraries, and after school organizations

Ecoschools provides a framework for advocacy: Access to programming that fits climate justice within the structures of curriculum are key to doing good work.

More and more curricular revisions are including learning about climate change through multiple courses and disciplines. In Canada and Ontario, the NGO Ecoschools offers certification of eco-based culture in schools regarding grounds, leadership, energy, waste, community, and pedagogy goes a long way to engage in defining a culture of eco advocacy in and beyond the boundaries of the school (Toronto District School Board 2019). If there is not a similar entity in your region, consider using this as a potential model for how to embed climate justice programming into your school. In an ideal world, this type of programming would be embedded in the school's mandate and practices and not done on a volunteer basis. Advocating for this increased status across the educational platform is still in its early stages.

Embed climate justice in classes and workshops: Climate justice work can be connected to virtually all parts of the school curriculum and to multiple intersections with parenting. This is not hard. It simply requires creativity, out of the box thinking, and cooperation. Finding allies within the system who can work with you or support your work will allow for multi-disciplinary reinforcement of learning: Do an English language poetry unit on nature, water, or pollution or create a blog or film. Photography classes can document natural processes on and around the school grounds. Social science, philosophy, environmental studies, science, and history can explore ethics around tar sands, climate migration, the economics of carbon mitigation, and re-thinking a history and a present system that is only beginning to acknowledge the experiences of Indigenous peoples. Music classes can create activist folk songs. Math classes can calculate carbon footprints. Physical education can include being active in healthy green spaces and advocating for clean air and water. Science classes can test the cleanliness of the air and water and energy output in the school. There are countless lesson and unit plans available online for free that can cater to specific curricular needs. These possibilities can translate easily into household discussions, field trips or weekend trips, and actions on road trips or hikes. In this way, curriculum to be an upstander can be rolled out as easily outside of school as part of life learning.

Outdoor education provides enriched learning (Bowers 2001): People will remember experiences outside far more easily than any particular classroom experience. Historical walking tours in the neighbourhood, English descriptive writing that establishes settings, photography in nature, social science field study, math quantifying observations about animals and plants observed over time, all of these are potential modes of learning that engage the sense of outdoors through experience (Dewey 1997). Community projects that stem from inquiries students have based on the place, history, and location of the school, community centre, or home also provides rich, community building, creative ways to learn about and advocate for justice.

Environmental education engages big picture thinking: Looking at climate justice issues by engaging in problem-based learning that arises when we experience change first-hand allows us to identify conflict as a result of climate change (Noddings 2002). This work uses systems thinking (Senge and Sterman 1992) that puts this kind of problem solving at the forefront of contexts for learning

(Russell 1999). When my son learned that his favourite tree was going to be torn down, he asked why. I didn't know. I engaged him and my high school class in finding out. We learned it was diseased, as a result of an infestation of a foreign bug into the neighbourhood that was killing many of the trees. We learned how this bug was introduced. We learned what couldn't be done about it. We also looked into finding ways to plant more replacement trees. At the time this was not available to us, but every person engaged in the work can now recognize the type of tree that was devastated, the progressive action that allowed for the destruction, and the lack of political initiative to engage in stopping the destruction and in replacing the trees. The young people got into the politics of the region by following a specific tree and the outcome of its journey. One day on the way to school, I encountered a city worker who rescued a baby duck lost on the concrete. As he waited for help to arrive, cradling the little duck for warmth, he offered to allow my students to come and see. I brought them out and they asked questions, some held the duck gently, and they spoke with the rescue team about what would happen to it once it was picked up. The students for the rest of their time at the school were able to identify the specific duck in the water, and watched out for the animals in the surrounding area. Some wrote about the experience as significant in their journals.

Popular education enables a discourse of restructuring power: This kind of approach to learning uses creativity and play to build communities and ground learning in understanding power and power structures and finding creative ways to practice resistance toward social and eco (in)justices (Freire 2013). Posters and tee-shirts advocating for a cause; drama works illustrating a power struggle connected with climate injustice; artworks like the ones mentioned above; mural projects that illustrate Indigenous connections to regional river systems all illustrate how communities can connect creatively to engage in advocating for social change. In a course learning to teach about Indigenous experiences in Canada, we were asked to sketch the story told of a woman's grandmother's experience in a residential school as told in a documentary. An artist in the class designed an extraordinary rendering of the story as a drawing. We later transferred this drawing onto a tee-shirt. Each of us now has this shirt that tells the story of residential schools. Our collective creative experience resulted in ongoing advocacy as we are retelling this story each time we each wear the shirt. Engaging young people in this type of learning is an effective way to spark justice work through building communities to decide collectively what action they wish to take.

Constructing formal justice-based pedagogy within the larger structure of the school can create a progressive, comprehensive learning context for justice: Much of our education comes from systemic, deeply embedded capitalist and neoclassical economics that are the basis of the school system structure (Klein 2008). Working within this system to dismantle its negative affects is going to garner pushback. Inroads can be made advocating for school-wide initiatives as well as working within individual classrooms.

Consider embedding within the school culture annual school-wide themes tackling issues of water, migration, energy and other such topics. These can become themes within all classes across all grade levels that are explored differently throughout the school year.

Consider different grade levels exploring these themes in the context of different relationships of power. Younger grades can focus on exploring the relationship between power and identity through various classes. The next grade level might see all of their courses through the lens of power conflicts. The following year students can look at their subjects through the paradigm of resistance to power. Finally, senior students can explore topics thinking creatively and innovatively to make room for new paradigms of power that navigate current systems. These options help to build communities of activists and systems thinking problem solvers (hooks 2010). As students mature, they are more able to refine their context of understanding, and can solve problems in different ways, allowing for creative opportunities. Problem-based (Hmelo-Silver 2004) and project-based learning (Bell 2010) helps to enhance critical thinking skills. Use of the arts (Seidel et al. n.d.) has proven that finding metaphor assists in seeing creative solutions in new ways (Bresler 2007).

Conclusion

In the face of lagging positive global change, it is essential to provide opportunities for young people to feel agency, inspiration, autonomy, and belonging in spaces that are becoming more scarce, vulnerable, and imperilled. The hope for constructive paradigm shifts is through grassroots efforts. Building communities that create and support allies who speak for and protect spaces and beings that inhabit them is key to their power to succeed. Ongoing consistent positive, creative, direct, and communal interaction is key to engaging young people in connecting to place. Outlined above are ideas individuals can incorporate into their daily lives to support their own justice work and more importantly to model and support this work in young people. There persists a disconnection between personal and widespread action and the need for systemic change to mitigate global impacts (Rand 2014). These guidelines navigate possibilities for individual and group activism. This is a time when action must happen now to ensure the security of our future generations. We live in an era where, like never before, the young are inheriting a world where we cannot currently envision solutions for climate conflicts we have created and continue to create. This is the paralyzing, anxiety-producing, overwhelming reality of our youth and future generations. Helping them to find hope and connection with each other and the systems they are tasked to try to protect, while reversing adversarial effects already in place, is essential to create the justice warriors needed to usher us into a new more sustainable paradigm.

References

Alexander, Bruce K. (2000) 'The Globalization of Addiction', *Addiction Research*, 8(6), pp. 501–526. [online]. Available at: https://doi.org/10.3109/16066350008998987 (Accessed: 16 March 2019).

Artnet. (2019) 'Andy Goldsworthy' Artnet [Online]. Available at: www.artnet.com/artists/andy-goldsworthy/ (Accessed: 16 March 2019).

Awâsis, Sâkihitowin. (2014) 'Chapter 23: Pipelines and Resistance across Turtle Island' in Black, T., Klein, N., McKibben, B., Russell, J.K., D'Arcy, S., Weis, A.J. (Eds.) *A Line in the Tar Sands: Struggles for Environmental Justice*. Toronto, ON; Oakland, CA: PM Press, pp. 253–266.

Beatty, V. (2013) *23 DIY Birdfeeders That Will Fill Your Garden With Birds* [Online]. DIY Crafts. Available at: www.diyncrafts.com/3515/home/23-diy-birdfeeders-will-fill-garden-birds (Accessed: 26 June 2017).

Bell, S. (2010) 'Project-Based Learning for the 21st Century: Skills for the Future'. *The Clearing House: A Journal of Educational Strategies, Issues and Ideas*, 83(2), pp. 39–43. [Online]. Available at: https://doi.org/10.1080/00098650903505415 (Accessed: 16 March 2019).

Bowers, C.A. (2001) 'How Language Limits Our Understanding of Environmental Education', *Environmental Education Research*, 7, pp. 141–151. [Online]. Available at: https://doi.org/10.1080/13504620120043144 (Accessed: 16 March 2019).

Bresler, L. (Ed.) (2007). *International Handbook of Research in Arts Education* Part 1, Dordrecht, The Netherlands: Springer.

Bruce Trail Conservancy. (n.d.) *Bruce Trail* [Online]. Available at: brucetrail.org (Accessed: 16 March 2019).

Canada's Historic Places. (2017) *Canada's Rivers Canada's History* [Online]. Available at: www.historicplaces.ca/en/pages/17_heritage_rivers-rivieres_patrimoniales.aspx (Accessed: 16 March 2019).

Centre for Social Innovation. (2017) *How to Be an Ally* [Online]. Available at: https://socialinnovation.org/how-to-be-an-ally/ (Accessed: 16 March 2019).

Centre for Social Justice. (2012) Centre for Social Justice [Online]. Available at: www.socialjustice.org/community/?f_cat=2w (Accessed: 16 March 2019).

Chiblow, S. (2015) "Little Turtle Walk" with Josephine Manadamin'. *Anishinabek Newsca*, 15 February [Online]. Available at: http://anishinabeknews.ca/2016/02/15/little-turtle-walk-with-josephine-manadamin/ (Accessed: 16 March 2019).

City of Toronto. (2017a) *Community and People* [Online]. Available at: www.toronto.ca/community-people/ (Accessed: 16 March 2019).

City of Toronto. (2017b) *Environment Days* [Online]. Available at: www.toronto.ca/services-payments/recycling-organics-garbage/community-environment-days/ (Accessed: 16 March 2019).

Cote, J., Bynner, J.M. (2008) 'Changes in the Transition to Adulthood in the UK and Canada: The Role of Structure and Agency in Emerging Adulthood', *Journal of Youth Studies*, 11(3), pp. 251–268. [Online]. Available at: https://doi.org/10.1080/13676260801946464 (Accessed: 16 March 2019).

Curthoys, L.P. (2007) 'Finding a Place of One's Own: Reflections on Teaching in and with Place', *Canadian Journal of Environmental Education*, 12(1), pp. 68–79.

David Suzuki Foundation. (2014) *Create a Pollinator-friendly Garden for Birds, Bees and Butterflies* [Online]. Available at: https://davidsuzuki.org/queen-of-green/create-pollinator-friendly-garden-birds-bees-butterflies/ (Accessed: 16 March 2019).

David Suzuki Foundation. (2019) 'Lip Balm Recipe' in Queen of Green [Online]. Available at: https://twitter.com/QueenofGreen/status/968274254402805760 (Accessed: 16 March 2019).

Depanneur. (2019) *Newcomer Kitchen* [Online]. Available at: https://thedepanneur.ca/newcomerkitchen/ (Accessed: 16 March 2019).

Dewey, J. (1997) *Experience and Education*. 1st edn. The Kappa Delta Pi Lecture Series. New York, NY: Simon & Schuster.

Doherty, T.J., Clayton, S. (2011) 'The Psychological Impacts of Global Climate Change', *American Psychologist*, 66(4), pp. 265–276. [Online]. Available at: https://doi.org/10.1037/a0023141 (Accessed: 16 March 2019).

Ehrenberg, R.G. (2009) *Tuition Rising: Why College Costs So Much*. Cambridge, MA: Harvard University Press.

Ernst, J. (2014) 'Early Childhood Educators' Use of Natural Outdoor Settings as Learning Environments: An Exploratory Study of Beliefs, Practices, and Barriers', *Environmental Education Research*, 20(6), pp. 735–752. [Online]. Available at: https://doi.org/10.1080/13504622.2013.833596 (Accessed: 16 March 2019).

Evergreen Brickworks. (2019) 'Our Projects' in Evergreen [Online]. Available at: www.evergreen.ca/our-impact/?gclid=Cj0KEQjw4cLKBRCZmNTvyovvj-4BEiQAl_sgQoaOPTjx88u2X2fDQadzklqy4cXJcvzncEDMP2rgsagaAgTp8P8HAQ (Accessed: 16 March 2019).

Firstpalette. (2019) 'Leaf Rubbings' in *First Palette—Your Step By Step Guide To Kids Crafts*. [Online]. Available at: www.firstpalette.com/craft/leaf-rubbings.html (Accessed: 16 March 2019).

First Story TO (2019) *First Story Toronto: Exploring the Aboriginal History of Toronto* [Online]. Available at: https://firststoryblog.wordpress.com/ (Accessed: 16 March 2019).

Flack, D. (2014) 'The top 5 DIY bike repair shops in Toronto', *BlogTO*, 14 May, [Online]. Available at: www.blogto.com/sports_play/2014/05/the_top_5_diy_bike_repair_shops_in_toronto/ (Accessed: 16 March 2019).

FLAP Canada (2019) 'What Should I Do If I Find an Injured Bird?', *Working to Safeguard Migratory Birds in Urban Environment through Education, Policy Development, Research, Rescue and Rehabilitation*, [Online]. Available at: www.flap.org/find-a-bird.php (Accessed: 16 March 2019).

Freire, P. (2013) *Education for Critical Consciousness*. London, UK; New York, NY: Bloomsbury Academic.

Friends of Algonquin Park. (2019) Algonquin Provincial Park [Online]. Available at: www.algonquinpark.on.ca/ (Accessed: 16 March 2019).

Gertten, F. (2015) *Bikes vs Cars* [documentary]. Denmark, Canada: Kino Lorber.

Gilpin, E.M. (2015) 'Eating Animals: Resurgent Relationships to What We Eat', *Guts Magazine*, 5, 5 November. [Online]. Available at: http://gutsmagazine.ca/eating-animals/ (Accessed: 16 March 2019).

Green Communities Canada. (2019) Green Communities Canada [Online]. Available at: http://greencommunitiescanada.org/ (Accessed: 16 March 2019).

Green Planet 4Kids. (2019) 'Why Reduce, Reuse and Recycle?', *Green Planet 4 Kids* [Online]. Available at: http://greenplanet4kids.com/comic/rrr/meaning-rrr?gclid=Cj0KEQjw4cLKBRCZmNTvyovvj-4BEiQAl_sgQlEOjeWrUuTZB0BtQYhiA3tPI22iTOJT4aw42i3EeXoaAsK48P8HAQ (Accessed: 16 March 2019).

Haluza-DeLay, R. (2001) 'Remystifying the City: Reawakening the Sense of Wonder in Our Own Backyards', *Thresholds in Education*, 27, pp. 36–40.

Harris, P. (2014) 'How many jobs do Canadians hold in a lifetime?', *Workopolis*, 12 April. [Online]. Available at: https://careers.workopolis.com/advice/how-many-jobs-do-canadians-hold-in-a-lifetime/ (Accessed: 16 March 2019).

Henderson, A., Webber, E. (n.d.) 'Ame Henderson + Evan Webber: performance encyclopaedia', Art Gallery of Ontario, [Online]. Available at: www.ago.net/performance-encyclopaedia (Accessed: 16 March 2019).

Hmelo-Silver, C.E. (2004) 'Problem-Based Learning: What and How Do Students Learn?', *Educational Psychology Review*, 16(3), pp. 235–266. [Online]. Available at: https://doi.org/10.1023/B:EDPR.0000034022.16470.f3 (Accessed: 16 March 2019).

Homesteading. (2019) 'How to Grow All the Food You Need in Your Backyard', *Homesteading*, [Online]. Available at: https://homesteading.com/homestead-handbook-grow-all-the-food-you-need-in-your-backyard/ (Accessed: 16 March 2019).

hooks, bell. (2010) *Teaching Critical Thinking: Practical Wisdom*. New York, NY: Routledge.

hooks, bell. (1994) *Teaching to Transgress: Education as the Practice of Freedom*. New York, NY: Routledge.

HotDocs. (2019) *Hot Docs International Documentary Festival* [Online]. Available at: http://hotdocs.ca/ (Accessed: 16 March 2019).

Inglehart, R.F., Norris, P. (2016) 'Trump, Brexit, and the Rise of Populism: Economic Have-Nots and Cultural Backlash' *Faculty Research Working Paper Series*, August. Cambridge, MA: Harvard Kennedy School. [Online]. Available at: file:///Users/barbarasniderman/Downloads/SSRN-id2818659.pdf (Accessed: 16 March 2019).

Instructables. (2019) 'Fastest Recycled T-Shirt Tote Bag', Instructables [Online]. Available at: www.instructables.com/id/FASTEST-RECYCLED-T-SHIRT-TOTE-BAG/ (Accessed: 16 March 2019).

IPPF. (2011) 'Want to change the world? Here's how... Young people as advocates', International Planned Parenthood Federation [Online]. Available at: www.ippf.org/resource/want-change-world-heres-how-young-people-advocates (Accessed: 16 March 2019).

Jobson, C. (2017) 'Support: Monumental Hands Rise from the Water in Venice to Highlight Climate Change', Colossal, [Online]. Available at: www.thisiscolossal.com/2017/05/support-hands-sculpture-venice/ (Accessed: 16 March 2019).

Kastner, M. (2019) *The Fairy Tree (age 11)*. [pencil on paper drawing 8.5x11]. Toronto, ON: personal collection.

Kastner, M., Sniderman, B. (2017) *What's it like to learn outside?* [interview]. Toronto, ON, 3 June.

Kiefer, J., Kemple, M. (1999) 'Stories From Our Common Roots: Strategies for Building an Ecologically Sustainable Way of Learning' in: Smith, G.A., Williams, D.R. (Eds.), *Ecological Education In Action: On Weaving Education, Culture, and the Environment*. Albany, NY: State University of New York Press, p. 21.

Klein, N. (2008) *The Shock Doctrine: The Rise of Disaster Capitalism*. New York, NY: Picador.

Klein, N. (2014) *This Changes Everything: Capitalism vs. the Climate*. New York, NY: Simon & Schuster.

Krapfel, P. (1999) 'Deepening Children's Participation through Logical Ecological Investigations' in: Smith, G.A., Williams, D.R. (Eds.), *Ecological Education in Action: On Weaving Education, Culture and the Environment*. Albany, NY: State University of New York Press.

LaFree, G., Dugan, L. (2007) 'Introducing the Global Terrorism Database', *Terrorism and Political Violence*, 19(2), pp. 181–204. [Online]. Available at: https://doi.org/10.1080/09546550701246817 (Accessed: 16 March, 2019).

LEAF. (2019) 'Schools', Leaf [Online]. Available at: www.yourleaf.org/schools (Accessed: 16 March 2019).

Leatherdale, S.T., Hammond, D., Ahmed, R. (2008) 'Alcohol, Marijuana, and Tobacco Use Patterns Among Youth in Canada', *Cancer Causes & Control*, 19(4), pp. 361–369. [Online]. Available at: https://doi.org/10.1007/s10552-007-9095-4 (Accessed: 16 March 2019).

Lifeline Syria. (2017) 'Call for Refugee Sponsorship' *Lifeline Syria* [Online]. Available at: http://lifelinesyria.ca/sponsor/ (Accessed: 16 March 2019).

Lourie, B., Smith, R. (2013) *Toxin Toxout: Getting Harmful Chemicals Out of Our Bodies and Our World*. Canada: Knopf Canada.

Marino, D. (1997) *Wild Garden: Art, Education, and the Culture of Resistance*. Toronto, ON: Between The Lines.

Monbiot, G. (2008) *Bring on the Apocalypse: Essays on Self-Destruction*. Toronto, ON: Anchor Canada.

Nabarro, D. (2010) 'Editorial Comment: Helping the Vulnerable', *The Journal of Nutrition*, 140(1), pp. 136S–137S.

NFB. (2019) National Film Board of Canada [Online]. Available at: www.nfb.ca/ (Accessed: 16 March 2019).

Noddings, N. (2002) *Educating Moral People: A Caring Alternative to Character Education*. New York, NY: Teachers College Press.

Ontario Heritage Trust. (2017) Doors Open Toronto [Online]. Available at: www.toronto.ca/explore-enjoy/festivals-events/doors-open-toronto/ (Accessed: 16 March 2019).

Ontario Provincial Parks. (2019a) 'Killarney', *Ontario Provincial Parks* [Online]. Available at: www.ontarioparks.com/park/killarney (Accessed: 16 March 2019).

Ontario Provincial Parks. (2019b) 'Ontario Parks', *Ontario Provincial Parks* [Online]. Available at: www.ontarioparks.com/en (Accessed: 16 March 2019).

Parks Canada. (2019) 'Georgian Bay Islands National Park' *Parks Canada* [Online]. Available at: www.pc.gc.ca/en/pn-np/on/georg (Accessed: 16 March 2019).

Perlin, R. (2012) *Intern Nation: Earning Nothing and Learning Little in the Brave New Economy*, Updated paperback ed. London; New York: Verso.

Petition Online Canada. (2019) 'Petition Online Canada', *Activism—Freedom in Sharing*, [Online]. Available at: www.activism.com/en_CA/petitiononlinecanada.com (Accessed: 16 March 2019).

Picchi, A. (2016) 'Young adults living with their parents hits a 75-year high', CBC News, 21 December. [Online]. Available at: www.cbsnews.com/news/percentage-of-young-americans-living-with-their-parents-is-40-percent-a-75-year-high/ (Accessed: 16 March 2019).

Planet In Focus. (2019) Planet In Focus Environmental Film Festival [Online]. Available at: http://planetinfocus.org/ (Accessed: 16 March 2019).

Rand, T. (2014) *Waking the Frog: Solutions for Our Climate Change Paralysis*. Toronto, ON: ECW Press.

Regent Park Film Festival. (2019) Regent Park Film Festival [Online]. Available at: http://regentparkfilmfestival.com/ (Accessed: 16 March 2019).

Roberts, J.T., Parks, B. (2006) *A Climate of Injustice: Global Inequality, North-South Politics, and Climate Policy*. MIT Press.

Russell, C.L. (1999) 'Problematizing Nature Experience' in Environmental Education: The Interrelationship of Experience and Story', *Journal of Experiential Education*, 22(3), pp. 123–137. [Online]. Available at: https://doi.org/10.1177/105382599902200304 (Accessed: 16 March 2019).

Schwartz, E.G. (1999) 'Exploring Children's Picture Books through Ecofeminist Literacy', in: Smith, G.A., Williams, D.R. (Eds.), *Ecological Education in Action: On Weaving Education, Culture and the Environment*. Albany, NY: State University of New York Press.

Seidel, S., Tishman, S., Winner, E., Hetland, L., Palmer, P. (n.d.) *Qualities of Quality: Excellence in Arts Education*, Cambridge, MA: Wallace Foundation and Arts Education Partnership, Project Zero Harvard Graduate School Education, [Online]. Available at: www.wallacefoundation.org/knowledge-center/Documents/Understanding-Excellence-in-Arts-Education.pdf (Accessed: 16 March 2019).

Senge, P.M., Sterman, J.D. (1992) 'Systems thinking and organizational learning: Acting locally and thinking globally in the organization of the future', *European Journal of Operational Research*, 59(1), pp. 137–150. [Online]. Available at: https://doi.org/10.1016/0377-2217(92)90011-W (Accessed: 16 March 2019)

Sierzputowski, K. (2016) 'Ai Weiwei Wraps the Columns of Berlin's Konzerthaus with 14,000 Salvaged Refugee Life Vests', *Colossal*, 16 February. [Online]. Available at: www.thisiscolossal.com/2016/02/ai-weiwei-konzerthaus-refugee-life-vests/ (Accessed: 16 March 2019).

Solnit, R. (2014) 'Call climate change what it is: violence', *The Guardian*, 7 April. [Online]. Available at: www.theguardian.com/commentisfree/2014/apr/07/climate-change-violence-occupy-earth (Accessed: 16 March 2019).

Sovacool, B.K., Sidortsov, R.V., Jones, B.R. (2014) *Energy Security, Equality and Justice*. Abingdon, Oxon, UK: Routledge, Taylor & Francis Group.

The Spruce. (2019) 'Free Birdhouse Plans—A List of the Best Free Bird House Plans Online', *The Spruce*, [Online]. Available at: www.thebalance.com/free-birdhouse-plans-1357100 (Accessed: 16 March 2019).

TIFF. (2019) *TIFF* [Online]. Available at: www.tiff.net/tiff/ (Accessed: 16 March 2019).

Toronto and Region Conservation for the Living City. (2016) 'Hike Our Natural Woodland Trails', Toronto Region and Conservation Authority, [Online]. Available at: https://trca.ca/activities/hiking/?gclid=Cj0KEQjw4cLKBRCZmNTvyovvj-4BEiQAl_sgQpCD3K3_wYOwO6_6uq2sy1gEsyrPCm6sQc7qt8UFKPAaAiE-8P8HAQ (Accessed: 16 March 2019).

Toronto District School Board. (2019) 'Eco Schools', Toronto District School Board, [Online]. Available at: www.tdsb.on.ca/ecoschools/Home.aspx (Accessed: 16 March 2019).

Toronto Humane Society. (2019) 'Adopt a Pet', Toronto Humane Society, [Online]. Available at: www.torontohumanesociety.com/adopt-a-pet (Accessed: 16 March 2019).

Tripadvisor Canada. (2019) 'Walking Tours in Toronto', Tripadvisor Canada [Online]. Available at: www.tripadvisor.ca/Attraction_Products-g155019-zfc12046-zfg11876-Toronto_Ontario.html (Accessed: 16 March 2019).

Turcotte, M. (2011) 'Commuting to work: Results of the 2010 general social survey' (Statistics No. 92), Canadian Social Trends, 24 August. [Online]. Available at: www.fonvca.org/agendas/sep2011/extras/statcan-commuting-to-work.pdf (Accessed: 16 March 2019).

Walker Leslie, C., Roth, C.E. (2003) *Keeping a Nature Journal: Discover a Whole New Way of Seeing the World Around You*. 2nd edn. North Adams, MA: Storey Publishing.

White, G. (2011) *Climate Change and Migration: Security and Borders in a Warming World*. Oxford; New York: Oxford University Press.

Young, K. (2009) 'Understanding Online Gaming Addiction and Treatment Issues for Adolescents', *The American Journal of Family Therapy*, 37(5), pp. 355–372. [Online]. Available at: https://doi.org/10.1080/01926180902942191 (Accessed: 16 March 2019).

19

RECONCILIATION IN THE WATERSHED

Strengthening relationships for climate justice

Elizabeth Lorimer

Wherever we live, we are part of a watershed – an interdependent ecosystem nested in a larger ecosystem, which is also a watershed. We all have a relationship with the bodies of water that sustain our lives and as living parts of a watershed we are affected when droughts, floods, and other extreme weather events impact our food, shelter, and livelihoods. This watershed perspective helps people make sense of the links between local and global weather, economies, and environmental change, and how people can work together for climate justice.

The term *watershed* can be understood in three different ways. First, as an ecosystem or area of land that collects precipitation and drains it through a network of streams and rivers into a common body of water (WWF, 2017). The word is used interchangeably with drainage basin, basin, or catchment area. Second, it can be understood as a basin of relationships between species and not just between humans and other living species, but also between humans themselves. How strong are our relationships with other humans and other communities within our local watersheds? What is the state of relations between Indigenous and non-Indigenous peoples in Canada and the US? How do we treat migrants and other vulnerable communities? Permaculturist Brock Dolman suggests that we perceive our watersheds as lifeboats and as threats arise, communities within that watershed lifeboat must work together to protect it. (Dolman & Lundquist, 2018). The third interpretation of *watershed* is more abstract. Watershed can be used to describe a historic moment or turning point, from which things will never be the same. The current climate crisis is a watershed moment for humanity. We have been presented with a turning point in which we can either take action or maintain the status quo and watch the destruction of our planet.

The impacts of climate change such as drought, floods, and extreme weather events and their impacts on food, habitat, and livelihoods are also shaped and defined by watersheds. These impacts are felt within the bioregional area of a watershed and therefore, building an understanding of this space and reconnecting to it in our daily lives is critical for our resilience to climate change and in transitioning the economy toward clean energy.

Watershed and identity

Despite these integral connections, we often do not think about our lives as they relate to the watershed. The watershed as a geographic location of origin does not typically shape our identities. Most people would identify themselves rather by political boundaries. For example, we say we are from a municipality, province, state, or country. These boundaries, however, are not connected to the natural flow of water across the land. In fact, in some cases, political boundaries have been drawn right down the middle of rivers or lakes, separating communities that are supported by that body of water. The Great Lakes provide a clear example of this division, separating Canada and the US and a number of states and provinces, but there are hundreds of other transboundary waterways like it throughout the world.

Maps are a great resource for learning about the world and where we live. They help shape our identity with the land. However, many of the maps that we use today are colonial artifacts and represent a limited way of seeing the world. Maps tend to simplify the history of the land. A typical political map of continental North America will use bright colours to illustrate the boundaries of provinces, territories, and states in Canada, the US, and Mexico (see Figure 19.1). The maps are often simplified and many of the boundaries separating one jurisdiction from the next are straight lines.

In contrast, a watershed map of continental North America (see Figure 19.2) shows larger variegated shapes, which follow the contours of the landscape – coasts, mountain ranges, tundra, arctic, prairies, and desert. These shapes are North America's oceanic watersheds and through them flow rivers and streams that pause in lakes or flow into larger rivers before pouring into the oceans that surround the continent. There are hundreds of smaller watersheds within these oceanic systems that feed and sustain life on the continent. In North America there are 20 major river basins and sub-basins with hundreds of smaller watersheds nested within them (CEC, 2010).

In North America, the health of these watersheds is threatened by a number of factors including pollution, alteration of flows (i.e., dams), resource extraction, agriculture, habitat loss, etc. As of 2017, WWF-Canada estimates that 53 of the 167 sub-watersheds are experiencing high or very high levels of stress (2017, p. 9). In the US, the Nature Conservancy estimates that 40–50% of waters are impaired or threatened. Climate change is considered one of these threats – with an estimated 21 sub-watersheds in Canada experiencing high impacts of climate change and a further 105 that are moderately impacted (WWF-Canada, 2017, p. 12). While climate change is on its own a stressor to watersheds, its impacts,

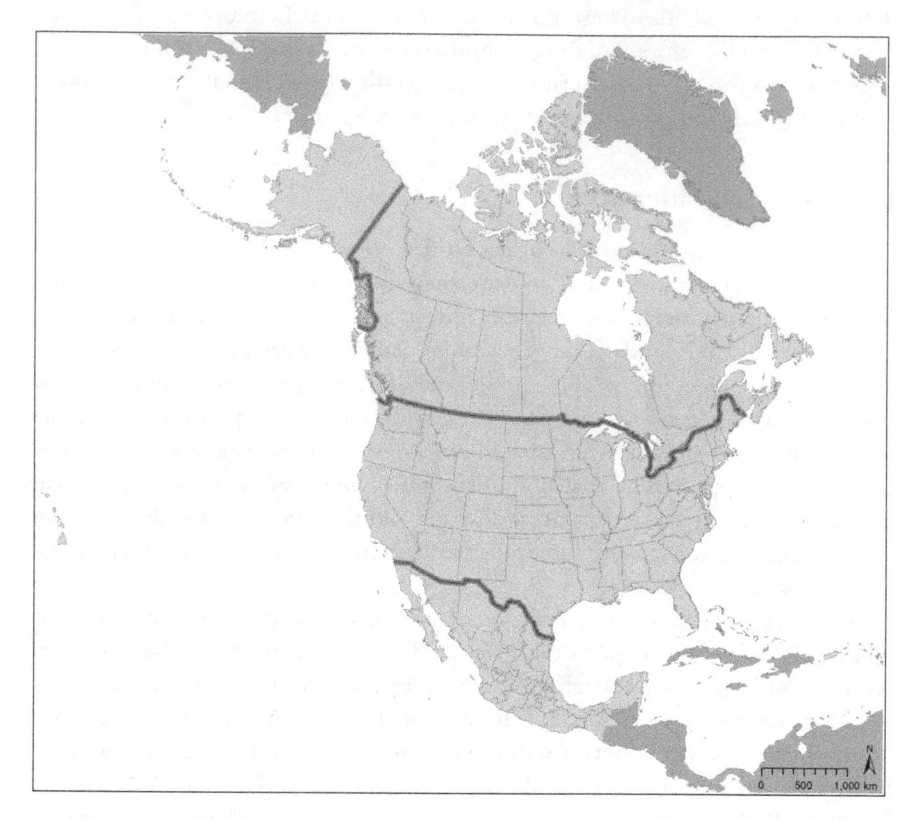

FIGURE 19.1 Political map of North America. The three countries located in North America, and many of the states and provinces within them, are often divided from each other by straight lines or along rivers. Map credit: CEC 2010, North American Environmental Atlas: Political Boundaries. Montreal, Canada: Commission for Environmental Cooperation. http://www.cec.org/tools-and-resources/north-american-environmental-atlas

such as glacier and ice cover loss, increased temperatures, drought, and forest fires also make watersheds less resilient and undermine their ability to mitigate the impacts of other threats.

The state of North America's watersheds is a reflection of our current development model, which is driven by extraction and commodification of natural resources. This model disconnects us from the watershed by placing value on its underlying fossil fuel and mineral deposits rather than its interconnected life-giving properties. William Jennings explains, "We have been taught to see the earth as commodified, territorialized, nationally bounded spaces. In fact it exists as a whole space joined together by water, dirt, sky, animal, plant, and microbe" (2017). Our identity with land and water has been shaped by what we have

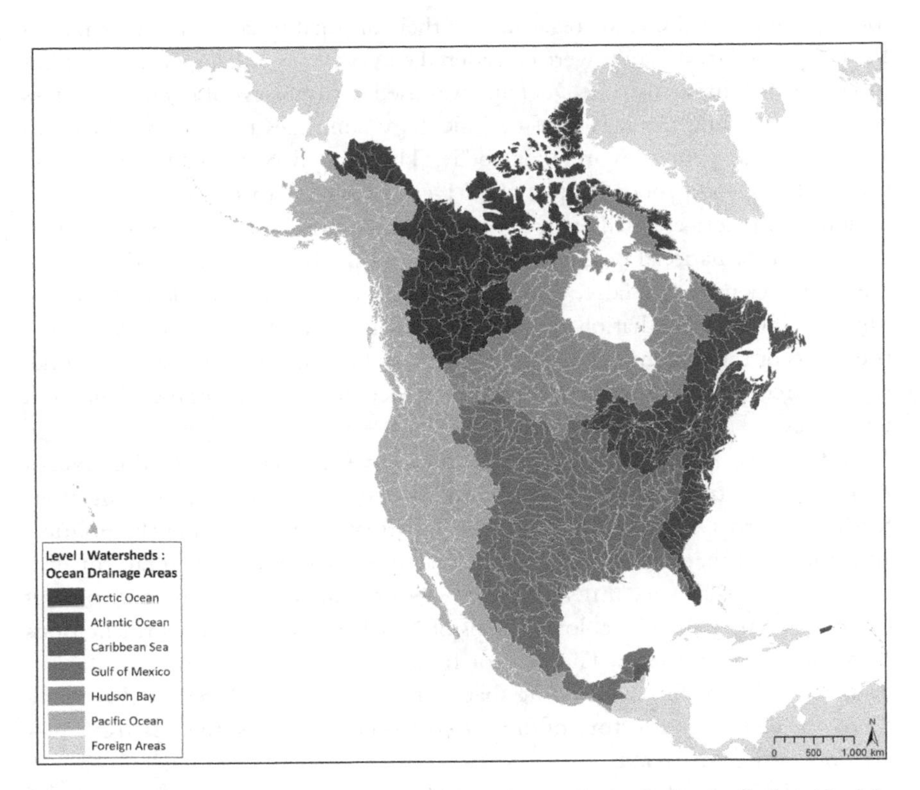

FIGURE 19.2 Oceanic watersheds of North America. Watersheds are distinguished by heights of land, since water flows downhill. They show the elevations and three-dimensionality of the landscape. Map credit: CEC 2010, *North American Environmental Atlas: Watersheds*. Montreal, Canada: Commission for Environmental Cooperation. http://www.cec.org/tools-and-resources/north-american-environmental-atlas

been taught but these teachings do not reflect the natural relationship between humans and this life-giving element.

Decolonizing the watershed

The current development model and the way in which we identify with and understand water have also been shaped by colonialism. Colonial powers created jurisdictions that ignored watershed boundaries. The political boundaries that were imposed on the landscape by colonial powers also ignored the way of life of Indigenous peoples, whose languages, cultures, and traditional practices are dependent on healthy lands and waters.

The maps that were drawn at the time of European arrival reflect the view held by colonizers about "discovery." In the 15th century, Papal Bulls established the Doctrine of Discovery, which provided legal and moral justification for explorers to claim lands regardless of their original inhabitants (Assembly of First Nations, 2018). Lands were considered empty or *terra nullius* if they were not inhabited by Christians. The Doctrine was used to dispossess Indigenous peoples of their rights and the legacy of this ideology continues to have ramifications for Indigenous peoples worldwide today. The United Nations Declaration on the Rights of Indigenous Peoples (UN Declaration) affirms that "all doctrines, policies and practices based on or advocating superiority of peoples or individuals on the basis of national origin or racial, religious, ethnic or cultural differences are racist, scientifically false, legally invalid, morally condemnable and socially unjust" (2007). Repudiation of these doctrines is foundational to recognizing the rights of Indigenous peoples, such as the right to self-determination and the right to free, prior and informed consent on decisions that affect their lands and territories.

Indigenous peoples understand that they have inhabited the land of North America since time immemorial. Figure 19.3 shows the extent of Indigenous territories across North America today. This map is much unlike the political and watershed maps discussed above. Created by Native Land Digital, this living map provides a very different view of North America than the one most of us are familiar with. The colonial maps of North America were representations of ownership and power. The map of Indigenous territories shows a different dynamic – the various overlapping shapes show diversity and suggest that while there was inevitably a history of disputes between territories, there is also a history of sharing and treaty.

Colonial maps effectively erased Indigenous presence from the land and neglected the traditional Indigenous knowledge about the watersheds that sustained life for thousands of years (Johnson, n.d.). As part of this process, many Indigenous names for locations and bodies of water were also rejected and replaced with European names. In some instances, the Indigenous name was kept but anglicized (or gallicized) to make it easier to pronounce by settlers.

Indigenous place names are intricately connected with Indigenous history, stories, and teachings and can carry valuable ecological knowledge and navigational information (Indigenous Corporate Training Inc., 2016). Some Indigenous names are mnemonic devices, in that they are meant to remind us of a place based on its specific characteristics, which can then be passed down orally throughout society to build common understanding of a place. For example, the Great Lakes are known in Anishinaabemowin as *Nayaano-nibiimaang Gichigamiin*, meaning the "Five Freshwater Seas." The Anishinaabemowin name for Georgian Bay is *Waaseyaagami-wiikwed*, meaning "Shining Waters Bay," and Chicago is an anglicized version of its Anishinaabemowin name *Gaa-zhigaagowanzhigokaag*, meaning "At the Place Abundant with Skunk-grass" (Lippert & Engel, 2015). These names impart specific characteristics of these places. Much of this understanding and identity of place has been lost today, however there are movements to reclaim Indigenous names across the continent. The official renaming of the

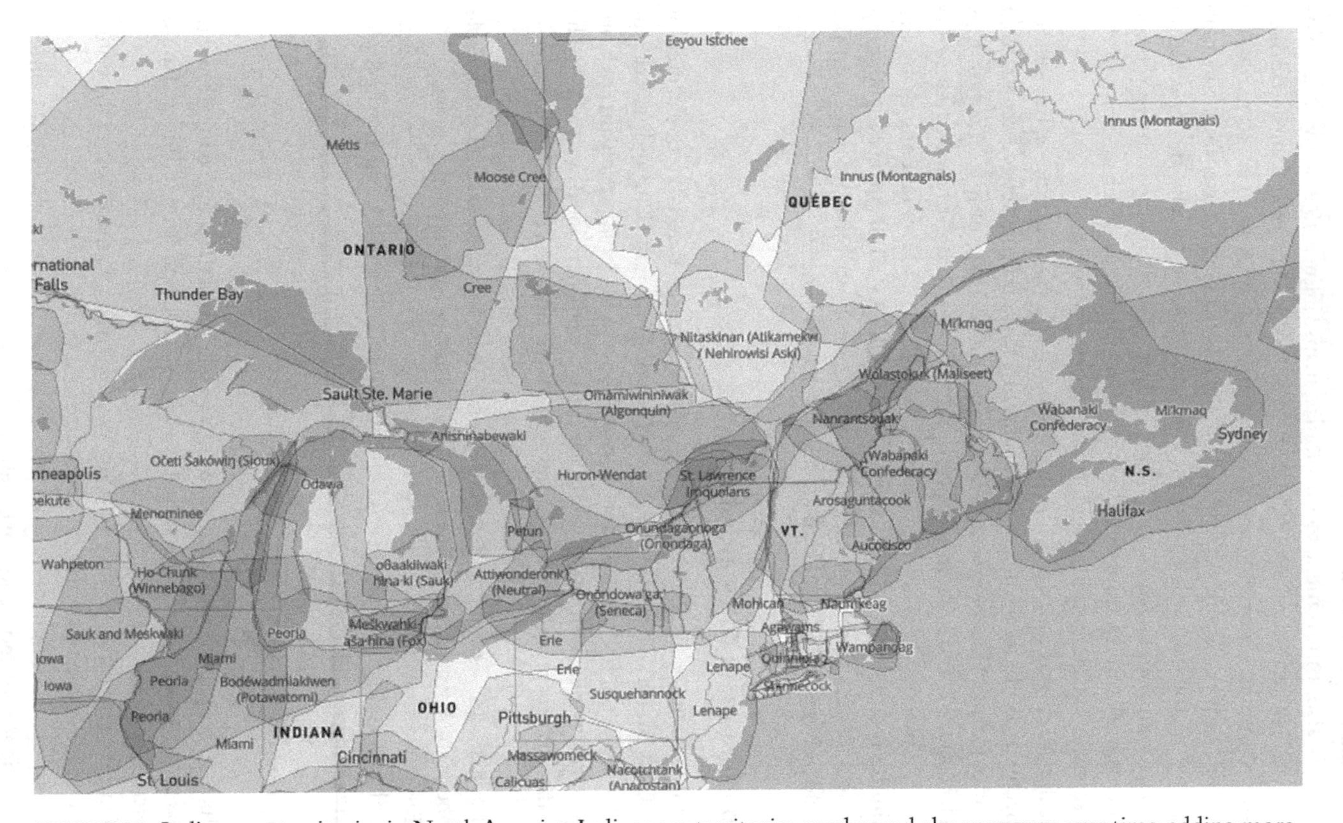

FIGURE 19.3 Indigenous territories in North America. Indigenous territories overlap and also may vary over time, adding more dimensions and relationships. Map credit: Native Land Digital, https://native-land.ca/

Queen Charlotte Islands in British Columbia to *Haida Gwaii* (meaning "Islands of the Haida people") is one example.

There is also a movement to decolonize maps. Figure 19.4 shows a decolonized map of the Great Lakes. Created by Charles Lippert and Jordan Engel, this map shows an Ojibwe perspective of the region with Indigenous names for places and bodies of water. At first glance, the map can be disorienting. The map is oriented to the east rather than the north, which is typically the orientation of colonial maps; and a Medicine Wheel has replaced the compass rose. Yet, the map offers just as much understanding of the region as any other map and in this case, captures Indigenous knowledge and presence in the region.

At the heart of reconciliation is decolonization: a re-centering of Indigenous perspectives in all areas of society, including the environment. Activist Syed Hussan explains, "Decolonization is a dramatic re-imagining of relationships with land, people and the states. Much of this requires study, it requires conversation, it is a practice, it is an unlearning" (Walia, 2012). Decolonizing maps and reclaiming Indigenous names is one way in which this can happen. Repudiating racist policies like the Doctrine of Discovery, which still manifests itself in the laws and policies of today in Canada and the US, is another.

Reconciliation in the watershed

It is this process and conversation of decolonization that is at the center of KAIROS Canada's *Reconciliation in the Watershed* program. Through a series of workshops hosted by local partners and Indigenous leaders, participants explore perspectives of decolonization while learning about their local watershed. The program and accompanying workshop series have two overarching objectives:

- Increase the number and diversity of Canadians who are knowledgeable about their watershed and able to identify issues related to its protection.
- Build relationships with Indigenous peoples and movements while highlighting Indigenous perspectives on water and the work Indigenous peoples are doing to prevent watershed destruction.

Each *Reconciliation in the Watershed* workshop:

- Introduces participants to their local watershed and the major issue related to its protection (for example, resource extraction, pollution, climate change, etc.).
- Equips participants to take part in decision-making and advocate for policy changes related to their local watershed and Indigenous rights.
- Incorporates Indigenous knowledge and perspectives, including the unique experiences of Indigenous women.
- Connects people with their local watershed through recreation, ceremony, and active reflection.

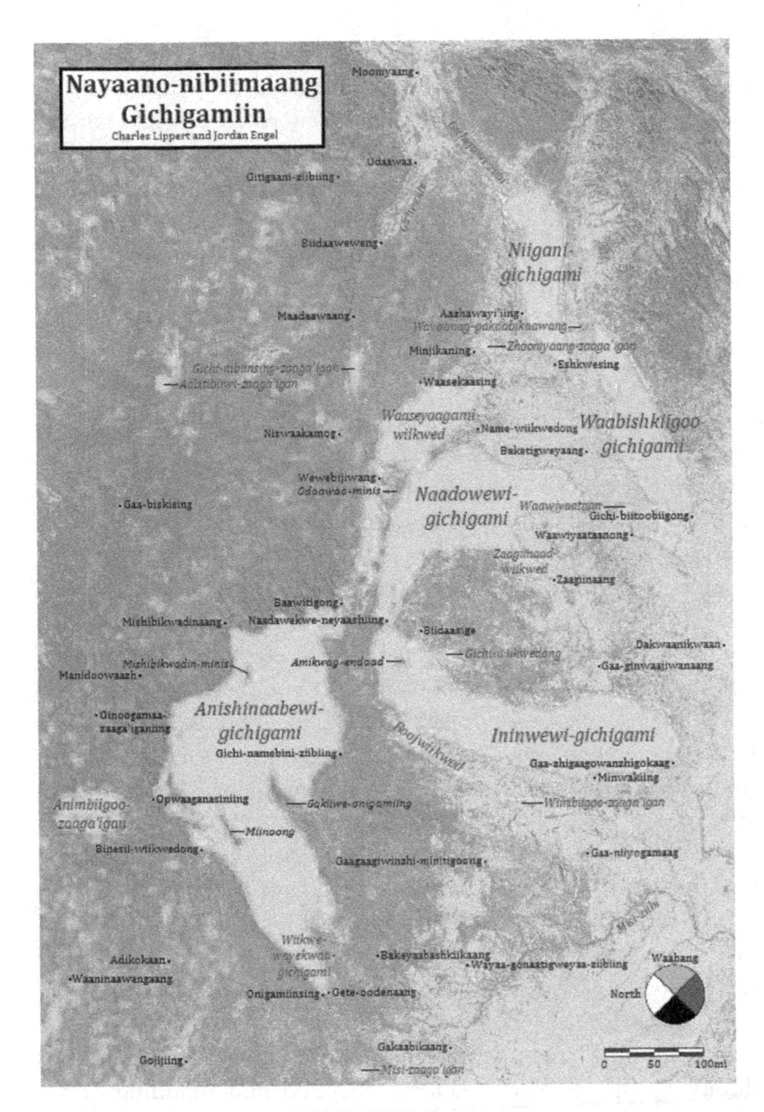

FIGURE 19.4 Ojibwe map of the Great Lakes. The Anishinaabe or Ojibwe traditionally orient themselves to the East, where the sun rises. Their language, *Anishinaabemowin*, is the most-spoken Indigenous language in the Great Lakes basin. The names of the five lakes in *Anishinaabemowin* are *Anishinaabewi-gichigami*: Lake Superior (Anishinaabe's Sea); *Ininwewi-gichigami*: Lake Michigan (Illinois' Sea); *Naadowewi-gichigami*: Lake Huron (Iroquois' Sea), also known as *Gichi-aazhoogami-gichigami* (Great Crosswaters Sea); *Waabishkiigoo-gichigami*: Lake Erie (Neutral's Sea), also known as *Aanikegamaa-gichigami* (Chain of Lakes Sea); and *Niigaani-gichigami*: Lake Ontario (Leading Sea), also known as *Gichi-zaaga'igan* (Big Lake). Map credit: Lippert and Engel, 2015; *The Decolonial Atlas*. https://decolonialatlas.wordpress.com/2015/04/14/the-great-lakes-in-ojibwe-v2/

The *Reconciliation in the Watershed* program builds on years of KAIROS' watershed education and Indigenous rights work, but is also influenced by the Watershed Discipleship movement. Coined by Ched Myers, Watershed Discipleship is a framework for ecological theology and practice, which recognizes that the world is at a critical (or watershed) moment of crisis and that this moment demands us to live out principles of social justice and sustainability in our local watersheds (2016). This theology acknowledges the role that colonialism has played in shaping the prevailing extractivist worldview, which has led to the climate crisis and has damaged relationships between Indigenous and non-Indigenous peoples. Myers (2016) explains,

> since the late medieval Doctrine of Discovery, a theology and/or politics of *entitlement* to land and resource – both in the colonizing and extractive sense – gave *carte blanche* first to imperial expansion and conquest, then to capitalist production and consumption without limits.

This has led to our own displacement and a breakdown in understanding that our local watersheds are part of our core identity. Followers of watershed discipleship contend that we need to re-place ourselves in our local watersheds in order to repair this relationship. Reconnecting people with their local watershed is a core objective of the *Reconciliation in the Watershed* program. Rebuilding this connection, through ceremony, recreation, and reflection, is intended to stimulate learning, and affection and ultimately a commitment to engage in actions to protect the water.

KAIROS launched the *Reconciliation in the Watershed* workshop series in 2017. Workshops in the wider Great Lakes watershed have addressed a number of issues, including groundwater depletion, industrial and agricultural pollution, impacts of urban sprawl, and more. Informed by local ecological and Indigenous rights issues, each workshop is unique and tailored to the local experience.

Similar to the beginning of this chapter, the workshop walks participants through the different maps of North America to build understanding of the colonial impacts on how we view the land and to introduce new ways of looking at the land through Indigenous perspective. One workshop participant summed up the exercise as follows,

> When viewing these five maps side by side one can see how the standard map of Canada is mostly disconnected from the landscape and waterways and disregards Indigenous sovereignty. Viewing the maps, and considering the new information they provide, left me with new questions. What would a new relationship between the Indigenous and settler population look like? How can I live a life in closer harmony to creation and daily give thanks for the gifts provided to me? What is my watershed and how can I connect to it in a deeper way?

Another core objective of the workshops is to build relationships between Indigenous and non-Indigenous peoples and one of the ways this is done is by highlighting the knowledge and leadership of Indigenous peoples, in particular Indigenous women. Water has significant cultural importance to Indigenous communities in Canada and Indigenous women share a sacred connection with water through their role as child bearers (Cave, 2016). They have particular responsibilities to protect and nurture water and for this reason, Indigenous women in Canada and around the world have been at the forefront of watershed protection and the Indigenous-led climate movement. Indigenous women have been largely left out of watershed governance and resource decisions, so showcasing their unique knowledge in the workshops has been a way of centering their perspective in the conversation.

Each component of the workshop focuses on some aspect of ceremony, learning, relationship-building, and action. Participants reflect on the role of the watershed in their identities and daily lives, the history of the watershed and its First Peoples, and the legacy of colonialism and industrialization on the watershed and on relationships between Indigenous peoples and settlers.

Ceremony is a key component and theologian Denise Nadeau (2016) explains its value as follows,

> The Anishinaabe way of life is centered on relationships, each of which carry responsibilities. The function of ceremony is to remind us about and to restore those relationships. This is why ceremony is critical to the work of protecting water and watersheds. If people feel a relational connection to the watershed in which they live, it is easier for them to act in an embodied way upon their responsibilities.

The workshop format is a living process. Each community is at a different stage in their understanding of their watershed and the issues related to their protection and each individual is at a different stage in their understanding of reconciliation and what it means to decolonize. As such, each community shapes the tone of the workshop and identifies what steps they are ready to take together to reconnect with water and repair relations in their local watershed.

Conclusion

The application of the *Reconciliation in the Watershed* program to the climate justice movement may not seem readily clear. Learning about your local watershed may seem elementary when one considers the stage we have reached in the climate crisis. Yet when considering this learning in the context of reconciliation and in rebuilding relationships between Indigenous and non-Indigenous peoples, its value becomes more evident. Climate justice requires human-centered approaches, which this learning process provides. Actions to address climate change and transition the economy toward clean energy will be informed by science but they must also be informed by other ways of knowing and living in relation with one another, as well. *Reconciliation in the Watershed* provides a platform to engage in this conversation, to shift our perspective, and to apply this learning in how we take action to protect life on earth.

References

Assembly of First Nations. (2018). *Dismantling the Doctrine of Discovery.* Ottawa: Assembly of First Nations.

Boucek, E. (2017 27-10). *Water flows, connects, and knows no borders.* Retrieved 2018 20-08 from KAIROS Canada: https://www.kairoscanada.org/water-flows-connects-knows-no-borders

Cave, K. (2016). Water Song: Indigenous Women and Water. *The Solutions Journal* , 7 (6), 64–73.

CEC. (2010 27-05). *North American Watersheds.* Retrieved 2019 25-02 from Commission for Environmental Cooperation: http://www.cec.org/content/newsletter-may-27-2010

Dolman, B., & Lundquist, K. (2018). *Basins of Relations – A Citizen's Guide to Protecting and Restoring Our Watersheds* (3rd ed.). Occidental, CA: Occidental Arts & Ecology Center.

Henry, J. (n.d.). *A Reflection on the United Nations Declaration on the Rights of Indigenous Peoples.* Retrieved 2019 24-02 from KAIROS Canada: https://www.kairoscanada.org/what-we-do/indigenous-rights/undrip-reflection

Indigenous Corporate Training Inc. (2016 16-02). *The Relationship between Indigenous Peoples and Place Names.* Retrieved 2018 24-02 from Working Effectively with Indigenous Peoples: https://www.ictinc.ca/blog/the-relationship-between-indigenous-peoples-and-place-names

Jennings, W. J. (2017 20-06). *July 23, Ordinary 16A (Genesis 28:10-19a).* Retrieved 2019 07-08 from The Christian Century: https://www.christiancentury.org/article/july-23-ordinary-16a-genesis-2810-19a

Johnson, S. (n.d.). *Teacher's Guide.* Retrieved 2019 26-02 from Native Land: https://native-land.ca/teachers-guide/

Lippert, C., & Engel, J. (2015 14-04). *The Great Lakes: An Ojibwe Perspective.* Retrieved 2019 24-02 from Decolonial Atlas: https://decolonialatlas.wordpress.com/2015/04/14/the-great-lakes-in-ojibwe-v2/

Myers, C. (2016). A Critical, Contextual, and Constructive Approach to Ecological Theology and Practice. In C. Myers, *Watershed Discipleship – Reinhabiting Bioregional Faith and Practice.* Eugene, OR: Wipf and Stock.

Nadeau, D. (2016). Foreword. In C. Myers, *Watershed Discipleship – Reinhabiting Bioregional Faith and Practice.* Eugene, OR: Wipf and Stock.

Native Land Digital. (2019). *Native Land.* Retrieved 2019 24-02 from https://native-land.ca/

Native Land Digital. (n.d.). *Teacher's Guide.* Retrieved 2019 26-02, from Native Land : https://native-land.ca/teachers-guide/

United Nations. (2007). *United Nations Declaration on the Rights of Indigenous Peoples.* New York: Department of Economic and Social Affairs, United Nations.

Walia, H. (2012 01-01). *Decolonizing Together – Moving Beyond a Politics of Solidarity Towards a Practice of Decolonization.* Retrieved 2018 24-02 from Briarpatch Magazine: https://briarpatchmagazine.com/articles/view/decolonizing-together

WWF. (2017). *Watershed 101.* Retrieved 2019 25-02 from WWF Watershed Reports: http://watershedreports.wwf.ca/#canada/by/threat-overall/profile/?page=watersheds

WWF-Canada. (2017). *A National Assessment of Canada's Freswhater – Watershed Reports.* Toronto: WWF-Canada.

20

CLIMATE JUSTICE MONTREAL

Who we are and what we do

Jen Gobby

March 15, 2019, the day of the worldwide Students' Climate Strike, was the biggest day of global climate action ever. One and a half million young people walked out of school and into the streets in over 2,000 places, in 125 countries on all continents, to demand real action on climate change.

On that day I marched with 150,000 other students through early spring streets in Montreal. I saw the joyful faces of a whole new generation of activists. For many of them, this was their first protest – their first taste of the thrill and power of collective action, of how good it feels to care deeply about something with thousands of other human beings. I also saw many familiar faces—many passionate, kind and hard-working climate activists I've come to know through my organizing work with Climate Justice Montreal (CJM) over the last five years.

CJM is a grassroots group pursuing environmental and climate justice through education, mobilization, and direct action in solidarity with frontline communities. We live and meet on unceded[1] Kanien'kehá:ka (Mohawk) territory on Tiotia:ke, the island called Montreal. In our organizing, we see the issues around climate justice as inherently linked to larger questions of capitalism and colonialism. Through our content and the way that we organize, we try to address and highlight these injustices. Our initiatives strive to take direction from and amplify the voices of those most affected by systemic oppression due to race, ethnicity, class, gender, age, sexuality, ability, illness, and mental health.

The climate justice movement in Canada has been recently reinvigorated by Powershift: Young and Rising. This conference, held most recently in Ottawa

1 "Unceded" means that a First Nation has not relinquished title to its land to the government, either by treaty or otherwise. See https://en.wikipedia.org/wiki/Wiikwemkoong_First_Nation; www.mironline.ca/kanienkehaka-sovereignty/ .

in February 2019, convened "hundreds of young people from across this land to build a powerful and intersectional youth climate justice movement" (Powershift 2019a, n.p.). I was there, along with other current and former CJM members, leading workshops on direct action techniques, movement theory, and climate justice organizing. I led two workshops, but spent the weekend feeling a powerful sense of having much more to learn than I had to teach. I was floored by the powerful words of keynote speakers such as Harsh Walia, Eriel Deranger, and Kanahus Manuel, madly scribbling notes as I listened, as they profoundly deepened my understandings of what is at stake, of what it will take to win. It was my first Powershift.

History of CJM

Climate Justice Montreal was formed at the first Power Shift Canada conference in 2009 in the lead-up to the Copenhagen climate negotiations. This was the first national gathering of Canadian youth climate change activists. After three days of workshops, discussions, and protests outside and inside Parliament in Ottawa, the Montreal contingent decided it was time to take action in Quebec. The vision for forming CJM was to start working together to build a bigger, better, stronger climate movement here in Montreal. After a Climate Camp to stop the construction of a tar sands oil pipeline pumping station in Dunham, Quebec, CJM lay dormant until it was rekindled at Powershift 2012. Since then, CJM has been ebbing and flowing, with a slow but constant turnover of members, but has managed to continue organizing. Much of our work has focused on anti-pipeline activism. We've been actively involved in opposing Kinder Morgan, Energy East, Line 9, and other proposed oil pipelines. Through all the work we do, we promote a decolonial approach to climate action in the mainstream environmental movement in Quebec. What we happen to be doing at any given moment in time depends on which members take initiative and what they are inspired to do. Sometimes the work we do is proactive, where we come up with an idea and make it happen. Other times we are more responsive, taking action to support projects and initiatives led by other groups and communities.

CJM structure and decision-making

We are a non-hierarchical collective, inspired and guided by the Jemez Principles for Democratic Organizing (Solis & Union, 1997). We make decisions through bi-weekly meetings, and rotate roles and responsibilities to distribute power between members, redistributing leadership roles to everyone in the group. In experimenting with new campaigns and actions and continuing a process of self-education and workshops, we aim to continuously increase our collective capacity to mobilize. The group stays in contact through an active-members listserv. We have a strong archive of past activities in our google docs folder, which is accessible to all members.

CJM makes decisions by consensus. Consensus means that a proposal does not pass until concerns about it expressed by meeting participants have been addressed to the satisfaction of all those present. We have chosen this approach because it helps ensure that a decision reflects the input of everyone present. It nurtures a spirit of openness, collaboration, and compromise. It helps us address power relationships within the group, with a view to equalize them. Consensus helps harness the creativity and lived experience of all participants. However, we understand that under some circumstances it is not possible to achieve consensus, but a decision needs nonetheless to be made. In extreme instances, the group can decide (with minimum 2/3 support from those present) to submit a proposal to a majority vote, in which case a super-duper-majority of 75% will be required for the proposal to pass. Before a decision is passed by vote, time will be allotted for discussion and dissenting views; abstentions and resignations from the group will also be recorded at this time. We feel that it's important to practice consensus, because as a non-hierarchical form of decision-making it prefigures alternative governance structures that embody the world we want to create; ones not based on systems of domination that prevail in current mainstream governance, but ones based on radical equality, inclusion, where everyone's voices and insights are valued. We continually work to undo the invisible hierarchies when they begin to develop amongst us.

CJM started out as a working group of the university-based organization Sustainable Concordia (SC). We had meetings in their offices and received funding through them, usually between $1,000 and $2,000 a year. When SC restructured in 2017 and no longer had working groups, CJM went out on our own and began to apply and receive money from other funding sources such as LUSH Cosmetics. The work we do is all volunteer based, with none of us getting paid for our work. The funds we manage to scrape together help us pay for meeting space, cover expenses for events and actions, and make direct donations to frontline Indigenous land defenders.

Climate Justice Montreal's Guiding Principles

Over the first few years of CJM, members worked together to hash out the group's Guiding Principles. Unlike more mainstream environmental groups and organizations in Montreal which focus mostly on environmental issues, CJM holds at the core of our work a commitment to solidarity, decolonization, and anti-oppression.

Solidarity approach

CJM is committed to:

- Establishing relationships with communities directly affected by any issues we take on
- Being transparent about our aims, interests, and decision-making process within those relationships

- Taking direction from affected communities and sharing the burden of their struggle through support work that amplifies their message and empowers the affected community within
- Considering and working for the well-being of affected communities in general
- Being transparent about the criteria we use to determine who we work with in a given community and take responsibility for the effects our interventions have
- being transparent about criteria for creating relationships, who receives CJM money, and how much they receive

Anti-oppression and decolonization

CJM is committed to:

- Understanding systemic oppression based on skin colour, ethnicity, class, gender, age, sexuality, ability, illness and mental health
- Understanding our own role in normalizing this oppression on a day to day level, and taking appropriate action to create equitable relationships
- Understanding our own history and the ways in which we benefit from historical and ongoing land theft
- Confronting said land theft and the systemic devaluing of certain groups of human beings that makes it possible
- Creating inclusive spaces by actively supporting and encouraging equal participation in meetings, roles, tasks, etc.
- Proactively sharing knowledge and skills to facilitate sharing of roles
- Noticing who is excluded from the group and seeking to understand why that is and take appropriate action

The movement ecosystem we are part of

CJM aims to centre the conversation on environmental activism around justice through a radical and anti-capitalist perspective. A big strength of CJM is our ability to help connect groups across the political spectrum in Quebec and across North America as a whole. Being part of this network means that we are well placed to introduce this perspective within initiatives that are already in progress, and gives us the credibility to put forward our own independent messaging through autonomous actions.

Our network provides links to the pulse of a few different worlds. As a bilingual, university-based grassroots group, we are connected to local student contexts, provincial NGOs, local anarchist organizing, and anglophone-Canadian and North American communities—many of whom have limited contact with each other. We are members of The Rising Tide Network, the Reseau Quécbecois des Groupes Écologistes, the Front Commun pour la transition énergetique, and others.

This year, several young organizers joined CJM and launched a Youth Mobilization sub-committee. The CJM Youth Mob are organizing actively as part of the Canada-wide Youth Rising network. As one of their first actions after joining CJM, they planned and carried out a day-long sit-in in Prime Minister Justin Trudeau's constituency office in Montreal to demand that the self-proclaimed "Youth Minister" stop buying pipelines and commit to leaving fossil fuels in the ground. This action was designed to pressure the Prime Minister to live up to his government's commitments to both climate mitigation and reconciliation with Indigenous nations. CJM's Youth Mob helped with planning of the March 15 largest ever climate strike and march in Montreal, and have been working hard to centre Indigenous rights and anti-racism in the action messaging. The older folks at CJM have been putting most of our time and efforts in the last six months into supporting the Youth Mob, providing trainings on direct action and security culture.

What we do: Education, mobilizing and direct action, and direct solidarity

We think of the work we do as spanning three categories of activism: *Education, mobilizing and direct action*, and *direct solidarity*. In our educational work, we host workshops, panel discussions, and other educational events to raise awareness and deepen analysis about climate justice and resistance to fossil fuel expansion. With direct action, we organize and participate in non-violent actions such as marches, demos, sit-ins, blockades, etc., to resist the expansion of fossil fuel projects. In our direct solidarity work, we strive to actively support Indigenous communities, particularly those on the front lines of resource extraction. We do this by amplifying Indigenous folks' voices of resistance, supporting community needs, and raising funds for land defenders.

Our education work over the last 10 years has included organizing a Climate Justice Skills Day, leading direct action trainings for new activists, and planning a series of screenings of the film "Kanehsatake: 270 years of Resistance" in communities around Montreal for the 25th anniversary of the Oka Crisis (a 78-day standoff in 1990 between Indigenous land protectors, the army, and police over threatened condominium and golf course development on an Indigenous burial ground). Last year we hosted a six-week Book Club series to read and discuss together with other Montreal activists Arthur Manuel's most recent book *The Reconciliation Manifesto* (Manuel 2017). We've given many workshops at schools and other organizations on topics such as Climate Justice 101, Climate Justice and Austerity, and Pipelines and Tar Sands. Currently, we are working on developing a Climate Justice podcast, and recently we hosted a symposium on environmental racism and land dispossession in Canada featuring important frontline voices Ellen Gabriel, Vanessa Gray, and Will Prosper as well as leading Canadian scholar of environmental racism Ingrid Waldon.

In 2015, we helped organize several protests against Energy East and Line 9 pipelines. At the National Energy Board (NEB) hearings for the Line 9 pipeline, activists interrupted the undemocratic and exclusive process with a story reflecting the experiences of marginalized communities along the pipeline. We have led sit-ins in MP and PM offices to pressure them to take a stance against fossil fuel development. When the Trudeau Liberals were elected in 2015, we helped out with the Climate Welcome Ottawa action demanding for a freeze of tar sands expansion.

In November 2016, we organized the #NODAPL Solidarity March to support the water protectors at Standing Rock. Nearly 1,000 participants marched together after a rally featuring Indigenous speakers. The march led to a busy downtown intersection where three different Canadian banks who were invested in the DAPL pipeline are located. The end of the march culminated in protestors entering their banks—bank cards in hand—and closing their accounts or threatening to close their accounts if the bank would not divest.

We've also been working in coalition with other groups in Montreal to organize other actions and rallies to pressure banks and other financial institutions such as the Caisse Desjardins and the Caisse de depot et placements du Quebec (CDPQ) to divest from fossil fuels.

FIGURE 20.1 No DAPL photo. CJM helped organize this march through Montreal in solidarity with Standing Rock on November 7, 2016. Over 1,000 people attended. Photo by Jen Gobby (chapter author).

We spend about one third of our annual budget donating money directly to frontline Indigenous land defenders. For example, this year we donated $400 to the Wet'suwet'en First Nation to support their legal defence fund related to arrests for blockading a planned gas pipeline on their territory in northern British Columbia. For several years in a row we helped organize transport from Montreal to Aawmjiwnaang First Nation for the annual Toxic Tour of Canada's Chemical Valley. In 2016, when Anishinaabe land defender Vanessa Gray was facing criminal charges for manually shutting down the Line 9 pipeline, we organized the event "*Defending Land Defenders—Solidarity with Vanessa Gray.*" This event brought together leading legal scholars to conduct a "public think tank" to brainstorm legal defence for Vanessa's case. The transcript of the event was sent to her lawyer to help with her court case. We also raised over $800 that evening to contribute to her legal costs. (Vanessa's charges were later dropped.)

In May 2017, Kanehsatà:ke, a Mohawk community close to Montreal, was hit by very heavy spring flooding. CJM members travelled there to move sandbags and provide other assistance to the community dealing with the rising waters. Kanehsatà:ke has also been facing unwanted housing development on their land in the last few years. In response, in 2018, we hosted a Phone Bank and Email Blitz, facilitating participants to contact the relevant Members of Parliament and their assistants to demand that they place a moratorium on development on Mohawk land until the conflict over land title has been resolved. After the phone bank was over, we went outside and staged this message with our CJM light brigade boards: Stop Oka Land Theft.

FIGURE 20.2 Stop Oka Land Theft. CJM Light Brigade Action on December 14, 2017. The Light Brigade message was assembled and displayed on the streets of Montreal to raise awareness of the ongoing struggle to defend Mohawk land rights and title on Kanehsatà:ke (near Oka, Quebec). Photo by Nicolas Chevalier.

This year we organized several events to help raise money for the "Tar Sands Trial." Beaver Lake Cree Nation is taking on the Alberta tar sands, which have poisoned water, eliminated whole forests, and decimated traditional food sources for the Beaver Lake Cree people. The Beaver Lake Cree Nation is the first ever to challenge and be granted a trial on the cumulative impacts of industrial development.

Currently we are working with organizers in Montreal Nord, a racialized and poor community facing environmental racism and other challenges. We are coordinating support donated by Montreal students to help research key questions that can help the Montreal Nord community address and transform conditions there.

I think all CJM members would agree that solidarity work is the most important work we do. We do this because we understand that the most transformative force for change in Canada is Indigenous resistance and resurgence and that as settlers, often the most powerful use of our time and other resources is in actively supporting Indigenous people and other frontline communities.

CJM's ongoing challenges

We're proud of what we've done and continue to do. But we also struggle with challenges and some fundamental questions on how to work towards climate justice effectively and concretely. Some of these challenges seem to have been with us throughout CJM's ten years.

One is that relationships are hard. And we're not always doing as well as we'd like in forging and taking care of the kind of relationships we know are required to build powerful movements. There are ongoing divisions and tensions between anglophone and francophone activist circles in Montreal. We often are working in isolation from each other. Keeping members is another challenge we face. We have regular membership turnover and sometimes have quite low capacity. This makes it hard to follow through on some of our long-term projects and commitments. It makes it hard to keep track of policies we've made, and build on solidarity relationships that we started. Like all settler groups, we are continually learning about how to be better allies with Indigenous people and communities. And we learn this largely by seeing where we've gone wrong. We debate amongst ourselves about how much to initiate our own projects and how much we should just focus on supporting Indigenous-led initiatives.

There are ongoing debates within our group—but also between CJM and other groups—over strategy. Much of our work has been focused on resisting pipelines. We've done much less work thinking about the alternatives and solutions. But as Naomi Klein made clear in one of her recent books, no is not enough. And we know that. But still we focus our educational and direct-action work on pointing out the problems. We debate the implications of this.

We've also had internal divisions over whether to focus our efforts by doing campaign-style organizing. Many members contend that long-term strategic campaigns are *how change happens*. Yet we don't seem to have a stable enough membership to work on long-term campaigns, and instead seem to inevitably do one-off events. These one-off events are determined, at any given moment, by who is present, who has time and energy, and what they are excited to do. Though it's probably not the most strategic approach, it seems to be how we operate.

Through CJM's history, we've been navigating tensions and conflicts in the Quebec environmental movement. Some environmental organizations in this province still espouse a narrow framing of environmentalism. Many do not centre Indigenous rights or social justice in their conceptions and some even argue that anti-racism, Indigenous rights, feminism, and other social justice struggles bog down the work of environmentalism. We have been told that doing land acknowledgements is "bad communication strategy in Quebec." This is the context in which CJM continues to try to argue that social justice and climate change are inextricable struggles, that you can't win one without the other.

This tension is alive today as Montreal's streets fill with young people mobilizing in unprecedented numbers, rising up and demanding action on climate change. The words "climate justice" were there, hand-painted on placards, and climate justice slogans were being chanted loudly and urgently. But this was also the first march I've taken part in that was not headed by Indigenous people. The front line of the march was white youth. I heard no land acknowledgement. At several points as I marched, I looked around me and all I could see was a sea of white faces chanting "Whose *future?? Our Future!!"* This is not CJM's vision of a decolonial, anti-racist climate movement that we've been trying to promote in Montreal for the last ten years.

This was the most successful climate mobilization in Montreal ever and I want to celebrate that. And I am. But I also feel disappointed at the narrow climate framing that did not centre Indigenous voices and Indigenous rights. And there was no mention of the mosque shooting in New Zealand that also happened on the same day.

I don't want to be one of those endlessly critical lefties for whom no win is ever good enough. At CJM we've discussed again and again the dynamics of the endless co-critiques on the left and the activist culture this creates. But at the same time, we are pretty sure that we will not win the climate fight until we undo systems of oppression between humans and that includes ongoing colonial relations in Canada. While we will not denounce the beautiful rising of youth we saw in Montreal today, CJM stands strong in our understanding that an intersectional approach is the only way we will win. It's the only way to bring about the kind of world we are working for.

Our convictions of a justice-focused, decolonial climate movement were strengthened and reinvigorated by the words of Ellen Gabriel and Vanessa Gray, two powerful Indigenous women leaders who spoke at the Symposium on Environmental Racism and Land Dispossession that CJM helped organize in February 2019.

At this event, Ellen[2] called on a new generation of activists to acknowledge the social injustice, environmental racism, and land dispossession at the heart of Canada's economy and culture, and understand deeply how these systems of oppression relate to climate change. She calls on us to let these connections deeply inform the kind of social change we pursue.

> We have never been more forced into the same melting pot as humans as we are today. As we move farther and farther away from the Original Instructions, that teach us to use the languages that our ancestors used and to nurture the relationship we have with the environment … As we move further away from that we are also moving further away from the solutions to surviving climate change.
>
> If Canada really was a leader in human rights then Indigenous rights would also be respected. We have the right to sovereignty, to protect our lands, our resources, our people, and not to be oppressed anymore. To protect our lands and our people as *we* see fit, not as how the authorities see fit—the authorities who are doing the dirty work of the corporate-controlled politicians and governments that we see today.
>
> We are getting sick from things we cannot see.
>
> For the millions of disenfranchised people around the world, because of the Trumps and Reagans of the world, for the millions, if not billions of people, we are looking for an answer. We are in a revolution. It's not starting a revolution, we are in a revolution. It's time for this revolution to get a little less quiet. To get a bit more rowdy.
>
> There is another generation that is interested in making change … the significant changes that are needed to take us out of this way that we are all living. I am convinced that the generation that is here, and your babies will be the ones who will make the change.
>
> I need you to have the courage and energy to speak out. In a respectful way. To educate your children, your families, your MP, your MNAs. To educate. Because we've been trying to change the curriculum but it's not working because we're Indigenous people, we're the ones who have been dehumanized for centuries and only now being recognized as having human rights.
>
> We are the people who have allowed (you) settlers to survive on this beautiful continent.
>
> As of last year, you are now the owner of the Trans Mountain pipeline. So, what are you going to do to make that change? We have been silent for too long. But I thank you for allowing us to speak now, for listening now.
>
> —*Ellen Gabriel, Tiotia:ke (Montreal), Feb 18, 2019*

2 Ellen Gabriel is a Mohawk activist and artist from Kanehsatà:ke Nation - Turtle Clan. She began her public activism during the 1990 Siege of Kanehsatà:ke (the 1990 "Oka" Crisis), and was chosen by the People of the Longhouse and her community of Kanehsatà:ke to be their spokesperson. Since 1990, Ellen has worked consistently and diligently as a human rights and environmental advocate for the collective and individual rights of Indigenous peoples.

Vanessa Gray[3] spoke at this event with a powerful message that the change that is needed will come when we all fulfil our respective responsibilities, to the earth and to each other.

> There is so much violence that comes with this proud Canadian [identity] that is portrayed over and over again, that is feeding into the system of taking land away, of dispossessing Indigenous people.
>
> The first oil well ever was in my territory, so I feel some responsibility to make those connections of where things started and where things are right now and where things will go.
>
> We had this land before Imperial Oil, before Shell, before Suncor, before Enbridge. This was a territory that we thrived on. It thrived with us because we shared ceremony and made a lot of sacrifice to live [in this place] called the Great Lakes.
>
> There is a future without Imperial Oil. There is a future without Suncor and Shell. I probably won't be around to see that, but I need to work on a path that will allow settler dispossession to end. That is up to you. That's up to the actions you can take.
>
> White privilege exists. I am not saying it is a good thing or a bad thing. It's up to you. You have a choice but if you don't [do something], you're sitting back and letting this happen. You have jobs, you have families, but this is also your responsibility to use that white privilege in effective ways. That's a thing I don't have. I would use it if I could.
>
> It's important to acknowledge responsibility. I acknowledge my responsibility as an Anishinaabe-kwe, and a Bear clan member, as my grandparents' granddaughter. There is probably violence in your family history, perpetrated on Indigenous people. That's your responsibility. Don't walk around as though you have nothing to do. You have a responsibility. That never goes away.
>
> Canadians should not only respect Indigenous rights, but also contribute to radical Indigenous resurgence. [The Indigenous people on the front lines] need your support.
>
> I can tell you that my ancestors are not a lie. My culture is not a lie. We are way more tangible and giving than Canada has ever been.
>
> The land that Canada continues to colonize holds the fighting spirit of my ancestors.
>
> —*Vanessa Gray, Tiotia:ke (Montreal), Feb 18, 2019*

3 Vanessa is an Anishinaabe-kwe (Anishinaabe woman) from the Aamjiwnaang First Nation, located in Canada's Chemical Valley. As a grassroots organizer, land defender, and educator, Vanessa works to decolonize environmental justice research by linking scholarly findings to traditional teachings. Vanessa is a co-founder of Aamjiwnaang & Sarnia Against Pipelines (ASAP), host of the annual Toxic Tour of Canada's Chemical Valley.

Climate change is part of the legacy of European expansion, settler colonialism, and the imposition of extractive, oppressive relations between people and between people and the land. Land was stolen from Indigenous people to grant access for the extraction of natural resources to accumulate wealth and power to settlers. Land was stolen, and cultural genocide was committed, to make room for settlers to live and create what is now called Canada. The ways of life and the economic systems that are continuing to cause climate change were imported to these lands by force. This colonial process is actively continuing today. Solving climate change means undoing colonial capitalism. In a video from Powershift 2019, Jose Two Crows Trembley says, "I don't know how this movement can go forward without centering Indigenous voices" (Powershift 2019b).

This decolonial climate analysis, which is at the heart of CJM, was all but missing from the climate strike today. The "climate justice" they were chanting for is intergenerational justice. And yes, that is of course real and important. But what about racial justice, economic justice, gender justice? What about justice for Indigenous peoples? When we at CJM chant "climate justice," we mean all those things and more.

CJM will continue to mill around the edges of these movements in Quebec, planting and nurturing seeds of radical change in the minds of the young people just waking up to this moment in history and to their power. Their signs and banners read "Systems Change, not Climate Change." Along with incredible Indigenous leaders and communities and radical climate organizers, like those who organize Powershift, we'll do our best to help the newly empowered youth think deeply about what system change means, about what systems we want to create, and about who gets to decide what a just climate future looks like.

References

Manuel, Arthur with R. Derrickson. 2017. *The Reconciliation Manifesto: Recovering the Land, Rebuilding the Economy*. Toronto, ON: Lorimer.

Powershift 2019a. Powershift: Young and Rising Website About. https://powershift-youngandrising.ca/, Accessed 22 March 2019.

Powershift 2019b. Powershift Young and Rising Highlights 2019. https://www.youtube.com/watch?v=dJ09pbUHZVk

Solís, Rubén, and with Southwest Public Workers Union. 1997. "Jemez Principles for democratic organizing." SouthWest Organizing Project.

21

LISTEN, THE YOUTH ARE SPEAKING

The Youth and Climate Justice Initiative of Western New York

Lynda H. Schneekloth, Rebekah A. Williams, and Emily Dyett

Young people will face the consequences of each decision made and not made regarding climate change. The youth, representing half of the global population, have been born into, and are inheriting, the climate crisis. This has been a defining story their entire lives, and it will frame the rest of their years. So it can be argued that there is no voice as morally powerful and persuasive as young people. And yet they, especially the young of frontline, marginalized communities, are not included in local and global decision-making processes. We need to listen to their voices, their hopes, and their demands for intergenerational justice (Hertsgaard 2019, accessed 29 March 2019).

The Youth and Climate Justice Initiative (YCJ) in Buffalo, NY, is one attempt to open the space of climate justice to hear, respect, and amplify the voice and actions of young people. The form and structure of this initiative is still evolving, but over the last 3 years, 15 young people, mostly from frontline and disadvantaged communities, have engaged in conversation, learning, teaching, lobbying, demonstrating, and participating in the climate justice movement through an initiative of the Western New York Environmental Alliance (WNYEA).

This chapter places this program in the context of the climate justice movement in Buffalo, NY; describes some of the actions, practices, collaboration, and leadership opportunities this program has offered young people; and speaks to this effort within the current worldwide explosion of youth-led climate justice action. The global/local youth climate movement has energized all who are currently working on climate justice, including people in Buffalo (at the eastern end of Lake Erie in the Great Lakes bioregion). Ilyas Khan, a participant in the Youth and Climate Justice Initiative, describes their perspective this way.

> What is climate change? According to the Oxford dictionary, "Climate Change is a change in global or regional climate patterns, in particular a

> change apparent from the mid to late 20th century onwards and attributed largely to the increased levels of atmospheric carbon dioxide produced by the use of fossil fuels." That's somewhat accurate, but what Climate Change really is, is the destruction of our environment, the displacement and death of people, and it is an injustice to the marginalized communities who will suffer the most from its effects. Climate Change is the product of corporate greed, the product of fossil fuel companies taking advantage of us, the citizens, destroying our environment, and destroying the lives of countless people.
>
> *Ilyas Xavier Khan, age 14*

The Youth and Climate Justice Initiative of the Western New York Environmental Alliance emerged within the climate justice movement of the region, starting around 2015. Buffalo has a strong civic culture, and many organizations work on issues of social, racial, environmental, gender, and economic justice. Over the last few decades, a range of local voices have emerged to respond to the myriad challenging conditions in this Great Lakes rustbelt city, a city that made its wealth in the early industrial revolution before there were any environmental protections. We have inherited a landscape of toxic waste and are the home of Love Canal, a place and an event often referred to as the first human-made environmental disaster (Brown 1979). The Buffalo Niagara region, a centre of manufacturing, fought for and built a strong union culture that has been weakened as jobs have disappeared. Buffalo remains a very segregated city, and is the third poorest city in the United States. Because we live on the Great Lakes at an international border, it is recognized that the health of our waters is critical, and we witness international commerce and immigration movement across the Niagara River between the United States and Canada. There are many reasons to organize, and the people of this region across sectors have gathered to fight for justice in their city and region (Magavern and Schneekloth 2018).

Since 2015, many Buffalo organizations have built and joined the climate justice movement, as it has become so apparent that climate change is here and that its effects impact different communities differently. The movement has been built by many diverse communities of activists. Environmental organizations such as the Western New York Environmental Alliance, the Sierra Club Niagara Group, and others, had been engaged in promoting renewable energy for some years. Peoples' justice advocate groups such as PUSH (People United for Sustainable Housing), Partnership for the Public Good, and the Clean Air Coalition, along with food justice groups such as the Massachusetts Avenue Project (MAP) and many others, were identifying the unequal impacts of climate change on frontline communities, not only across the globe, but right here in Buffalo (see organizations' websites in the Reference list).

The conversations and actions converged when various groups joined forces in the campaign to keep deep horizontal fracking out of the state and were successful! Governor Andrew Cuomo issued an order to prohibit fracking in

December 2015. This campaign was a crash course in climate change for many New Yorkers and served to organize diverse groups into a movement reflected in the energy of the New York City's People's Climate March on September 21, 2014: "To change everything, we need everyone" (People's Climate March 2014).

The year 2015 was also significant in galvanizing the climate justice movement in the Buffalo Niagara region and across the planet, with the lead-up to the Paris Climate Meeting (COP 21) in December of that year. Young people and Indigenous folks led with deep values about the injustice of this crisis. That same year, justice and climate change were also lifted up by Pope Francis with the papal encyclical *Laudato Sí—On Care for Our Common Home* (Laudato Sí 2015). It was becoming clearer to so many that "[T]he rights of human beings and the rights of nature are two names for the same dignity" (Galeano 2010). The sensibility of this declaration by Uruguayan writer and activist Eduardo Galeano opened the possibility of many new alliances.

In 2015, the Buffalo environmental community organized a campaign, Rise up for Climate Justice, to educate the community on the importance of the upcoming Paris talks and to encourage folks to push President Obama and other elected officials for action. At the same time, seven of the most significant social/racial/economic/environmental justice organizations joined together in a coalition, called the Crossroads Collective, and were funded by a significant grant from the Chorus Foundation to focus on climate justice in Buffalo. In December 2015, Crossroads and the environmental groups jointly sponsored a "Party for the Planet" in Buffalo. The silos that had for years fragmented various justice and rights movements were weakening and a conversation had begun.

All of these organizations were aware of, and had ascribed to, the Jemez Principles for Democratic Organizing which state that those most impacted by any action should be the voices that are heard (Jemez Principles 1996). Our community was experimenting with the challenging principles of inclusivity, building right relationships among ourselves, trying hard to hold onto a vision of a just transition, not only to a new form of producing and managing energy/electricity, but to a vision of a democratic society where work is transformed and the Earth and each other are held sacred.

WNYEA had a policy of adopting platforms to guide the work of the Alliance. In early 2015, the board adopted climate justice as a major campaign platform. The Alliance's environmentally focused member groups had worked for some years on climate change, but the clarity of the message of real and unequal impacts on people and communities made it apparent that this was a much broader cultural issue that had to be addressed in addition to the science-based work.

It was at this moment that Rebekah Williams, Youth Education Director of the MAP who oversaw their Leadership and Healthy Youth programs, and Derek Nichols of Grassroots Gardens—both groups focused on food justice—looked around and saw that there were almost no youth, especially frontline youth, in the climate justice movement. If any group will be impacted by the coming injustices brought about by climate change, it is the youth. MAP and Grassroots Gardens had strong youth programs that focused on climate justice through the lens of food, but were not ready to expand at that moment. Williams, who had close ties with the Western New York Environmental Alliance, initiated a conversation with WNYEA about their interest in developing a program to engage youth in the local and global climate justice struggle. The members of the Alliance found that this effort would support their mission, and organized to seek support and to frame ways of engaging young people.

To better understand what young people knew about climate change along with its impact on their lives and on the planet, a collection of groups from MAP, Grassroots Gardens, Sierra Club, WNYEA, the Ujima theatre group, and others held a Youth Summit on April 12, 2016, where 70 young people gathered. Using popular education methods, the young people engaged in interactive conversations and exercises to uncover their knowledge and interest in the climate crisis. The participants' knowledge varied widely, from well-informed, to thinking primarily about recycling, to almost nothing. One of the questions that emerged during the final skits the students created, prompted by Lorna Hill from Ujima, was "where are the grownups?" Haunting.

Initially, the young people who joined the resulting Youth and Climate Justice Initiative were the students who were already engaged with MAP and other environmental organizations. They knew the value of these leadership experiences, and were eager to broaden their knowledge and opportunities to impact climate change. At this point, we had no formal mechanism to select Fellows, but gladly embraced these young people as our inaugural Fellows to help us frame an initiative for which we could seek specific funding. In 2016, we called the first formal gathering of the eight inaugural Fellows and the adult advisory group met to develop an agenda and program.

In early 2017, the Fellows travelled with a delegation from WNYEA to New York City to meet with the Overbrook Foundation to seek funding for the coming year. The young people were an impressive group by that time; they had clear opinions about climate change, and were articulate about the impacts on frontline communities. Later that spring, with WNYEA support, the young people travelled to the Climate March in Washington, D.C. where they also attended the Powershift Youth Summit. We had begun, we obtained funding for the next year, and WNYEA asked Rebekah Williams to be the consultant for this effort. The process and program took shape.

Also in 2017, students participated on a panel at the Western New York Youth Climate Action Summit sponsored by the New York State Master Teacher Program (a professional network for schoolteachers), and in September they made a presentation at a Buffalo Humanities Institute symposium where climate activist Bill McKibben was the featured speaker. Eleven of the youth Fellows had a two-hour special meeting with McKibben who affirmed their engagement and deeply encouraged them to continue. One of the Fellows was featured in a press conference with McKibben, advocating for state-wide climate justice legislation. Later that day, the students offered a session on advocacy and our Youth and Climate Justice Initiative at the symposium. Gabe Cohen speaks of his experience this way.

> The focus of our work was to ready us all as young leaders, and take on the task of promoting safe and sustainable practices for the earth and to engage other youth to take on that task as well. My time in the Alliance has been fun, informative, and has helped me grow into a better leader.
>
> *Gabe Cohen, age 18*

The social justice Crossroads Collective, shortly after it was formed in 2015, began to work closely with a justice group in Oakland, California—the Movement Generation Justice and Ecology Project (MG). Several Crossroads members went to California for training, including two youth members. In the fall of 2017, MG facilitated a three-day retreat near Buffalo for 35 Buffalo-based activists. Movement Generation's curriculum entwines ecological and social justice. It is a provocative experience and conversation. The youth coordinator and advisory group engaged in the Youth and Climate Justice Initiative decided to offer a shorter version of the retreat for the Fellows and others affiliated with WNYEA in the community who work with young people. We held this one-and-a-half-day retreat in January 2018 on a snowy cold day in Buffalo, with 30 participants. It was led by Crossroads Collective members and one of the first Fellows, Dillon Hill, who had participated in the fall 2017 retreat.

This Justice and Ecology retreat was an immersive/interactive experience that sought to explore the deep systematic causes of our current crisis and envision a just transition, to "shift to an economy that is ecologically sustainable, equitable and just for all its members" (Movement Generation n.d.,). This framework challenges activists to take action that is both strategic and visionary. How do we move away from the exploitive and extractive economy we live in to a more democratic, decentralized and diverse economy that respects people and the earth? This is a powerful conversation, and the youth who participated were fully engaged and eager to continue in the climate justice space. It has had an impact, as Skylar Moffett explains.

> Even though I do not live in an urban area, being a part of the Youth Climate Justice Fellowship has opened my eyes to the 'justice' aspects of climate change and how much it affect people living in frontline communities. Without the fellowship and all of the opportunities it has brought me, I would not have as much knowledge on the topic as I do now, and my confidence would be as strong. For now I truly feel my voice as a youth matters and what I do can make a difference, which is the same with every other youth who has a passion.
>
> *Skylar Moffett, age 18*

FIGURE 21.1 The Justice and Ecology retreat brought together Fellows from the Youth and Climate Justice Initiative with leaders of environmental organizations where they enacted the "web of life" and the implications of cutting strands to viscerally understand the harm of human actions. Photo used with permission of the Western New York Environmental Alliance.

Other opportunities for the Fellows that spring included working with a research group at the University of Buffalo, CoRE, whose goal is "Inclusive Innovation," i.e. to find solutions to some of the most challenging societal problems. A workshop with CoRE faculty and affiliates questioned our current form of production of solar panels, and opened a conversation around "what is the factory of the future?" The Fellows also engaged in a planning retreat to outline their expectations for the coming year, and planned and then led a workshop at the 2017 Western New York Climate Action Summit on "Talking to People in Positions of Power." They also participated in the 2018 Buffalo Humanities Institute Symposium in collaboration with the Ujima climate justice play, *Free Fred Brown*. Other opportunities for Fellows, often in small groups, included advocating for climate justice at the offices of Congressman Brian Higgins, tabling at the Holiday Market on behalf of the Erie County BYO Bag Coalition, tabling for the Electric Bus Campaign with the Sierra Club, participating in People's Education with the Energy Democracy Committee of People United for Sustainable Housing (PUSH) Buffalo, and other opportunities—too many to list!

FIGURE 21.2 Fellows from the Youth and Climate Justice Initiative of the Western New York Environmental Alliance attending the Justice and Ecology Retreat were asked to draw and share the story they have been told about their origins. Photo used with permission from the Western New York Environmental Alliance.

In the fall of 2018, applications were opened for new Fellows, as some of the young people were graduating and moving on. Leadership changed as well; Emily Dyett was hired as the new Youth Climate Justice Coordinator through a continuing grant from the Overbrook Foundation, as Rebekah Williams left for new work. The new school year opened with five new Fellows integrated into the group of five that were continuing. Since fall, the Youth Fellows have participated in promoting, lobbying, rallying, and letter-writing on behalf of the New York State climate justice coalition group, New York Renews, to advance the state CCPA Legislation (Climate and Community Protection Act), a comprehensive climate justice bill to move to renewables with a focus on investing in frontline communities, with transition supports for impacted communities and workers. The movement in support of climate legislation was successful and in June 2019, the Governor signed into law a modified version of the CCPA, the most advanced climate bill to date in the U.S. In February 2019, the youth Fellows also played a key role in an event called the WNY Climate Conversations that included a visit and talk from the Zero Hour Youth Climate Movement Global Outreach Director, Sohayla Eldeeb.

As this chapter is being written in April 2019, the youth are about to participate in a full-day activity, "Urban Ecological Regeneration" training at the Lyceum at Silo City, a regionally based ecological education group focused on the regeneration

FIGURE 21.3 Over 50 young people participated in the "Strike School for Climate" in Buffalo, New York on March 15, 2019. There were an estimated 1.6 million students from over 120 countries walking out of school that day to protest adult inaction on climate change (Strike School for Climate 2019). Photo by William Jacobson, used with permission.

of the contaminated landscapes that surround us in Buffalo. They will also continue their relationship with CoRE with a workshop on "Solar, Clean Production, Data and Advocacy," and will present at the 2019 WNY Youth Climate Action Summit, a continuing project of the NY State Master Teachers Program.

The Fellows have been deeply inspired by the worldwide climate justice movement of young people. They ask why they should bother to go to school when they are not even sure there will be a future. All of these "kids" demand to be taken seriously, as Swedish activist Greta Thunberg so articulately said when a patronizing voice suggested that we need to give young people hope. Thunberg (2019) responded: "I don't want your hope. I want you to act as if the house is on fire. Because it is."

Buffalo youth organized in their schools to participate in the student strike on March 15, 2019, and used the communication network of the WNYEA Youth and Climate Justice Initiative, and their own social media platforms, to recruit. They were able to reach students in the region, and young people from six different schools participated. Approximately 50 "kids" and a few grownups marched in front of the Buffalo City Hall on a cold, blustery Buffalo afternoon chanting, "No more coal, no more oil, keep the carbon in the soil."

Buffalo youth have enthusiastically embraced emerging youth climate organizations such as #thisiszerohour, the Sunrise Movement, Fridays for the Future, and others. The young people who run climate justice groups are very articulate—not only about intergenerational injustice that is impacting them and the lives of 50% of the earth's human population, but also about species extinction and the profound inequalities globally and locally caused by the changing climate. The youth who have been truth-speakers since COP21 in Paris are moving toward a just transition to end the extractive-based economy and move to a livable, healthy future on this planet (Carrington 2019).

> March 15th is both a beginning and an end. An answer to the appalling damage that we have inflicted upon the earth—and a beginning to treating the environment the way each and every one of us wants to be treated.
>
> *Ariel Braverman, age 11, a youth organizer and future Fellow*

The Fellows, many of whom come from frontline communities, are well aware of the injustices of climate change and our current economic structure on their families and communities. They are aware that they live on the Great Lakes, in an economy responsible historically and currently for significant greenhouse gas emissions and a bioregion that contains 21% of the world's fresh surface water. They know this water is precious and is endangered by climate change, with Lake Erie already 2°F warmer than it was in 1970. This rapid change is stressing animals and plants, decreasing winter ice cover, exacerbating pollution, and generating enormous storms. The students, knowing that local work must be done, are setting up meetings with their Buffalo Common Council members demanding a Climate Action Plan that includes their voices. These efforts are being supported and uplifted by the Youth and Climate Justice Campaign as significant youth-initiated activities.

> I am a Youth Fellow in the Climate Justice movement. What motivates me to be a part of the movement is the work I see people putting in and how other youth fellows and I feel about climate change, trying to help educate others of how serious climate change is. My experience so far has been amazing. I have had the chance to express my thoughts about every topic regarding the environment we learned about. I have had a chance to teach young children about the importance of water. And not least, I have felt like a contributor to the climate change movement.
>
> *Ingabire Adams, age 17*

As we enter into the third year of the Youth and Climate Justice Initiative, we are considering some of the changes that may be necessary to meet with a new group of fellows. The Fellows that started three years ago are almost all graduating from high school and heading off to college. Many of these young people came out of the food justice community and therefore entered the Youth and Climate Justice Initiative with knowledge, experience, and a critique of the ways in which our culture is based on racial, economic, gender, and class discrimination.

This background enabled them to quickly grasp the concept of climate justice beyond the scientific discourse of climate change. They were ready.

We will continue to reach out to youth to participate in the Fellows program through our affiliated organizations in Crossroads and WNYEA, and encourage our current Fellows to talk to their peers and friends about joining. Although we don't yet know what the new group of applicants will be and where they will start on their knowledge journey, we do know that they will have a world movement to join in addition to our local actions, and they will come with different backgrounds and experiences. Although the number of Fellows is limited to 10–15 per year, these young people literally reach hundreds of other young people in the region through their presentations, workshops, symposium, and rallying. This is the network we hope to continue to engage.

As we plan the 2019–2020 year, we maintain the goals of the Youth and Climate Justice Initiative:

(1) to lift up and amplify youth voices in the climate justice space for education and leadership, and
(2) to build and support a youth activist network to get involved in local solutions and to connect with community activists.

We are also adding a stipend for the Youth Fellows' participation in the coming year to clearly place value on their work and to enable them to engage more fully.

We plan to spend the first part of the year in a "Youth Climate Justice Leader Certificate Program" where the Fellows will help frame and participate in a series of interactive sessions based on the model of popular education on climate change, climate justice, and just transitions. We will offer opportunities to ground them in our bioregion and their home space, and give them tools to understand the structure of government for advocacy, lobbying, and the use of social media. And in the second part of the year, we will work with the Fellows to find spaces for them to move out into the community and to offer their insights and sense of climate justice to other youth, expanding circles of young advocates. This account describes the Youth and Climate Initiative of the WNY Environmental Alliance as of April 2019, including our aspirations for the coming year. But already we are working with the young people to evolve the program as the context changes, opportunities emerge, new Fellows are selected, and national calls for youth action provide platforms for larger groups of young people to participate. Young people want to be engaged, which is made so clear in the *Open Letter from the Global Coordination Group of the Youth-Led Climate Strike.*

> We, the young, are deeply concerned about our future. Humanity is currently causing the sixth mass extinction of species and the global climate system is at the brink of a catastrophic crisis. Its devastating impacts are already felt by millions of people around the globe. Our generation grew up with the climate crisis and we will have to deal with it for the rest of our

> lives. Despite that fact, most of us are not included in the local and global decision-making process. We are the voiceless future of humanity. We will no longer accept this injustice … so we are rising up. We are going to change the fate of humanity.
>
> *(Open Letter from the Global Coordination Group of the Youth-Led Climate Strike 2019).*

This is zero hour; there is a fierce urgency to bring about climate justice. Those of us who are "adults" and have been so slow to take action or to push for action, carry pain in our hearts for what we are leaving to our children. We have not been good ancestors. David Orr's words have resonance: "We will stand before whoever is able and willing to judge, or perhaps the silence of extinction, as a generation that willfully and unnecessarily imposed egregious wrongs on all future generations, depriving them of liberty, property, and life" (Orr, 2019 p. 73). Bless the youth who have the courage to look directly in the eye of the climate crisis, see what needs to be done, and have shouted their demands for their right to life and equity. We can't change what has or hasn't happened, but we can stand for climate justice with the youth.

References

Brown, Michael H. (1979). "Love Canal and the Poisoning of America," *The Atlantic*, December 1. http://chej.org/about-us/story/love-canal/and www.theatlantic.com/magazine/archive/1979/12/love-canal-and-the-poisoning-of-america/376297/, accessed 29 March 2019.

Buffalo Humanities Institute. https://humanitiesinstitute.buffalo.edu/ and https://humanitiesinstitute.buffalo.edu/event/buffalo-humanities-festival-environments-spotlight-speaker-bill-mckibben/

Carrington, D. (2019). "School climate strikes: 1.4 million people took part, say campaigners." *The Guardian*, March 19. www.theguardian.com/environment/2019/mar/19/school-climate-strikes-more-than-1-million-took-part-say-campaigners-greta-thunberg, accessed 29 March 2019.

CoRE. Collaboratory for a Regenerative Economy. Retrieved from http://engineering.buffalo.edu/materials-design-innovation/research.html, accessed 29 March 2019.

Crossroads Collective Buffalo. https://openbuffalo.org/issues/issue:crossroads-collective/ and www.facebook.com/CrossroadsBuffalo/

Cuomo, A. (2015). https://www.nytimes.com/2014/12/18/nyregion/cuomo-to-ban-fracking-in-new-york-state-citing-health-risks.html, accessed 29 March 2019.

Galeano, E. (2010). https://climateandcapitalism.com/2010/04/21/eduardo-galeano-message-to-the-mother-earth-summit/, accessed 29 March 2019.

Grassroots Gardens WNY. www.grassrootsgardens.org/

Hertsgaard, M. (2019). "The climate kids are coming." *The Nation*, March 25.

Jemez Principles (1996). https://www.ejnet.org/ej/jemez.pdf, accessed 29 March 2019.

Laudato Sí (2015). http://w2.vatican.va/content/francesco/en/encyclicals/documents/papa-francesco_20150524_enciclica-laudato-si.html, accessed 29 March 2019.

Lyceum at Silo City. www.lyceumsilo.city/

Magavern, S. and L. Schneekloth (2018). *The Climate Justice Movement in Western New York*, 24 Buff. Envtl. L.J. 59. (2016–2018). Provided by: Charles B. Sears Law Library. Content downloaded/printed from *HeinOnline*

Massachusetts Avenue Project (MAP). http://mass-ave.org/

Movement Generation Justice and Ecology Project. https://movementgeneration.org/

Movement Generation (n.d.). https://movementgeneration.org/our-work/training-analysis/retreats/, accessed 29 March 2019.

NY RENEWS. www.nyrenews.org/

Open Letter from the Global Coordination Group of the Youth-Led Climate Strike (2019). https://wattsupwiththat.com/2019/03/02/we-the-young-open-letter-from-the-student-climate-change-strikers/, accessed 29 March, 2019.

Orr, D.W. (2009). *Down to the Wire: Confronting Climate Collapse.* London, UK: Oxford University Press. p. 73.

People's Climate March (2014). https://reporter.rit.edu/news/change-everything-we-need-everyone

Powershift. https://powershift.org/, accessed 29 March, 2019.

Strike School for Climate (2019). https://en.wikipedia.org/wiki/School_strike_for_climate and www.theguardian.com/environment/2019/mar/19/school-climate-strikes-more-than-1-million-took-part-say-campaigners-greta-thunberg, accessed 29 March 2019.

This is Zero Hour (2019). https://thisiszerohour.org, accessed July 2019.

Thunberg, Greta (2019). https://me.me/i/adults-keep-saying-we-owe-it-to-people-to-give-443f542b5f0e4f2ba60d04890fd045e0, accessed 29 March 2019.

Ujima Co., Inc. www.ujimacoinc.org/

Western New York Youth Climate Summit. https://wnyyouthclimatesummit.org/

Western New York Environmental Alliance. www.growwny.org/

We the Young … Open Letter from the Global Coordination Group of the Youth-Led Climate Strike (2019). https://wattsupwiththat.com/2019/03/02/we-the-young-open-letter-from-the-student-climate-change-strikers/

22

EDUCATION REFORM IN THE STRUGGLE FOR CLIMATE JUSTICE

Gabriel Yahya Haage and Natália Britto dos Santos

To understand the interdisciplinary and ethical challenges posed by climate change, education is key. Schools can model grassroots agency and ethical values even in marginal circumstances, helping students see the possibilities for local action to understand and mitigate climate change.

This chapter explores how the Child-Friendly School system can be used as an inspiration for creating a better education system. Child-Friendly Schools are schools set up by charities, including the United Nations International Children's Fund (UNICEF), in developing nations, often during crisis situations. These schools' ability to thrive in the most difficult of circumstances and their Climate Change and Environmental Education programs make them a good model for reforming education systems worldwide. In addition, we discuss other education initiatives, both local and international, which can serve as inspirations for developing a robust climate justice education program.

Child-Friendly Schools

Based directly on the United Nations Convention on the Rights of the Child (1989), Child-Friendly Schools are established and maintained in crisis regions by charities like UNICEF, often working in conjunction with multiple governmental bodies. Over 100 countries have incorporated the principles of Child-Friendly Schools, with many schools being set up in severely impoverished and war-torn regions (Miske 2010). For instance, in 2015 alone, UNICEF's efforts resulted in nearly 140,000 out-of-school children in West and Central African countries receiving access to Child-Friendly Schools (UNICEF 2016). UNICEF is promoting the Child-Friendly Schools model worldwide.

Three key components of the Child-Friendly School system are particularly interesting: Its interdisciplinary approach to climate change and environmental issues, its emphasis on developing moral fortitude, and its determination to empower and increase the agency of its students. All three aspects are not only important to the success of Child-Friendly Schools, but are in fact necessities, when we consider where and under what circumstances they are constructed and must thrive.

Interdisciplinary education in Child-Friendly Schools

A major benefit of Child-Friendly Schools is their interdisciplinarity and their emphasis on holistic education. Their Climate Change and Environmental Education program is meant to be part of every aspect of the curriculum, from math to creative writing (Iltus 2013). This can be achieved by creating ecological activities that connect these disciplines. Through activity, this curriculum can instill both a deep understanding of and respect for nature. The holistic curriculum also encourages seeing all forms of injustice as interrelated. For instance, this school system emphasizes how an environmental worldview can help decrease gender inequalities (Iltus 2013). Several successful applications of this holistic approach to injustice are evident. For instance, by encouraging the creation of school gardens in Child-Friendly Schools, students not only learn about the environment and gain a sense of empowerment but also have ready access to nutritious food (Iltus 2013). When one considers that food insecurity is often one of the reasons girls are kept home from school, this holistic approach can also strike a blow against gender inequality (Iltus 2013).

Such concepts are applicable beyond the crisis areas in which Child-Friendly Schools are normally set up. Indeed, such interdisciplinary and holistic perspectives could help improve the mainstream education system with which most students and educators are familiar. While what is considered "mainstream education" varies, here we use the term to refer to the education framework that has been dominant in much of the world for the last several decades, one in which strictly defined, seldom-interacting disciplines are the norm (Elshof 2011; Bigelow, Kelly and McKenna 2015). Many education systems also emphasize standardization, inflexible curricula, and ranking as a way of evaluating success.

By adopting the interdisciplinary and holistic spirit of Child-Friendly Schools, in contrast, educational structures can be improved, particularly when it comes to their approach to the topic of climate change. Disciplines that are normally kept apart could interact in more meaningful ways. For instance, history textbooks could benefit by not only increasing the discussion of climate change, but also explaining its strong link to the historical industrialization of nations (Bigelow, Kelly and McKenna 2015). This could result in students who are better informed both about the realities of climate change and its link to the current market system (Bigelow, Kelly and McKenna 2015). Fortunately, certain activist organizations, like Rethinking Schools, are actively working towards such goals (Bigelow, Kelly and McKenna 2015).

As discussed above, Child-Friendly Schools have found great success by considering the interrelated nature of most issues. By incorporating such systems thinking into mainstream education, students could become better equipped to understand the complex, non-linear components in the climate crisis (Elshof 2011). More specifically, by encouraging the notion that solving the climate crisis cannot be done without considering the interrelated nature of social, political, and scientific components, educators may feel more comfortable discussing issues like differential responsibility in the classroom (Elshof 2011).

Moral development and Child-Friendly Schools

The teaching of moral development in schools is a contentious topic. Historically, controversy has arisen regarding which values to emphasize and who should be responsible for teaching morality to students (Cragg 1978; Santor 2000). And yet, more than mathematics or literature, the concepts of empathy, moral responsibility, and the desire to remedy global injustice are key to the development of a generation capable of dealing with climate change and climate injustice (Weil 2004; Assadourian 2017). Teachers who wish to encourage such moral instruction often find it difficult in the current education system due to its emphasis on standardized tests and inflexible curricula (Sondel 2015). In fact, a "hidden" ethical system is being implicitly imparted to students by the mainstream education framework itself. This "hidden" curriculum supports the current highly unequal economic system, rather than emphasizing compassion, social justice, and moral responsibility (Sondel 2015; Portelli and Konecny 2013). The current focus in education on standardization, profit, and rankings demonstrates this. Similarly, the mainstream education system generally fails to express the moral dimension of climate change as a dangerous and urgent crisis that is "directly related to our assumptions about economic growth, globalism and consumption" (Houghton 2010). Certainly, textbooks, by determining what is discussed and what is omitted, can send a strong moral message. Many textbooks

> fail to help students think systemically, and to name the role of the capitalist system in pushing the climate crisis forward. The texts may talk in terms of the role of 'industrialized nations' and 'developing countries,' but they fail to mention, let alone interrogate, the nature of the global economic system.
>
> *(Bigelow, Kelly and McKenna 2015)*

Textbooks can also have a troubling message when it comes to the responsibilities of developing nations for alleviating climate change (Bigelow, Kelly and McKenna 2015). It is therefore understandable that students develop a moral compass that aligns with the current, detrimental, consumerist lifestyle.

In contrast, Child-Friendly Schools explicitly include a justice-oriented moral component. The desired outcome of training in such schools is not just knowledge, but moral fortitude. Amongst these goals are the fostering of an environmental ethics and a respect for local and Indigenous knowledge (Iltus 2013).

These moral sentiments can be developed through their ubiquitous Climate Change and Environmental Education program (Iltus 2013). Through activities that connect students to nature and foster cooperation amongst students of diverse backgrounds, they can develop a respect for nature and other students and break down detrimental social taboos (Iltus 2013). Such a curriculum is often a necessity, as social taboos, particularly regarding gender and socioeconomic inequalities, can greatly impact young people, and climate change will almost certainly lead to an exacerbation of these issues (Back and Cameron 2007).

Agency and Child-Friendly Schools

Unfortunately, the strict curricula and ubiquitous standardization in many schools can lead to inflexibility and can discourage innovation both at the school and classroom level (Portelli and Konecny 2013; Hursh 2005). Such a restrictive system can make it difficult to encourage activism in children (Sondel 2015). This is troubling, as the fight against climate injustice will require strong activism from the younger generation.

In contrast, Child-Friendly Schools encourage student agency and decision-making power over their schools and community. Children are considered agents of change and when "empowered and educated on climate change [they] can reduce the vulnerability of themselves and their communities to risk and contribute to sustainable development" (Iltus 2013). Child-Friendly Schools encourage students to form groups, petition the government, and set up student-run programs and activities. An illustrative activity is "risk-mapping," in which children are asked to identify the locations they believe are safe and unsafe in their schools and communities (Wright, Mannathoko and Pasic 2009). As Child-Friendly Schools are often set up in crisis areas prone to natural disasters, gathering local information is vital in maintaining a safe school. Such activities increase empowerment while "addressing environmental risks and vulnerabilities to climate change within schools and their surrounding communities" (Wright, Mannathoko and Pasic 2009).

In the end, the beneficial components of Child-Friendly Schools, which help address the interrelatedness of climate change and socioeconomic vulnerabilities, stem from a concerted effort to break down barriers. These include the barriers between traditionally formal and non-formal education, both in terms of academic subjects and moral development. Similarly, the barrier between community and school are blurred, with students actively encouraged to participate and become leaders in both realms.

Closer to home

Several North American school programs have adopted ideas similar to those included in Child-Friendly Schools. For instance, in British Columbia, Canada, a group of teachers developed a Climate Justice package in 2014, with 8 modules addressed to students between grades 8 and 12 (Cho 2014). The toolkit explores climate justice within the local context, and provides a framework to discuss many social and environmental issues such as consumerism, the industrial food system, energy, and waste. Although it is aligned with prescribed learning outcomes already in the province's scholarly curriculum, the Climate Justice package is not yet mandatory.

In Portland, Oregon, the local school board passed a resolution in May 2016 to incorporate climate justice in its schools' curriculum (Hellmann 2017). A pilot program was implemented at Lincoln High School for the 2016/2017 school year, comprising climate education and environmental activism lessons. Students participated in both in-class and external activities, such as a march to City Hall and advocating for environmental justice in public meetings. These examples show that it is possible to develop awareness about climate justice while also giving children agency to participate on local forums, empowering them to contribute positively in their communities.

In addition to these two cases that explicitly aim to include climate justice in the curriculum, there are many other ecological literacy and community-building initiatives in formal education. For instance, in Canada, the Ontario EcoSchools is a voluntary certification program that encourages schools to create an EcoTeam of students and develop environmental learning projects and actions with the whole community (Ontario EcoSchools 2009). In the 2015/2016 school year, there were 1,720 EcoSchools across the province, impacting more than 880,000 students and their communities. This program and similar initiatives could expand to include climate justice issues as either the core of potential projects or as an essential component of other activities, since socio-environmental topics are deeply interrelated.

These examples can serve as an inspiration to develop similar endeavours in other school boards, and thus incorporate climate change and climate justice into the curriculum of a greater number of students.

Non-formal education initiatives

In the realm of non-formal education (outside schools), flexibility often results in more practical learning through games, challenges, group work, and time outdoors. In addition, children usually have more agency to decide what they want to do and learn, given their interests. Consequently, children usually become more involved throughout the learning process and more interested in the outcomes.

Moreover, non-formal education can reach a great number of children. For example, the Scouts Movement is a voluntary non-political education movement present in 224 countries and territories, with over 40 million members (World Organization of the Scouts Movement 2018). Its main purpose is to encourage children to engage in their own development and become good citizens that serve their communities. The Scouts education method is based on learning by doing, nurturing moral values, group work aligned with personal development, and outdoor activities. The World Organization of the Scouts Movement (WOSM) coordinates an Environment Program, with resources to support environmental education, and the Messengers of Peace, an initiative to recognize and inspire scouts' projects that bring positive change in local communities.

The Climate Change Challenge Badge (Youth and United Nations Global Alliance 2015), developed by the Youth and United Nations Global Alliance (YUNGA) in collaboration with United Nations agencies, civil society, and other organizations (including the WOSM and the World Association of Girl Guides and Girl Scouts—WAGGGS), is part of YUNGA's learning and action series, which also includes challenge badges on other important social and environmental issues. These badges are based on the Sustainable Development Goals adopted by the UN, which are also a driving force behind the Child-Friendly Schools framework. The Climate Change Challenge Badge aims to develop awareness about why climate change matters and motivate children and young people to take action and become agents of change in their own communities. The associated booklet can be used by educators and young leaders, adapting the suggested activities to local realities and priorities (Youth and United Nations Global Alliance 2015).

Similar initiatives exist around the globe, and educators everywhere are developing climate change and climate justice activities and action with children, through active learning and community engagement.

Ways forward

By helping children develop the necessary values and principles, parents and educators can foster a generation that fights for global justice and understands the dangers of the currently prevalent consumerist way of life. Such education reform must include a concerted effort to give children agency. Children must be seen as current agents of change, not only "future leaders." Care must also be taken in how the subject of climate change and climate justice is incorporated into the school curriculum. Ideally, it should not simply be "another topic to cover," but should pervade all aspects of the curriculum.

The Child-Friendly School framework models an education system that focuses on children's agency, experiential knowledge, and moral values. By showing that agency and democratic participation can lead to education improvements even in the regions where climate injustice is greatest, Child Friendly-Schools are a lesson for education reformers and environmentalists in general, that cooperation and collective action have powerful effects.

Formal and non-formal education groups in North America are currently building climate justice into their programs. From the Eco-schools in Ontario to the international Climate Change Challenge Badge, many inspirational examples exist. One child at a time, one school at a time, or one community at a time, the push for education reform must continue as an essential contributor towards climate justice.

References

Assadourian, E. (2017) 'Introduction. EarthEd: Rethinking Education on a Changing Planet' in Assadourian, E. and Mastny, L. (eds.) *EarthEd: Rethinking Education on a Changing Planet*. Washington, D.C.: Island Press, pp. 3–20.

Back, E. and Cameron, C. (2007) *Our Climate, Our Children, Our Responsibility: The Implications of Climate Change for the World's Children*. London, UK: UNICEF UK.

Bigelow, B., Kelly, A. and McKenna, K. (2015) Bringing Climate into the Classroom: Inside a Teaching Retreat around Naomi Klein's This Changes Everything. *Radical Teacher*, 1(102), pp. 35–42.

Cho, R. (2014) *Climate Justice in BC: Lessons for transformation* [Online]. Available at: http://teachclimatejustice.ca (Accessed: 7 June 2017).

Cragg, A.W. (1978) Moral Education in the Schools: The Hidden Values Argument. *Interchange*, 10(1), pp. 12–19.

Elshof, L. (2011) Can Education Overcome Climate Change Inactivism. *Journal for Activist Science and Technology Education*, 3(1), pp. 15–51.

Hellmann, M. (2017) 'Portland Public Schools First to Put Global Climate Justice in Classroom', Yes Magazine [online], 6 April 2017. Available at: www.yesmagazine.org/planet/portland-public-schools-first-to-put-global-climate-justice-in-classroom-20170406 (Accessed: 7 June 2017).

Houghton, E. (2010) Green School Hubs for a Transition to Sustainability. *Our Schools/Our Selves*, 19(100), pp. 205–236.

Hursh, D. (2005) Neo-Liberalism, Markets and Accountability: Transforming Education and Undermining Democracy in the United States and England. *Policy Futures in Education*, 3(1), pp. 3–15.

Iltus, S. (2013) *Climate Change and Environmental Education: A Companion to the Child Friendly Schools Manual*. Geneva, Switzerland: UNICEF.

Miske, S.J. (2010) Child-Friendly Schools-Safe Schools, paper presented at the: *Second International Symposium on Children at Risk and in Need of Protection*. Ankara, Turkey, 24 April.

Ontario EcoSchools. (2009) Ontario EcoSchools/ÉcoÉcoles de l'Ontario [Online]. Available at: www.ontarioecoschools.org (Accessed: 15 June 2017).

Portelli, J.P. and Konecny, C.P. (2013) Neoliberalism, Subversion and Democracy in Education. *Encounters on Education*, 14, pp. 87–97.

Santor, D. (2000) 'A Canadian Experience: Transcending Pluralism' in Gardner, R., Cairns, J., and Lawton, D. (eds.) *Education for Values: Morals, Ethics and Citizenship in Contemporary Teaching*. London, UK; Sterling, VA: Stylus Pub, pp. 323–337.

Sondel, B. (2015) Raising Citizens or Raising Test Scores? Teach for America, "No Excuses" Charters, and the Development of the Neoliberal Citizen. *Theory & Research in Social Education*, 43(3), pp. 289–313.

UNICEF. (2016) *Annual Results Report 2015: Education*. New York, NY: UNICEF.

Weil, Z. (2004) *The Power and Promise of Humane Education*. Gabriola Island, BC: New Society Publishers.

World Organization of the Scouts Movement (WOSM). (2018) World Scouting [Online]. Available at: www.scout.org (Accessed: 20 June 2019).

Wright, C., Mannathoko, C., and Pasic, M. (2009) *Child Friendly Schools Manual*. New York, NY: UNICEF, Division of Communication.

Youth and United Nations Global Alliance (YUNGA). (2015) *Climate Change Challenge Badge* [Online]. Available at: www.fao.org/yunga/resources/challengebadges/climate-change/en/ (Accessed: 8 June 2017).

23

PHOTOGRAPHS, PERFORMANCE, AND PROTEST

The fight for climate justice through art

Alison Adams

In March 2003, Senator Barbara Boxer of California brought a series of photographs to the US Senate floor (*March 19, 2003 Senate Session Recording*, 2003). The photographs, from the book *Arctic National Wildlife Refuge: Seasons of Life and Land* by photographer Subhankar Banerjee, depicted the Arctic National Wildlife Refuge (ANWR) (Carter and Banerjee 2003). The Senate had been debating whether to open the Refuge to oil drilling, and there were strong opinions both in favor of and against the proposal. Boxer urged her fellow senators to read the book and visit Banerjee's upcoming exhibition of photographs at the Smithsonian National Museum of Natural History.

Banerjee had taken the photographs over the course of two years, travelling 4,000 miles by kayak and snowmobile, on foot and on raft, with Iñupiat guide Robert Thompson. His book was the first comprehensive photographic portrait of the Arctic through all four seasons, and the photographs served as a powerful counterargument to the assertions of the oil industry and its supporters that the ANWR was a "flat white nothingness" (US House, Committee on Resources 2003, pp. 8–9) and a "frozen wasteland" ('Interview with Senator Ted Stevens' 2005). Instead, the images portrayed the Arctic as alive: the open wintry expanses were the migration grounds of Caribou, as well as the hunting grounds of polar bears and of the Gwich'in people of Alaska; they were the cemeteries of the Iñupiat people, and home to numerous species of birds. Banerjee's photographs powerfully portrayed the shifting light, colours, and life of the Arctic as winter turned to spring, spring to summer, and summer to fall.

Though Boxer's statements generated significant controversy regarding Banerjee's Smithsonian exhibition, which was ultimately moved from a space near the Natural History Museum's rotunda to a less prominent downstairs gallery, the proposal to open the ANWR to drilling was rejected in a 52–48 vote (Banerjee 2004).

Art can be a powerful tool in the fight for climate justice. Unlike newspaper articles, presentations, research reports, and lectures, art conveys ideas by creating an experience, rather than a specific message to be heard or read and then assimilated (Blücher 1951). Viewers must interact with artworks; this interaction can be in the form of the visual perusal of an image (an experience not unlike exploring a landscape, albeit faster), listening to music, participating in the lives of others by watching their stories enacted on stage, or wandering around a sculpture and sizing up its particular occupation of the viewer's space. Experience is known to be an effective educational tool (Lieberman and Hoody 1998) and is a central tenet of foundational philosophies of learning (Dewey 1938). The most powerful works of art create experiences that move their audiences, drawing viewers psychologically closer to the subject matter at hand, and bringing to the fore feelings that may previously have been dormant. Olafur Eliasson, a prominent environmental artist, wrote:

> I believe that one of the major responsibilities of artists [...] is to help people not only get to know and understand something with their minds but also to feel it emotionally and physically. By doing this, art can mitigate the numbing effect created by the glut of information we are faced with today, and motivate people to turn thinking into doing.
>
> *(Eliasson 2016)*

Artists have a unique role to play in social movements because they create emotional and physical experiences that can motivate people to act.

Art-making and art-viewing are available to everyone. Though certain forms of art are limited to privileged environments, art can—and does—originate from and exist in democratic spaces. Art can be created by professional artists, but it can also be created by someone without title or training. Art-making and appreciation are components of human society that are shared across cultures and communities with different ways of knowing. Art, thus conceived, can provide a bridge between—or even within—communities (Rathwell and Armitage 2016). It can transcend language and cultural barriers, because it need not rely on written or spoken word, or on cultural norms. Art can be a powerful way to initiate conversation and spark action between and among communities that are pushed to the margins of political institutions and decision-making processes.

Visual art contributes to the climate justice movement in at least three ways: it can be a missive from communities that are and have been marginalized in decision-making processes about climate governance; it is a rallying point and information-sharing tool within the movement; and it serves as a way to fight extractive industry in places deemed by those in power as "not valuable." While other art forms, including music, poetry, and literature, can be equally effective forces in the movement, I focus exclusively on visual and interactive performance art in this chapter. The examples discussed here are primarily from North America, although there are, of course, artists working within the climate justice movement all over the world.

Art as a voice for marginalized and more-than-human communities

Art can be a critical way for marginalized communities to participate in the climate change conversation. Although women, People of Colour, Indigenous people, and people from the Global South are the most vulnerable to the effects of a changing climate (Figueiredo and Perkins 2013), they are underrepresented in decision-making processes regarding responses to climate change (Macchi et al. 2008). Procedural justice, one of a handful of types of justice discussed by scholars and ethicists, is the idea that decision-making processes themselves should be fair; in the context of climate change, procedural justice requires meaningful inclusion of the most-affected groups in those processes that determine climate responses, policies, and agreements (Tomlinson 2015). Although it is imperative that the most vulnerable populations are afforded a significant and consequential role in climate change negotiations and decision-making from local to global scales, art can serve as an additional avenue—not mediated by government structures—for these groups to participate in the climate conversation.

Art played such a role in the Standing Rock Sioux's fight against Energy Transfer Partner's construction of the Dakota Access Pipeline (DAPL). Energy Transfer Partners had plans to construct an oil pipeline from North Dakota to central Illinois, routing through disputed Sioux territory and across Lake Oahe just outside the edge of the Standing Rock Sioux reservation, where the Sioux said it would threaten sacred water supplies and damage or destroy culturally and historically significant sites. For months, a physical standoff took place between law enforcement and the Standing Rock Sioux, as well as representatives from hundreds of other tribes and non-Indigenous supporters of the Standing Rock tribe. Led by Standing Rock Sioux organizers, those who opposed the pipeline stood in its path, blocking Energy Transfer Partners from continuing construction.

The "Art Action" camp at Standing Rock housed Sioux artists, artists from other Indigenous tribes and the surrounding communities, and well-known artists who travelled to the camp to support the protestors. Most of the pieces created in the camp were visually arresting banners and patches emblazoned with phrases such as "Protect the sacred," or "Mni Wiconi," Lakota for "Water is life." The banners and patches were eye-catching, and guaranteed that any photograph of the front lines contained the message the protestors wanted to communicate. In a February 2017 interview, Sarain Fox, the host of the documentary series *RISE*, commented on the importance of art in what came to be known as the "#NoDAPL" movement:

> I sat down with a lot of creators [of #NoDAPL art] and they explained that art is just as important as bodies on the front line. The way they explained it is so simple: The media will always try and silence Indigenous people. They will always try and silence the voices amplifying justice. But when

> you control the messaging and when you can put art on the front line, no matter what angle they take photos from, no matter who the media outlet is, the only photo they can publish has the art in it, and the art contains the messaging. You can't be silenced as long as art is part of the movement.
> *(Weisenstein 2017)*

In the case of the Standing Rock protests, art was a literal vehicle for communication in the climate justice movement. Although the pipeline was completed and began transporting oil shortly after President Donald Trump took office in January 2017, in June of the same year a federal judge ruled that the US Army Corps of Engineers did not adequately assess the environmental impacts of the project during the permitting process. Subsequently, arguments were submitted to determine whether the pipeline would be shut down. At the time of the writing of this chapter, that portion of the case had yet to be decided.

The use of art as a communication medium for those excluded from climate change conversations can extend to include non-human organisms as well. Many scholars, activists, and ethicists argue that humans have a moral responsibility to allow non-human species the space, resources, and conditions they need to flourish (Berry 1999). Since plants, animals, fungi, and microorganisms cannot participate in climate change negotiations and decision-making processes through traditional channels, art serves as a way—albeit human-mediated—to highlight the impacts of climate change on these individuals, populations, and ecosystems. Maya Lin's piece *What Is Missing?* is a memorial to environmental change (Lin 2009). The artwork, which appeared first as a DVD and sound sculpture in 2009, and later as an online interactive database, chronicles the myriad ways in which environmental sights, sounds, and species are changing or—in many cases—disappearing. The film version was shown at the California Academy of Sciences on Earth Day in 2009, and again a year later in New York City's Times Square on a large screen alongside the Square's typical multitude of brightly coloured advertisement billboards and videos. It ran for approximately two weeks at both showings, slowly and deliberately presenting viewers with 12 kinds of anthropogenic environmental loss, including the loss of biodiversity and the loss of mountain tops, among others. Representative text laid over video imagery was displayed as the film rotated through each type of loss. The most recent version of *What Is Missing?* is an interactive online map (whatismissing.net), where users can submit stories and images of lost species or experiences, and browse others' submissions. When Lin created the website, she consulted many of the earliest written descriptions of landscapes and species, so that she could include stories of historical losses as well (*Maya Lin's Memorial to Vanishing Nature* no date).

What Is Missing? creates a unique multi-sensory experience of the scale and ubiquity of the effects of anthropogenic environmental change on non-human organisms, while maintaining micro-scale details about each of those changes. Furthermore, the piece includes multiple avenues through which the viewer can connect with the issues personally. Like the film, the website highlights the loss of non-human species and environmental sights and sounds, but it also provides a way for ordinary citizens, including the most marginalized individuals, to tell their own stories about environmental change on a public platform. Additionally, by displaying the film in one of the most powerful cities in the world, amid rampant consumerism, heavy traffic, and dense crowds, Lin juxtaposes her message of environmental (in)justice with drivers of climate change in a place where those who *can* participate in formal negotiations might see it. "Carry these messages," the film seemed to suggest to viewers in Times Square. "Take them with you to your communities, to COP, to your legislators." *What is Missing?* represents a dynamic insertion of the effects of anthropogenic environmental change on non-human life into the climate change conversation.

Activists at the 2015 COP21 climate change conference in Paris took this idea to its literal conclusion by embodying non-human organisms. During the conference, a French group calling themselves the "Zoological Ensemble for the Liberation of Nature" invaded a Volkswagen dealership dressed as plants and animals, throwing leaves and vegetables, pretending to scratch tires, and placing banana peels under cars. Before they left the building, the activists shouted: "We are nature defending itself!" (Demos 2015).

Art as a tool to power the movement

Art can also operate as a tool *within* the movement. It can unify activists and organizers, and share information within or between communities of resistance. Two projects that were part of the People's Climate March in New York City in 2014 emphasize the role art can play in unifying climate justice activists. Artist and activist Rachel Schragis created a 6-inch-by-90-foot canvas that can be unrolled like a scroll. The piece, titled *The Same Thing*, depicts various facets of the climate justice movement based on stories and information collected from more than 100 activists involved in the march (Schragis 2016). The scroll, which highlights the complexity and interconnectedness of different parts of the climate movement, is intended to be unfolded by a large group of people arranged in a circle. As each image on the scroll emerges, the facilitator sings a song written by Schragis; the song's message is to find hope in the connectedness of our struggles. Schragis created the piece as part of a residency at 350.org, in response to an identified need for artwork that unified individuals in the movement (Vázquez 2016). Ryan Camero, an arts activist who helped organize the first unfolding of the scroll—at the 2013 Gabfestry Festival for artist activists in Machias, Maine—commented: "You could really feel a sense of collective empathy on all of those different struggles that encapsulate the climate movement" (E 2016). Since then, the scroll has travelled around the world, serving as the focal point of unifying experiences in different locations and events, including the COP21 conference in Paris.

The Climate Ribbon project (Ledorze et al. 2014) was the culminating art piece of the 2014 People's Climate March, and like *The Same Thing*, it has since travelled around the world. In the piece, a Tree of Life sculpture is erected and draped with thousands of ribbons upon which individuals write descriptions of the things they most fear losing due to climate change. Things that people have written include "The smell of wet grass at dawn," "That tree that I planted in my grandma's garden when I was four," and "Being able to drink water from the tap." Fears of a broader scope are reflected on the tree as well: one woman wrote "For the happiness and safety of all people now and in the future, for all living things, and for the systems on which we depend" (Ledorze et al. 2014). People who visit the tree add a ribbon of their own, and take another ribbon as a commitment to protect the thing written on it. The project was inspired by Northeastern Native American wampum belts that signify the importance of agreements made between people. In this way, the ribbons on the tree represent, collectively, a "people's treaty"—a unifying agreement between individuals to continue to fight for climate justice. The project's website explains: "Together our commitments [to fight climate change] weave a giant thread connecting all of us as we work for a healthy, sustainable planet" (Ledorze et al. 2014). To those who participate in the project, the ribbons on the tree serve as a visual reminder of the number of people working to address climate change; for each fear of loss reflected on the tree, there is a corresponding person committed to working to avoid that loss. Furthermore, the piece requires individuals to consider the effects of climate change in the context of their own lives, an exercise that provides a point of entry to an otherwise dauntingly large problem. As in Schragis's *The Same Thing*, the *Climate Ribbon* tree requires people's participation to become a complete work of art. This participation heightens the experiential aspect of the artwork, making both *The Climate Ribbon* project and *The Same Thing* particularly powerful unifying tools within the climate justice movement.

In some communities, art can be a means of communication about a changing climate. A recent study of 30 professional artists in the Cape Dorset and Pangnirtung Inuit communities in Nunavut, Canada found that art served, in part, as a means of communication about environmental change (Rathwell and Armitage 2016). One artist who participated in the study, Tim Pitsiulak, creates prints depicting Inuit people, animals, and the interactions between them. Pitsiulak told the researchers: "I try to do the best work I can … talking through the art about the climate changing … the ice breaking up on the flow edge and people boating much later" (Rathwell and Armitage 2016). Another artist, Elisapee Ishulutaq, commented:

> When I was young the ice was not dangerous … now it's getting dangerous and through art, artists can get it out there […] Now a days hunters don't really listen to elders for their knowledge or wisdom and having some sort of visual aid like art would really put a clearer picture into what the elders are trying to say.
>
> *(Rathwell and Armitage 2016)*

FIGURE 23.1 Elisapee Ishulutaq artwork. Elisapee Ishulutaq. *Iqluviaq (snow house)*. 2009. Reproduced in greyscale. Ishulutaq's works, like this oil stick drawing, depict traditional Inuit ways of life before the transition to permanent settlements. Image courtesy of Marion Scott Gallery.

Ishulutaq's oilstick-on-canvas drawings are reflections of traditional Inuit life, and the ways that life is changing can be traced in her artworks. In these communities, art can be an effective way to share observations, stories, and knowledge about the effects of climate change (see Figure 23.1).

Art as a demonstration of the life in "wastelands"

Finally, art can be a conceptual tool used to demonstrate that spaces that are described as empty or of little value by those in power are actually alive, owned, occupied, or home for people and other living beings. This is a matter of urgent importance in the climate justice movement: throughout history, places described by governments, corporations, and the elite as empty or worthless, in those words or others, have been sites of environmental and social injustices. Danika Cooper, an Assistant Professor of Landscape Architecture and Environmental Planning at the University of California, Berkeley, has argued that the desert of the American Southwest is such a place (Cooper 2017). The deserts are unmapped: on many maps, they are literally depicted as empty space, suggesting that nothing there is worth noting. This implication has allowed the southwestern desert to be the site of myriad injustices, including internment camps, where Japanese people were forced to stay during World War II, and the largest of which was built on the Colorado River Indian Reservation despite the Tribal Council's objections (Fox and Nubile 2008); the testing of atomic weapons, which has had lasting health impacts on the environment and human health, particularly that of Indigenous people, in the Southwest (Kuletz 2001); and the storage of nuclear waste (Zhang 2016). Valerie Kuletz writes about nuclear weapons testing sites in the Southwest:

> Boundaries here are politically and socially constructed. An invisible line marks off the national sacrifice zone of [the] China Lake [weapons research centre] from the national treasure of Death Valley. This line is arbitrary, but it works to construct the way most Euro-Americans see these places—one is perceived as desolate and the other as mystically beautiful.
>
> *(Kuletz 2001, p.247)*

The way governments, and others in power, define and assign value to spaces permeates society, such that one space can be a site of unfathomable destruction while a neighbouring space one of treasured beauty.

The forces causing and resulting from climate change are not immune from these patterns. This is particularly evident in the use of so-called "vacant" or "worthless" land for fossil fuel extraction. As mentioned previously, descriptions of the Arctic National Wildlife Refuge (ANWR) as a "flat white nothingness" and a "frozen wasteland" were central to the arguments of proponents of opening the ANWR for drilling in 2003, and these arguments re-emerge each time the question comes up for debate in Congress. When Senator Boxer brought them to the Senate floor, Banerjee's photographs provided a direct counterpoint to these depictions, making immediately evident to anyone who saw them that that Arctic is the traditional home of thriving human communities and the feeding, breeding, and migration grounds of diverse fauna.

Garth Lenz's photographs serve a similar purpose. They show the stunning beauty of ecosystems juxtaposed with the environmental devastation wrought by extractive fossil fuel industries. His photographs of the boreal forest portray the people, animals, and plants that call the region home, along with images of the massive destruction caused by bitumen extraction in the tar sands. Lenz's work has received significant public attention: he has given a TED Talk, and his photographs have appeared in *TIME* magazine and *The New York Times*, among others (Lenz no date). These photographs are much more explicitly political than Banerjee's, although they serve a similar purpose: demonstrating that there is already life—and value—in the places where fossil fuel industries operate.

The fact that Banerjee and Lenz are privileged outsiders in the ANWR and the boreal forest, respectively, highlights a tension of this particular role for art in the climate justice movement. The people who live and have lived in these areas for centuries or, in many cases, millennia, do assert and have asserted—by their very existence and by their protests—that there is life and value in these landscapes. One of the many effects of colonialism is that the opinions and values of these individuals and communities are not afforded the same respect in settler societies and governments as those of settler communities and corporations. The influence Banerjee's and Lenz's photographs have had cannot be separated from the photographers' respective roles as settlers. While the work of Banerjee and Lenz is a powerful tool in the movement, it is important to recognize and afford more respect, weight, and power to the art and voices emanating directly from frontline communities in the climate justice struggle.

Incorporating art in climate justice work

Art is a crucial component of the climate justice movement. Unlike academic and formal modes of communication, art has the ability to create an experience for viewers, even across language and cultural barriers, thus connecting more people—more powerfully—to the issues at hand. Art can serve as an alternative platform for messages from communities that are underrepresented in climate change negotiations and decision-making; as a tool to unify and relay messages within the movement; and as a way to correct conceptions of certain landscapes as unoccupied or not valuable.

Art-making is not limited to professional artists. Anyone can start a climate justice art project in their community; host *The Same Thing* scrolls, the *Climate Ribbon* project, or another similar interactive piece; or highlight artwork made by people or groups from frontline communities by bringing it to policymakers or posting about it on social media. Those who are able can pay for art, show it, and donate money to artists who create pieces that can influence the struggle for climate justice. Action on climate is not limited to the halls of academia and the floors of Congress—art is action, too.

References

Banerjee, S. (2004) 'A Conversation with Subhankar Banerjee'. Available at: www.subhankarbanerjee.org/PDF/BanerjeeGPGInterview.pdf (Accessed: 4 June 2017).

Berry, T. (1999) 'Chapter 9: Ethics and Ecology', in Berry, T., *The Great Work: Our Way into the Future*. New York, NY: Crown/Broadway Books, pp. 100–106.

Blücher, H. (1951) 'Fundamentals of a Philosophy of Art'. Available at: www.bard.edu/library/archive/bluecher/lectures/phil_art/philart.php (Accessed: 27 June 2017).

Carter, J. and Banerjee, S. (2003) *Arctic National Wildlife Refuge: Seasons of Life and Land*. 1st edition. Seattle, WA: Mountaineers.

Cooper, D. (2017) 'Mapping the Desert'. American Association of Geographers Annual Meeting, Boston, MA, 8 April.

Demos, T. J. (2015) *Playful Protesters Use Art to Draw Attention to Inadequacy of* Paris *Climate Talks, Truthout*. Available at: www.truth-out.org/news/item/34006-playful-protesters-use-art-to-draw-attention-to-inadequacy-of-paris-climate-talks (Accessed: 28 June 2017).

Dewey, J. (1938) *Experience and Education*. Indianapolis, IN: Kappa Delta Pi.
E, C. (2016) 'Art for Climate Justice', *SustainUS*, September. Available at: http://sustainus.org/2016/09/art-for-climate-justice/ (Accessed: 29 June 2017).
Eliasson, O. (2016) *Why art has the power to change the world, World Economic Forum*. Available at: www.weforum.org/agenda/2016/01/why-art-has-the-power-to-change-the-world/ (Accessed: 19 June 2017).
Figueiredo, P. and Perkins, P. E. (2013) 'Women and water management in times of climate change: participatory and inclusive processes', *Journal of Cleaner Production*. (Special Volume: Water, Women, Waste, Wisdom and Wealth), 60, pp. 188–194. doi: 10.1016/j.jclepro.2012.02.025.
Fox, J. and Nubile, J. (2008) *Passing Poston: An American Story* [Documentary film].
Kuletz, V. (2001) 'Invisible Spaces, Violent Places: Cold War Nuclear and Militarized Landscapes', in Peluso, N. L. and Watts, M. (eds) *Violent Environments*. Cornell University Press.
Ledorze, D. et al. (2014) *The Climate Ribbon Project*. Available at: http://theclimateribbon.org
Lenz, G. (n.d.) *Artist website*. Available at: www.garthlenz.com/ (Accessed: 4 July 2017).
Lieberman, G. A. and Hoody, L. L. (1998) 'Closing the Achievement Gap: Using the Environment as an Integrating Context for Learning. Executive Summary'. Available at: https://eric.ed.gov/?id=ED428942 (Accessed: 21 June 2017).
Lin, M. (2009) *What Is Missing?* [Multimedia, website]. Available at: http://whatismissing.net
Macchi, M. et al. (2008) *Indigenous and Traditional Peoples and Climate Change*. IUCN. Available at: https://cmsdata.iucn.org/downloads/indigenous_peoples_climate_change.pdf (Accessed: 28 June 2017).
March 19, 2003 Senate Session Recording (2003) Washington, D.C.: C-SPAN.
Maya Lin's Memorial to Vanishing Nature (n.d.) Available at: http://e360.yale.edu/features/maya_lin_a_memorial_to_a_vanishing_natural_world (Accessed: 28 June 2017).
PBS. 'Interview with Senator Ted Stevens' (2005) *PBS News Hours with Jim Lehrer*.
Rathwell, K. and Armitage, D. (2016) 'Art and artistic processes bridge knowledge systems about social-ecological change: An empirical examination with Inuit artists from Nunavut, Canada', *Ecology and Society*, 21(2), p. 21. doi: 10.5751/ES-08369-210221.
Schragis, R. (2016) *The Same Thing* [Multimedia]. Artist's personal collection.
Tomlinson, L. (2015) *Procedural Justice in the United Nations Framework Convention: Negotiating Fairness*. Springer International Publishing. Available at: www.springer.com/us/book/9783319171838 (Accessed: 21 June 2017).
US House, Committee on Resources (2003) *H.R. 39, Arctic Coastal Plain Domestic Energy Security Act*. Hearing, 12 March. Washington, D.C.: Government Printing Office.
Vázquez, D. H. (2016) 'The Same Thing, A Climate Human Powered Scroll by Rachel Schragis', *When We Fight We Win!*, 8 March. Available at: www.whenwefightwewin.com/the-same-thing-a-climate-human-powered-scroll-by-rachel-schragis/ (Accessed: 29 June 2017).
Weisenstein, K. (2017) 'How Art Immortalized #NoDAPL Protests at Standing Rock: Interview with Sarain Fox', *Creators*, 28 February. Available at: https://creators.vice.com/en_us/article/gv3pg3/how-art-immortalized-nodapl-protests-at-standing-rock (Accessed: 28 June 2017).
Zhang, S. (2016) 'America's Nuclear-Waste Plan Is a Giant Mess', *The Atlantic*, 2 November. Available at: www.theatlantic.com/technology/archive/2016/11/nuclear-waste-wipp-new-mexico/506117/ (Accessed: 28 June 2017).

24

CONCLUSION

Moving ahead for climate justice

Patricia E. Perkins

To conclude this book, I would like to revisit the broad global imperative of climate justice—the most urgent and morally compelling challenge of our time, in my view and the view of many people. How is it that the activist initiatives outlined in this book, which are described as furthering climate justice, really address global redistribution and the pressing, extremely unjust livelihood threats that millions of human beings, who did almost nothing to cause climate change and are ill-equipped to face it, are now suffering?

Many of the chapter authors have answered this question explicitly. For example, greater awareness of the effects of rising sea levels and tropical storms in Bangladesh and elsewhere (Chapters 6, 7, 8) will hopefully lead North Americans to press their governments to act more decisively to stop climate change and assist its victims. Better understanding of the costs at home and abroad of extraction and fossil fuel transport and processing (Chapters 2, 3, 4, 13, 14, 15, 16) will hasten the energy transition by reducing social license and support for fossil fuels. Educating people in North America about colonialism and its ongoing impacts for Indigenous peoples and for everyone (Chapters 2, 6, 15, 16, 17, 20) contributes to the governance transitions and acceptance of other ways of organizing society that are needed overall to stop climate change. Educational, artistic, and community organizing initiatives (Chapters 2, 4, 5, 9, 14, 15, 16, 17, 18, 19, 20, 21, 22, 23) are also part of paving the way and facilitating this shift.

Three themes stand out in this discussion:

1) **More global solidarity and international climate justice activism are sorely needed**. While it is undoubtedly true that "starting close to home" to learn the motivations and skills for climate justice activism is a good strategy, it is also possible to leave the hardest global distribution questions for too late. International justice priorities demand creative new

approaches in policy, individual action, and collective organizing. These relate to such issues as trade agreements which facilitate consumption and obfuscate/shift the burdens and ecological footprint of over-consumption by the wealthy; foot-dragging with regard to necessary changes in transportation including personal vehicles, air travel, public transit, etc.; the global food industry and whether it meets humanity's needs efficiently and sustainably; and the imperative of facilitating migration on humanitarian grounds by those who must move, while also assisting and defending people's right to remain. Climate justice is a global problem which requires an ongoing search for global solutions. (Chapters 2, 3, 7, 8, 10, 11, and 12 discuss these issues in more detail.)

2) **Wealthy fossil-fuel-producing and high-consuming countries must cut extraction and drastically reduce their emissions immediately.** This includes the grossly inefficient Canadian tar sands extraction megaproject, as well as fracking, coal extraction, and diesel burning. The ecological effects of nuclear and hydroelectric power generation and storage also must be reconsidered. Emissions reductions also are vital in relation to the extraction and processing of metals needed for the green transition, which have huge environmental and atmospheric impacts. Since the bulk of global mining capital is organized from Toronto, this is an area where local and national actions can have global effects. There is no excuse for hesitation; the science and policy conclusions are clear. Comparing the carbon footprint of consumption in rich and poor countries, and by rich and poor people, reveals the huge extent of these injustices. The impacts worldwide are immense and the moral burden of wealthy countries' failure to act is crushing. These are actions that we can and must take, lead, and advocate for. (See examples of how to do this in Chapters 2, 9, 10, 11, 12, and 20).
3) **Decolonization—including self-education by settlers to counter systematic learned ignorance about this and disrupt, restructure, and replace colonial systems of power—is an urgent prerequisite for climate justice.** It is also, as noted in several chapters, an integral part of changing cultural and governance traditions to prevent the greed, rapacity, and over-consumption by some that are inherent to the capitalist, colonial economic system which has produced the climate crisis. Indigenous resurgence is growing worldwide; and in Canada in particular, with attention and care, new traditions of uneasy but at times respectful collaboration are emerging between settlers and Indigenous people. Learning and sharing the skills for building these collaborations is a central part of climate justice activism. (This is discussed further in Chapters 2, 16, 17, 18, and 20.)

As youth climate activists are forcefully saying, and Indigenous leaders are modelling: we know what's needed; let's just get on with it!

ACTION GLOSSARY

Anti-Oppression (See Chapter 20): Work and practices that recognize the power imbalances and other forms of oppression that exist in society (including racism, sexism, homophobia, and other forms of discrimination), and attempt to mitigate and remove them.

Carbon Budget (See Chapters 3, 9, 10): The total amount of carbon that can be emitted into the Earth's atmosphere without global temperatures rising by a certain amount; a direct relationship between carbon emissions and global warming, based on best scientific estimates.

Carbon Footprint (See Chapter 24): An estimated measure of the total greenhouse gas emissions caused by a person, group, event, or product. Comparing carbon footprints globally gives an idea of climate injustices related to consumption levels. http://css.umich.edu/factsheets/carbon-footprint-factsheet

Climate Debt (See Chapters 3, 6, 9, 10): The amount by which a country's carbon emissions have exceeded a calculated per capita global average, over time. In other words, the extent to which a person, group, or country has emitted more than its fair share of greenhouse gases.

Climate Refugees (See Chapters 6, 7, 8): A contested term for people forced to migrate due to climate change and related political and economic factors.

Commons (See Chapters 12, 17): Cultural and environmental amenities, resources, or materials that are accessible to all members of a group, and are managed collectively for shared benefit. If not well managed, they can become open-access resources subject to privatization.

Community Art (See Chapters 17, 20, 23): When a group of people document and share their experiences by making art—murals, painting, sculpture, performances, music, etc.—this is community art! It can be a method for community organizing, doing participatory research, recording local ecological knowledge, and registering/communicating protest as well as representing the experiences and feelings of the artists in relation to their environment.

Community Energy Plan (See Chapter 4): A document setting out how a community intends to integrate local energy and land use and infrastructure to reduce greenhouse gas emissions and improve the community.

Community Engagement (See Chapters 5, 6, 13, 14, 15, 17, 19, 20, 23): A relational process that facilitates communication, interaction, involvement, and exchange among people through dialogue, facilitated information-sharing, collective planning, and strategizing.

Consensus (See Chapter 20): General agreement reached by discussion rather than by voting.

Decarbonization (See Chapter 3): The reduction of carbon intensity, and therefore of average CO2 emissions, per unit of consumption or electricity generated, via increased use of low-carbon energy sources such as renewables, nuclear, and shifts to lower-emissions fuels.

Decolonization (See Chapters 1, 6, 16, 17, 19, 20): The undoing of colonialism; ongoing critique and contestation of the effects and continuing manifestations of colonial privilege, political structures, laws, research processes, cultural dominance, economic and political rights, etc., and in particular their impacts on Indigenous peoples.

Diaspora (See Chapter 7): A dispersed population whose origin lies in a separate geographic location. Some diasporas were created by the slave trade, colonialism, famines, or political conflict; many diasporas maintain ties to their homelands.

Direct Action (See Chapters 16, 20, 23): An activist term for taking immediate steps to reach political goals by using nonviolent civil resistance, sit-ins, strikes, occupations, blockades, or more violent political actions.

Education Reform (See Chapters 17, 18, 22): The goal of changing public education systems, curricula, goals, pedagogy, indicators of success, etc.

Extraction (See Chapter 6, 16, 20): For environmental activists, the mining of metals and fossil fuels and the resulting harmful impacts. (See F. Dubois, M. Tessier and D. Widgington, *Extraction! Comix Reportage*. Montreal: Cumulus Press. https://issuu.com/chesterrhoder/docs/extraction_comix_reportage.)

Food Sovereignty (See Chapters 6, 11, 12): People's right to healthy and culturally appropriate food produced using ecologically sound and sustainable methods, and their right to define their own food and agriculture systems. (See https://foodsecurecanada.org/who-we-are/what-food-sovereignty.)

Fossil Divestment (See Chapters 10, 20): Selling stocks, bonds, and investment funds that imply partial ownership of or help to finance oil and gas companies.

Fossil Fuel Subsidies (See Chapters 3, 9, 10): Government benefits, payments or tax exemptions provided to coal, oil, natural gas or related infrastructure companies for extraction, fracking, pipelines, transportation or port construction, or any other aspect of getting fossil fuels to market.

Free Prior and Informed Consent (FPIC) (See Chapter 2): A requirement central to the United Nations Declaration of the Rights of Indigenous Peoples that all affected by development projects, especially Indigenous peoples, must understand and agree in advance to the developments along with their expected impacts, or else the developments should not be undertaken.

Frontline Actions (See Chapters 16, 20, 21): Demonstrations, blockades, and protests to physically stand in the way and prevent unwanted extraction, forestry, exploration, construction or development projects from taking place.

Frontline Communities (See Chapter 5, 6, 7, 15, 16, 21): Areas that are particularly severely impacted by climate change and environmental injustice. The climate justice organization Ecotrust says, "Frontline communities are those that experience 'first and worst' the consequences of climate change. These are communities of colour and low-income, whose neighbourhoods often lack basic infrastructure to support them and who will be increasingly vulnerable as our climate deteriorates. These are Native communities, whose resources have been exploited, and labourers whose daily work or living environments are polluted or toxic" (https://ecotrust.org/centering-frontline-communities/). See "Sacrifice Zones."

Green Finance (See Chapters 3, 9, 10): Making investment capital available for renewable energy supplies or production, energy transition infrastructure, or other non-carbon-generating development.

Green Infrastructure (See Chapter 5): Water management installations that protect, restore, and/or mimic the natural water cycle by planting trees, restoring wetlands, installing green roofs, and minimizing rainfall and snow-melt runoff by facilitating water infiltration into the ground and its slow dispersal. Corollary benefits include temperature modulation due to shade cover, food production options, insect habitat, soil health, etc.

Green Jobs (See Chapters 2, 3): Employment opportunities in renewable power supplies or generation, public transit, renewable resource production, waste management, environmental infrastructures, and other non-extractive, non-fossil fuel related sectors.

Green New Deal (See Chapter 2, 9): A package of government policies designed to use public expenditures to create safe low-carbon jobs, including in renewable energy production and services, expand green industries, and retrain workers to speed the energy transition away from fossil fuels. These can be financed at least in part through removal of subsidies and tax breaks for carbon-intensive industries and firms.

Intersectionality (See Chapters 1, 6, 20): The acknowledgement that different injustices linked to gender, race, ethnicity, sexuality, dis/ability, etc. are interrelated, through social forces that share root causes. See P.H. Collins and S. Bilge (2016) *Intersectionality*. Cambridge: Polity Press.

Just Transition (See Chapters 2, 3, 21): An approach to the post-fossil fuel energy transition that prioritizes justice, particularly for unionized workers. Like Green New Deal proposals, a Just Transition involves speeding the energy transition beyond fossil fuels through retraining and job creation for workers in renewable industries and power production.

Land Acknowledgement (See Chapter 1, 19, 20): A formal statement recognizing the unique and enduring relationship that exists between Indigenous peoples and their traditional territories, and naming and thanking the Indigenous peoples of the local area, often made at the beginning of events, ceremonies, lectures and meetings, especially in Canada. (See S. Mills, "What are land acknowledgements and why do they matter?",

Local Love, https://locallove.ca/issues/what-are-land-acknowledgements-and-why-do-they-matter/#.XJ62hhNKhsY.)

Land Protectors or Land Defenders (See Chapter 6, 16, 20): Environmental activists, usually Indigenous people, who protect territories by occupying and physically blocking other people and their vehicles from entering an area threatened with exploration, construction, deforestation or extraction. (See H. Rollman, "Protesters? Or land protectors?" *The Independent*, Oct. 28, 2016, http://theindependent.ca/2016/10/28/protesters-or-land-protectors/.)

Leap Manifesto (See Chapters 17, 20): A political document issued by a coalition of Canadian authors, leaders, artists, and activists during the federal election campaign in September 2015. It calls for a restructuring of the Canadian economy to rapidly reduce climate change by ending dependence on fossil fuels (see leapmanifesto.org).

Networking (See Chapters 7, 8, 12, 13, 14, 15, 18, 20, 21): Exchange of information, contacts, perspectives, and ideas among people or groups, to share knowledge and build broader social ties and approaches.

Participatory Action Research (Chapters 14, 17): A research approach that emphasizes collective inquiry, experimentation, and the development of shared actions to address community priorities through collaboration between researchers and community members.

Participatory Eco-Cultural Mapping (Chapter 6): An activation and community-focusing method in which community members map geographic space with its ecological and cultural meanings for local people—including elders, and people with various positions in the community—in the past, present, and future.

Photo-Voice (See Chapter 23): A research and community engagement method in which diverse participants take photographs in their communities and share their commentary on the photos, narrating what they find interesting, problematic, or beautiful in the local area, thus allowing others to see things from their viewpoint.

Reconciliation (See Chapters 19, 20): The restoration of friendly, respectful relations. In Canada, the Truth and Reconciliation Commission (2008–2015) held hearings across the country to document the history and lasting impacts of the Indian Residential School system, and issued a report with 94 "calls to action" for building reconciliation in Canada. As of March 2018, 10 of the calls were considered by the Canadian Broadcasting Corporation to be complete, 15 were in progress, 25 had projects proposed, and 44 were unmet (CBC News, "Beyond 94: Where is Canada at with reconciliation?" March 19, 2018).

Right to Remain (See Chapter 6, 7, 8): The right not to be displaced by economic development, climate change, gentrification, or wealthier people's expansionist, profit-driven desires.

Sacrifice Zone (See Chapter 7, 15, 16): An area or territory subject to extreme negative environmental impacts from extraction, industrial development, air/water/land emissions, pollution, deforestation, and/or climate change, resulting in grave environmental, health, and livelihood impacts on local populations.

Settlers (See Chapters 16, 17, 19, 20): Non-Indigenous occupants of territories previously and still caretaken by Indigenous peoples. See Ruth

Koleszar-Green (2018), "What is a Guest? What is a Settler? *Cultural and Pedagogical Inquiry* 10(2), https://journals.library.ualberta.ca/cpi/index.php/cpi/article/view/29452.

Social Capital (See Chapters 13, 14, 15, 17, 23): A measure of social interactions, interpersonal relationships, and goodwill within a community, which allows people to rely on and help each other and increases well-being.

Social License (See Chapters 3, 10): Society's acceptance and implicit approval of corporate actions or ways of doing business.

Solidarity (See Chapters 2, 5, 6, 7, 8, 11, 12, 13, 14, 15, 16, 17, 18, 19, 20): Understanding and unity of interests, objectives, and sympathies among different groups in society, often across differences such as those of geography, race, religion, or ethnicity.

Toxic Tours (See Chapter 16, 20): Visits organized by environmental or community organizations to highlight pollution, contamination, emissions, and their socioeconomic and health impacts on their local area and its people. A technique for raising awareness, used worldwide in campaigns for environmental justice.

Turtle Island (See Chapter 6, 16, 17): Indigenous name for North America, referring to many Indigenous legends telling the creation story of a turtle that holds the world on its back.

SUGGESTED FURTHER READING ON CLIMATE JUSTICE

Adger, W. N., Paavola, J., Huq, S. and Mace, M. J. (eds.) (2006). *Fairness in Adaptation to Climate Change*. Cambridge, MA: MIT Press.

Angus, I. (ed.) (2010). *The Global Fight for Climate Justice: Anticapitalist Responses to Global Warming and Environmental Destruction*. Black Point, Nova Scotia: Fernwood Publishing and Resistance Books.

Atleo, E. R. (2011). *Principles of Tsawalk: An Indigenous Approach to Global Crisis*. Vancouver, BC: UBC Press.

Bond, P. (2018). "Social movements for climate justice during the decline of global governance: from international NGOs to local communities." In S. Lele et al. (eds.), *Rethinking Environmentalism: Linking Justice, Sustainability, and Diversity*. Cambridge, MA: MIT Press, Strüngmann Forum Reports, vol. 23, pp. 153–182.

Cappello, S. and Harcourt, W. (2009). 'Gender and climate justice', *International Journal of Green Economics*, 3(3–4), pp. 343–350.

Cohen, M. G. (ed.) (2017). *Climate Change and Gender in Rich Countries: Work, Public Policy and Action*. London/New York, NY: Routledge/Earthscan.

D'Arcy, S., Black, T., Weis, T. and Russell, J. K. (2014). *A Line in the Tar Sands: Struggles for Environmental Justice*. Toronto, ON: Between the Lines.

Druckman A. and Jackson, T. (2010). 'The bare necessities: how much household carbon do we really need?', *Ecological Economics*, 69(9), pp. 1794–1804.

Ensor, J. and Berger, R. (2009). *Understanding Climate Change Adaptation: Lessons from Community-Based Approaches*. Rugby/Warwickshire: Practical Action Publishing.

Estes, N. (2019). *Our History Is the Future: Standing Rock Versus the Dakota Access Pipeline, and the Long Tradition of Indigenous Resistance*. London/New York, NY: Verso.

Godfrey, P. and Torres, D. (eds.) (2016). *Systemic Crises of Global Climate Change: Intersections of Race, Class and Gender*. London/New York, NY: Routledge.

Haji-Abdi, A., Okot, B. and Wade, W. (2008). "Diaspora communities: global networks and local learning." In J. Blewitt (ed.), *Community, Empowerment and Sustainable Development*. Totnes, UK: Green Books, pp. 139–154.

Jacobsen, S. G. (ed.) (2018). *Climate Justice and the Economy: Social Mobilization, Knowledge and the Political*. London/New York, NY: Routledge.

Jafry, T. (ed.) (2018). *Routledge Handbook of Climate Justice.* London/New York, NY: Routledge/Earthscan.

Klein, N. (2014). *This Changes Everything: Capitalism vs. the Climate.* New York, NY: Simon and Schuster.

Macgregor, S. (2014). 'Only resist: feminist ecological citizenship and the post-politics of climate change', *Hypatia*, 29(3), pp. 617–633.

McGregor, D. (2004). 'Coming full circle: indigenous knowledge, environment, and our future', *American Indian Quarterly*, 28(3/4), pp. 385–410.

Mearns, R. and Norton, A. (eds.) (2010). *Social Dimensions of Climate Change: Equity and Vulnerability in a Warming World.* Washington, DC: The World Bank.

Mickelson, K. (2009). 'Beyond a politics of the possible? South-North relations and climate justice', *Melbourne Journal of International Law*, 10(2), pp. 411–423.

Nagel, J. (2016). *Gender and Climate Change: Impacts, Science, Policy.* London/New York, NY: Routledge.

Page, E. A. (2006). *Climate Change, Justice and Future Generations.* Cheltenham, UK: Edward Elgar.

Roberts, J. T. and Parks, B. C. (2007). *A Climate of Injustice: Global Inequality, North-South Politics, and Climate Policy.* Cambridge, MA: MIT Press.

Robinson, M. (2018). *Climate Justice: Hope, Resilience, and the Fight for a Sustainable Future.* New York, NY/London/Oxford/New Delhi/Sydney: Bloomsbury Publishing.

Rosenzweig, C. and Wilbanks, T. J. (2010). 'The state of climate change vulnerability, impacts, and adaptation research: Strengthening knowledge base and community', *Climatic Change*, 100(1), pp. 103–106.

Schlosberg, D. and Collins, L. B. (2014). 'From environmental to climate justice: climate change and the discourse of environmental justice', *Wiley Interdisciplinary Reviews: Climate Change*, 5, pp. 359–374.

Singer, M. (2019). *Climate Change and Social Inequality: The Health and Social Costs of Global Warming.* London/New York, NY: Routledge/Earthscan.

Whyte, K. P. (2014). 'Indigenous women, climate change impacts, and collective action', *Hypatia*, 29(3), pp. 599–616.

Whyte, K. P. (2017a). "Is it colonial déja vu? Indigenous peoples and climate injustice." In J. Adamson and M. Davis (eds.), *Humanities for the Environment: Integrating Knowledges, Forging New Constellations of Practice.* London/New York, NY: Earthscan Publications, pp. 88–104.

Whyte, K. P. (2017b). 'Indigenous climate change studies: indigenizing futures, decolonizing the anthropocene', *English Language Notes*, 55(1–2), pp. 153–162.

Whyte, K. P. (2019). "Way beyond the lifeboat: an indigenous allegory of climate justice." In D. Munshi, K. Bhavnani, J. Foran and P. Kurian (eds.), *Climate Futures: Reimagining Global Climate Justice.* Berkeley, CA: University of California Press.

Wu, F. (2018). "China: climate justice without a social movement?" In D. Munshi, K. Bhavnani, J. Foran and P. Kurian (eds.), *Climate Futures: Reimagining Global Climate Justice.* Berkeley, CA: University of California Press.

INDEX